Funtional methods in differential equations

CHAPMAN & HALL/CRC MATHEMATICS

OTHER CHAPMAN & HALL/CRC MATHEMATICS TEXTS:

Functions of Two Variables, Second edition
S. Dineen

Network Optimization
V. K. Balakrishnan

Sets, Functions, and Logic: An introduction to abstract mathematics, Third edition
Kevin Devlin

Algebraic Numbers and Algebraic Functions
P. M. Cohn

Computing with Maple
Francis Wright

Dynamical Systems: Differential equations, maps, and chaotic behaviour
D. K. Arrowsmith and C. M. Place

Control and Optimization
B. D. Craven

Elements of Linear Algebra
P. M. Cohn

Error-Correcting Codes
D. J. Bayliss

Introduction to the Calculus of Variations
U. Brechtken-Manderscheid

Integration Theory
W. Filter and K. Weber

Algebraic Combinatorics
C. D. Godsil

An Introduction to Abstract Analysis-PB
W. A. Light

The Dynamic Cosmos
M. Madsen

Algorithms for Approximation II
J. C. Mason and M. G. Cox

Introduction to Combinatorics
A. Slomson

Elements of Algebraic Coding Theory
L. R. Vermani

Linear Algebra: A geometric approach
E. Sernesi

A Concise Introduction to Pure Mathematics
M. W. Liebeck

Geometry of Curves
J. W. Rutter

Experimental Mathematics with Maple
Franco Vivaldi

Solution Techniques for Elementary Partial Differential Equations
Christian Constanda

Basic Matrix Algebra with Algorithms and Applications
Robert A. Liebler

Computability Theory
S. Barry Cooper

Full information on the complete range of Chapman & Hall/CRC Mathematics books is available from the publishers.

Veli-Matti Hokkanen
University of Jyväskylä

Gheorghe Moroşanu
University of Stuttgart

Functional methods in differential equations

CHAPMAN & HALL/CRC

A CRC Press Company
Boca Raton   London   New York   Washington, D.C.

Library of Congress Cataloging-in-Publication Data

Hokkanen, Veli-Matti.
    Functional methods in differential equations / Veli-Matti Hokkanen, Gheorghe Morosanu.
       p. cm. — (Research notes in mathematics ; 432)
    Includes bibliographical references and index.
    ISBN 1-58488-283-2
    1. Boundary value problems. 2. Functional analysis. I. Morosanu, Gheorghe II. Title.
    III. Series.

QA379 .H65 2002
515′.35—dc21
                                                     2002018853

Dedicated to Professor Wolfgang L. Wendland on the occasion of his 65th birthday

Contents

Introduction

In recent years, functional methods have become central to the study of theoretical and applied mathematical problems. An advantage of such an approach is its generality and its potential unifying effect of particular results and techniques.

Functional analysis emerged as an independent discipline in the first half of the 20th century, primarily as a result of contributions of S. Banach, D. Hilbert, and F. Riesz. Significant advances have been made in different fields, such as spectral theory, linear semigroup theory (developed by E. Hille, R.S. Phillips, and K. Yosida), the variational theory of linear boundary value problems, etc. At the same time, the study of nonlinear physical models led to the development of nonlinear functional analysis. Today, this includes various independent subfields, such as convex analysis (where H. Brézis, J.J. Moreau, and R.T. Rockafellar have been major contributors), the Leray-Schauder topological degree theory, the theory of accretive and monotone operators (founded by G. Minty, F. Browder, and H. Brézis), and the nonlinear semigroup theory (developed by Y. Komura, T. Kato, H. Brézis, M.G. Grandall, A. Pazy, etc.).

As a consequence, there has been significant progress in the study of nonlinear evolution equations associated with monotone or accretive operators (see, e.g., the monographs by H. Brézis [Brézis1], and V. Barbu [Barbu1]). The most important applications of this theory are concerned with boundary value problems for partial differential systems and functional differential equations, including Volterra integral equations. The use of functional methods leads, in some concrete cases, to better results as compared to the ones obtained by classical techniques. In this context, it is essential to choose an appropriate functional framework. As a byproduct of this approach, we will sometimes arrive at mathematical models that are more general than the classical ones, and better describe concrete physical phenomena; in particular, we shall reach a concordance between the physical sense and the mathematical sense for the solution of a concrete problem.

The purpose of this monograph is to emphasize the importance of functional methods in the study of a broad range of boundary value problems, as well as that of various classes of abstract differential equations.

Chapter 1 is dedicated to a review of basic concepts and results that are used throughout the book. Most of the results are listed without proofs. In some instances, however, the proofs are included, particularly when we could not identify an appropriate reference in literature.

Chapters 2 through 6 are concerned with concrete elliptic, parabolic, or hyperbolic boundary value problems that can be treated by appropriate functional methods.

In Chapter 2, we investigate various classes of, mainly one-dimensional, elliptic boundary value problems. The first section deals with nonlinear nondegenerate boundary value problems, both in variational and non-variational cases. The approach relies on convex analysis and the monotone operator theory. In the second section, we start with a two-dimensional capillarity problem. In the special case of a circular tube, we obtain a degenerate one-dimensional problem. A more general, doubly nonlinear multivalued variant of this problem is thoroughly analyzed under minimal restrictions on the data.

Chapter 3 is concerned with nonlinear parabolic problems. We consider a so-called algebraic boundary condition that includes, as special cases, conditions of Dirichlet, Neumann, and Robin-Steklov type, as well as space periodic boundary conditions. The term "algebraic" indicates that the boundary condition is an algebraic relation involving the values of the unknown and its space derivative on the boundary. The theory covers various physical models, such as heat propagation in a linear conductor and diffusion phenomena. We treat the cases of homogeneous and nonhomogeneous boundary conditions separately, since in the second case we have a time-dependent problem. The basic idea of our approach is to represent our boundary value problem as a Cauchy problem for an ordinary differential equation in the L^2-space. As a special topic, we investigate in the last section of this chapter, the problem of the higher regularity of solutions.

In Chapter 4 we consider the same nonlinear parabolic equation as in Chapter 3, but with algebraic-differential boundary conditions. This means that we have an algebraic boundary condition as in the previous chapter, as well as a differential boundary condition that involves the time derivative of the unknown. This problem is essentially different from the one in Chapter 3, and a new framework is needed in order to solve it. Specifically, we arrive at a Cauchy problem in the space $L^2(0,1) \times \mathbb{R}$ (see (4.1.6)-(4.1.7)). Actually, this Cauchy problem is a more general model, since it describes physical situations that are not covered by the classical theory. More precisely, if the Cauchy problem has a strong solution (u, ξ), then necessarily $\xi(t) = u(1,t)$; in other words, the second component of the solution is the trace of the first one on the boundary. Otherwise, $\xi(t) \neq u(1,t)$, but it still describes an evolution on the boundary. This is important in concrete cases, such as dispersion or diffusion in chemical substances. As in the preceding chapter, we study the case of a homogeneous algebraic boundary condition separately from the nonhomogeneous one. The higher regularity of solutions is also discussed.

Chapter 5 is dedicated to a class of semilinear hyperbolic partial differential systems with a general nonlinear algebraic boundary condition. We first study the existence, uniqueness, and asymptotic behavior of solutions as $t \to \infty$, by using the product space $L^2(0,1)^2$ as a basic functional setup. The theory has applications in physics and engineering (e.g., unsteady fluid flow with nonlinear pipe friction, electrical transmission phenomena, etc.). Unlike the parabolic case, we do not separate the homogeneous and nonhomogeneous cases, since we can always homogenize the problem. Although this leads to a time-dependent system, we can easily handle it by appealing to classical results on nonlinear nonautonomous evolution equations. In the second section of this chapter, we discuss the higher regularity of solutions. This is important, for instance, for the singular perturbation analysis of such problems. The natural functional framework for this theory seems to be the C^k-space. It is also worth noting that the method we use to obtain regularity results is different from the one in Chapters 3 and 4, and involves some classical tools such as D'Alembert type formulae, and fixed point arguments.

In Chapter 6, we consider the same hyperbolic partial differential systems as in the preceding chapter, but with algebraic-differential boundary conditions. Such conditions are suggested by some applications arising in electrical engineering. As before, we restrict our attention to the homogeneous case only. This problem has distinct features, as compared to the one involving just algebraic boundary conditions. We now consider a Cauchy problem in the product space $L^2(0,1)^2 \times \mathbb{R}$. In the case of strong solutions, we recover the original problem, but in general, this incorporates a wider range of applications. Moreover, the weak solution of this Cauchy problem can be viewed as a generalized solution of the original model.

The remainder of the book is dedicated to abstract differential and integro-differential equations to which functional methods can be applied.

In Chapter 7, the classical Fourier method is used in the study of first and second order linear differential equations in a Hilbert space H. The operator appearing in the equations is assumed to be linear, symmetric, and coercive. In order to use a more general concept of solution, we replace the abstract operator in the equation by its "energetic" extension. A basic assumption is that the corresponding energetic space is compactly embedded into H. This guarantees the existence of orthonormal bases of eigenvectors, and enables us to employ Fourier type methods. Existence and regularity results for the solution are established. In the case of partial differential equations, our solutions reduce to generalized (Sobolev) solutions. Finally, nonlinear functional perturbations are handled by a fixed-point approach. As applications various parabolic and hyperbolic partial differential equations are considered. Since the perturbations are functional, a large class of integro-differential equations is also covered.

In Chapter 8, we discuss the existence and regularity of solutions for first order linear differential equations in Banach spaces with nonlinear functional perturbations. The main methods are the variation of constants formula for linear semigroups and the Banach fixed-point theorem. The theory is applied to the study of a class of hyperbolic partial differential equations with nonlinear boundary conditions.

In Chapter 9, we consider first order nonlinear, nonautonomous differential equations in Hilbert spaces. The equations involve a time-dependent unbounded subdifferential with time-dependent domain, perturbed by time-dependent maximal monotone operators and functionals that can be typically integrals of the unknown function. The treatment of the problem without functional perturbation relies on the methods of H. Brézis [Brézis1]; the problem with functional perturbation is handled by a fixed-point reasoning. As an application, a nonlinear parabolic partial differential equation with nonlinear boundary conditions is studied.

Chapter 10 is concerned with implicit differential equations in Hilbert spaces. Results on the existence, uniqueness, and continuous dependence of solutions for related initial value problems are presented. The study of implicit differential equations is motivated by the two phase Stefan problem, which has recently attracted attention because of its importance for the optimal control of continuous casting of steel.

We continue with some general remarks regarding the structure of the book. The material is divided into chapters, which, in turn, are divided into sections. The main definitions, theorems, propositions, etc. are denoted by three digits: the first indicates the chapter, the second the corresponding section, and the third the position of the respective item in the section. For example, Proposition 1.2.3 denotes the third proposition of Section 2 in Chapter 1. Each chapter has its own bibliography but the labels are unique throughout the book.

We also note that many results are only sketched, in order to keep the book length within reasonable limits. On the other hand, this requires an active participation of the reader.

With the exception of Chapter 1, the book contains material mainly due to the authors, as considerably revised or expanded versions of earlier works. An earlier book by one of the authors must be here quoted [Moro6].

We would like to mention that the contribution of the former author was partly accomplished at Ohio University in Athens, Ohio, USA, in the winter of 2001. The work of the latter author was completed during his visits at Ohio University in Athens, Ohio, USA (fall 2000) and the University of Stuttgart, Germany (2001).

We are grateful to Professor Klaus Kirchgässner (University of Stuttgart) and Dr. Alexandru Murgu (University of Jyväskylä) for their numerous comments on the manuscript of our book. Special thanks are due to Professor Sergiu Aizicovici (Ohio University, Athens) for reading a large part of the manuscript and for helpful discussions.

We also express our gratitude to Professors Haïm Brézis, Eduard Feireisl, Jerome A. Goldstein, Weimin Han, Andreas M. Hinz, Jon Kyu Kim, Enzo Mitidieri, Dumitru Motreanu, Rainer Nagel, Eckart Schnack, Wolfgang L. Wendland and many others, for their kind appreciation of our book.

Last but not least, we dearly acknowledge the kind cooperation of Alina Moroşanu who has contributed to the improvement of the language style.

References

[Barbu1] V. Barbu, *Nonlinear Semigroups and Differential Equations in Banach Spaces*, Noordhoff, Leyden, 1976.

[Brézis1] H. Brézis, *Opérateurs maximaux monotones et semi-groupes de contractions dans les espaces de Hilbert*, North-Holland, Amsterdam, 1973.

[Moro6] Gh. Moroşanu, *Metode funcţionale în studiul problemelor la limită*, Editura Universităţii de Vest din Timişoara, Timişoara, 1999.

Chapter 1

Preliminaries

This chapter has an introductory character. Its aim is to remind the reader of some basic concepts, notations, and results that will be used in the next chapters. In general, we shall not insist very much on the notations and concepts because they are well known. Also, the proofs of most of the theorems will be omitted, the appropriate references being indicated. However, there are a few exceptions, namely Propositions 1.2.1, 1.2.2, and 1.2.3, which might be known, but we could not identify them in literature. The material of this chapter is divided into several sections and subsections.

1.1 Function and distribution spaces

The L^p-spaces

We denote $\mathbb{R} = (-\infty, \infty)$, $\mathbb{N} = \{0, 1, 2, \ldots\}$, and $\mathbb{N}^* = \{1, 2, \ldots\}$. Let X be a real Banach space with norm $\|\cdot\|_X$. If $\Omega \subset \mathbb{R}^N$, $N \in \mathbb{N}^*$, is a Lebesgue measurable set, we denote, as usual, by $L^p(\Omega; X)$, $1 \leq p < \infty$, the space of all equivalence classes (with respect to the equality a.e. in Ω) of (strongly) measurable functions $f: \Omega \to X$ such that $x \mapsto \|f(x)\|_X^p$ is Lebesgue integrable over Ω. In general, every class of $L^p(\Omega; X)$ is identified with one of its representatives. $L^p(\Omega; X)$ is a real Banach space with the norm

$$\|u\|_{L^p(\Omega;X)} = \left(\int_\Omega \|u(x)\|_X^p \, dx \right)^{\frac{1}{p}}.$$

We shall denote by $L^\infty(\Omega; X)$ the space of all equivalence classes of measurable functions $f: \Omega \to X$ such that $x \mapsto \|f(x)\|_X$ are essentially bounded in Ω. Again, every class of $L^\infty(\Omega; X)$ is identified with one of its representatives. $L^\infty(\Omega; X)$ is a real Banach space with the norm

$$\|u\|_{L^\infty(\Omega;X)} = \operatorname*{ess\,sup}_{x \in \Omega} \|u(x)\|_X.$$

In the case $X = \mathbb{R}$ we shall simply write $L^p(\Omega)$ instead of $L^p(\Omega; \mathbb{R})$, for every $1 \leq p \leq \infty$. On the other hand, if Ω is an interval of real numbers, say $\Omega = (a, b)$ where $-\infty \leq a < b \leq \infty$, then we shall write $L^p(a, b; X)$ instead of $L^p\big((a, b); X\big)$. We shall also denote by $L^p_{loc}(\mathbb{R}; X)$, $1 \leq p \leq \infty$, the space of all (equivalence classes of) measurable functions $u: \mathbb{R} \to X$ such that the restriction of u to every bounded interval $(a, b) \subset \mathbb{R}$ is in $L^p(a, b; X)$. If $X = \mathbb{R}$, then this space will be denoted by $L^p_{loc}(\mathbb{R})$.

The theory of L^p-spaces is well known. So, classical results, such as Fatou's lemma, the Lebesgue Dominated Convergence Theorem, etc., will be used in the text without recalling them here.

Scalar distributions. Sobolev spaces

In the following we assume that Ω is a nonempty open subset of $\mathbb{R}^N$. Denote, as usual, by $C^k(\Omega)$, $k \in \mathbb{N}$, the space of all functions $f: \Omega \to \mathbb{R}$ that are continuous on Ω, and their partial derivatives up to the order k exist and are all continuous on Ω. Of course, $C^0(\Omega)$ will simply be denoted by $C(\Omega)$. In addition, we shall need the following spaces

$$C^\infty(\Omega) = \{\phi \in C(\Omega) \mid \phi \text{ has continuous partial derivatives of any order}\},$$
$$C_0^\infty(\Omega) = \{\phi \in C^\infty(\Omega) \mid \text{supp } \phi \text{ is a compact set included in } \Omega\},$$

where supp ϕ is the support of ϕ, i.e., the closure of the set $\{x \in \Omega \mid \phi(x) \neq 0\}$. When $C_0^\infty(\Omega)$ is endowed with the usual inductive limit topology, then it is denoted by $\mathcal{D}(\Omega)$.

DEFINITION 1.1.1   *A linear continuous functional $u: \mathcal{D}(\Omega) \to \mathbb{R}$ is said to be a distribution on Ω. The linear space of distributions on Ω is denoted by $\mathcal{D}'(\Omega)$.*

Actually, $\mathcal{D}'(\Omega)$ is nothing else but the dual of $\mathcal{D}(\Omega)$. Notice that if $u \in L^1_{loc}(\Omega)$ (i.e., u is Lebesgue integrable on every compact subset of Ω), then the functional defined by

$$\mathcal{D}(\Omega) \ni \phi \mapsto \int_\Omega u(x)\phi(x)\, dx$$

is a distribution on Ω. Such a distribution will always be identified with the corresponding function u and so it will be denoted by u.

Now, recall that the partial derivative of a distribution $u \in D'(\Omega)$ with respect to x_j is defined by

$$\frac{\partial u}{\partial x_j}(\phi) = -u\Big(\frac{\partial \phi}{\partial x_j}\Big) \text{ for all } \phi \in D(\Omega),$$

and the higher order partial derivatives of u are defined iteratively, i.e.,

$$D^\alpha u(\phi) = (-1)^{|\alpha|} u(D^\alpha \phi) \text{ for all } \phi \in \mathcal{D}(\Omega),$$

where $\alpha := (\alpha_1, \alpha_2, \ldots, \alpha_N) \in \mathbb{N}^N$ is a so-called multiindex and $|\alpha| = \alpha_1 + \alpha_2 + \ldots + \alpha_N$. If $\alpha = (0, 0, \ldots, 0)$, then, by convention, $D^\alpha u = u$.

DEFINITION 1.1.2 *Let $1 \le p \le \infty$ and let $k \in \mathbb{N}^*$ be fixed. Then, the set*

$$W^{k,p}(\Omega) = \{u \colon \Omega \to \mathbb{R} \mid D^\alpha u \in L^p(\Omega) \text{ for all } \alpha \in \mathbb{N}^N \text{ with } |\alpha| \le k\}$$

(where $D^\alpha u$ are the derivatives of u in the sense of distributions) is said to be a Sobolev space of order k.

Recall that, for each $1 \le p < \infty$ and $k \in \mathbb{N}^*$, $W^{k,p}(\Omega)$ is a real Banach space with the norm

$$\|u\|_{W^{k,p}(\Omega)} = \left(\sum_{|\alpha| \le k} \|D^\alpha u\|^p_{L^p(\Omega)} \right)^{\frac{1}{p}}.$$

$W^{k,\infty}(\Omega)$ is also a real Banach space with the norm

$$\|u\|_{W^{k,\infty}(\Omega)} = \max_{|\alpha| \le k} \|D^\alpha u\|_{L^\infty(\Omega)}.$$

The completion of $\mathcal{D}(\Omega)$ with respect to the norm of $W^{k,p}(\Omega)$ is denoted by $W_0^{k,p}(\Omega)$. In general, $W_0^{k,p}(\Omega)$ is strictly included in $W^{k,p}(\Omega)$. In the case $p = 2$ we have the notation

$$H^k(\Omega) := W^{k,2}(\Omega), \ H_0^k(\Omega) := W_0^{k,2}(\Omega).$$

These are both Hilbert spaces with respect to the scalar product

$$(u, v)_k := \sum_{|\alpha| \le k} \int_\Omega D^\alpha u(x) D^\alpha v(x) \, dx.$$

The dual of $H_0^k(\Omega)$ is denoted by $H^{-k}(\Omega)$. If Ω is an open bounded subset of $\mathbb{R}^N$, with a sufficiently smooth boundary $\partial\Omega$, then

$$H_0^1(\Omega) = \{u \in H^1(\Omega) \mid \text{ the trace of } u \text{ on } \partial\Omega \text{ vanishes }\}.$$

If, in particular, Ω is an interval of real numbers, say $\Omega = (a, b)$ with $a < b$, then we shall write $C_0^\infty(a, b)$, $W^{k,p}(a, b)$, $H^k(a, b)$, and $W_0^{k,p}(a, b)$ instead of $C_0^\infty\big((a, b)\big)$, $W^{k,p}\big((a, b)\big)$, $H^k((a, b))$, and $W_0^{k,p}((a, b))$, respectively. If a, b are finite numbers, then every element of $W^{k,p}(a, b)$, $k \in \mathbb{N}^*$, $1 \le p \le \infty$, can be identified with an absolutely continuous function $f \colon [a, b] \to \mathbb{R}$ such that $d^j f / dt^j$, $1 \leq j \leq k-1$, exist and are all absolutely continuous on $[a, b]$, and $d^k f / dt^k$ (that obviously exists a.e. in (a, b)) belongs to $L^p(a, b)$ (more precisely, the equivalence class of $d^k f / dt^k$, with respect to the equality a.e. on (a, b) belongs to $L^p(a, b)$). Moreover, every element of $W_0^{k,p}(a, b)$ can be identified with such a function f, which also satisfies the conditions

$$\frac{d^j f}{dt^j}(a) = \frac{d^j f}{dt^j}(b) = 0, \ 0 \leq j \leq k - 1.$$

Recall that if $-\infty < a < b < \infty$ and $k \in \mathbb{N}^*$, then $W^{k,1}(a, b)$ is continuously embedded into $C^{k-1}[a, b]$ (in particular, $W^{1,1}(a, b)$ is continuously embedded in $C[a, b]$). Finally, we set for $k \in \mathbb{N}^*$ and $1 \leq p \leq \infty$,

$$W_{loc}^{k,p}(\mathbb{R}) = \{u \colon \mathbb{R} \to \mathbb{R} \mid D^\alpha u \in L_{loc}^p(\mathbb{R}) \text{ for all } \alpha \in \mathbb{N} \text{ with } \alpha \leq k\}.$$

Vectorial distributions. The spaces $W^{k,p}(a, b; X)$

Let Ω be an open interval (a, b) with $-\infty \leq a < b \leq \infty$ and denote by $\mathcal{D}'(a, b; X)$ the space of all continuous linear operators from $\mathcal{D}(a, b) := \mathcal{D}((a, b))$ to X. The elements of $\mathcal{D}'(a, b; X)$ are called *vectorial distributions* on (a, b) with values in X. If $u \colon (a, b) \to X$ is integrable (in the sense of Bochner) over every bounded interval $I \subset (a, b)$ (i.e., equivalently, $t \mapsto \|u(t)\|_X$ belongs to $L^1(I)$, for every bounded subinterval I), then u defines a vectorial distribution, again denoted by u, as follows,

$$u(\phi) := \int_a^b \phi(t) u(t) \, dt \text{ for all } \phi \in \mathcal{D}(a, b).$$

The distributional derivative of order $j \in \mathbb{N}$ of $u \in \mathcal{D}'(a, b; X)$ is the distribution defined by

$$u^{(j)}(\phi) := (-1)^j u\left(\frac{d^j \phi}{dt^j}\right), \text{ for all } \phi \in \mathcal{D}(a, b),$$

where $d^j \phi / dt^j$ is the j-th ordinary derivative of ϕ. By convention, $u^{(0)} = u$.

Now, for $k \in \mathbb{N}^*$ and $1 \leq p \leq \infty$, we set

$$W^{k,p}(a, b; X) = \{u \in L^p(a, b; X) \mid u^{(j)} \in L^p(a, b; X), \ j = 1, 2, \ldots, k\},$$

where $u^{(j)}$ is the j-th distributional derivative of u. For each $k \in \mathbb{N}^*$ and $1 \leq p < \infty$, $W^{k,p}(a, b; X)$ is a Banach space with the norm

$$\|u\|_{W^{k,p}(a,b;X)} = \left(\sum_{j=0}^k \|u^{(j)}\|_{L^p(a,b;X)}^p\right)^{\frac{1}{p}}.$$

Also, for each $k \in \mathbb{N}^*$, $W^{k,\infty}(a, b; X)$ is a Banach space with the norm

$$\|u\|_{W^{k,\infty}(a,b;X)} = \max_{0 \le j \le k} \|u^{(j)}\|_{L^\infty(a,b;X)}.$$

As in the scalar case, for $p = 2$, we may use the notation $H^k(a, b; X)$ instead of $W^{k,2}(a, b; X)$. Recall that, if X is a real Hilbert space with its scalar product denoted by $(\cdot, \cdot)_X$, then for each $k \in \mathbb{N}^*$, $H^k(a, b; X)$ is also a Hilbert space with respect to the scalar product

$$(u, v)_{H^k(a,b;X)} = \sum_{j=0}^{k} \int_a^b \left(u^{(j)}(t), v^{(j)}(t)\right)_X dt.$$

As usual, for $k \in \mathbb{N}^*$ and $1 \le p \le \infty$, we set

$$W^{k,p}_{loc}(a, b; X) = \left\{ u \in \mathcal{D}'(a, b; X) \mid u \in W^{k,p}(t_1, t_2; X), \right.$$
$$\left. \text{for every } t_1, t_2 \in (a, b) \text{ with } t_1 < t_2 \right\}.$$

In what follows, we shall assume that $-\infty < a < b < \infty$. For $k \in \mathbb{N}^*$ and $1 \le p \le \infty$, denote by $A^{k,p}([a, b]; X)$ the space of all absolutely continuous functions $f: [a, b] \to X$ for which $d^j f / dt^j$, $1 \le j \le k - 1$, exist, are all absolutely continuous, and (the class of) $d^k f / dt^k \in L^p(a, b; X)$.

If X is a reflexive Banach space and $v: [a, b] \to X$ is absolutely continuous, then v is differentiable a.e. on (a, b), $dv/dt \in L^1(a, b; X)$, and

$$v(t) = v(a) + \int_a^t \frac{dv}{ds}(s)\, ds, \ a \le t \le b.$$

Therefore, if X is reflexive, then $A^{1,1}(a, b; X)$ coincides with the space of all absolutely continuous functions $v: [a, b] \to X$, i.e.,

$$A^{1,1}([a, b]; X) = AC([a, b]; X).$$

We also recall the following result.

THEOREM 1.1.1
Let $1 \le p \le \infty$ and $k \in \mathbb{N}^*$ be fixed and let $u \in L^p(a, b; X)$ with $-\infty < a < b < \infty$. Then $u \in W^{k,p}(a, b; X)$ if and only if u has a representative in $A^{k,p}([a, b]; X)$.

So, $W^{k,p}(a, b; X)$ will be identified with $A^{k,p}([a, b]; X)$. If X is reflexive, then $W^{1,1}(a, b; X)$ can be identified with $AC([a, b]; X)$, while $W^{1,\infty}(a, b; X)$ can be identified with $Lip([a, b]; X)$ (the space of all Lipschitz continuous functions $v: [a, b] \to X$).

THEOREM 1.1.2
Let X be a real reflexive Banach space and let $u \in L^p(a, b; X)$ with $-\infty < a < b < \infty$ and $1 < p < \infty$. Then, the following two conditions are equivalent:

(i) $u \in W^{1,p}(a, b; X)$;

(ii) There exists a constant $C > 0$ such that

$$\int_a^{b-\delta} \|u(t+\delta) - u(t)\|_X^p \, dt \leq C\delta^p \text{ for all } \delta \in (0, b-a].$$

Moreover, if $p = 1$ then (i) implies (ii) (actually, (ii) is true for $p = 1$ if one representative of $u \in L^1(a, b, X)$ is of bounded variation on $[a, b]$, where X is a general Banach space, not necessarily reflexive).

Now, let V and H be two real Hilbert spaces such that V is densely and continuously embedded in H. If H is identified with its own dual, then we have $V \subset H \subset V^*$, algebraically and topologically, where V^* denotes the dual of V. Denote by $\langle \cdot, \cdot \rangle$ the dual pairing between V and V^*, i.e., $\langle v, v^* \rangle = v^*(v)$, $v \in V$, $v^* \in V^*$. For $v^* \in H^* = H$, $\langle v, v^* \rangle$ reduces to the scalar product in H of v and v^*.

Now, for some $-\infty < a < b < \infty$, we set

$$W(a, b) := \{u \in L^2(a, b; V) \mid u' \in L^2(0, T; V^*)\},$$

where u' is the distributional derivative of u. Obviously, every $u \in W(a, b)$ has a representative $u_1 \in A^{1,2}([a, b]; V^*)$ and so u is identified with u_1. Moreover, we have:

THEOREM 1.1.3
Every $u \in W(a, b)$ has a representative $u_1 \in C([a, b]; H)$ and so u can be identified with such a function. Furthermore, if $u, \tilde{u} \in W(a, b)$, then the function $t \mapsto (u(t), \tilde{u}(t))_H$ is absolutely continuous on $[a, b]$ and

$$\frac{d}{dt}\left(u(t), \tilde{u}(t)\right)_H = \langle u(t), \tilde{u}'(t) \rangle + \langle \tilde{u}(t), u'(t) \rangle \text{ for a.a. } t \in (a, b).$$

Hence, in particular,

$$\frac{d}{dt}\|u(t)\|_H^2 = 2\langle u(t), u'(t) \rangle \text{ for a.a. } t \in (a, b).$$

Compactness results

We recall here a couple of useful compactness results. We begin with a general version of Ascoli's theorem (see [Dieudo, p. 143]).

THEOREM 1.1.4

Let $\mathcal{F}$ be a family of mappings from a separable topological space X to a compact metric space M. If $\mathcal{F}$ is equi-continuous, then every sequence of $\mathcal{F}$ has a subsequence, which converges uniformly on compact sets of X.

In particular, let $S \subset \mathbb{R}$ be a nonvoid bounded set. If $\mathcal{F}$ is equi-continuous, then there exist a subsequence (f_n) of elements of $\mathcal{F}$ and a continuous $f\colon S \to M$ such that $f_n(t) \to f(t)$ uniformly on S, as $n \to \infty$.

The next important compactess result is proved in [Lions, p. 58].

THEOREM 1.1.5

Let $T > 0$, $p_0, p_1 \in (1, \infty)$, the sets B_0, B, and B_1 be real Banach spaces, and $\Lambda_0\colon B_0 \to B$, $\Lambda_1\colon B \to B_1$ be continuous linear injections such that

(i) B_0 and B_1 are reflexive;

(ii) $\Lambda_0 B_0$ is dense in B and $\Lambda_1 B$ is dense in B_1;

(ii) $\Lambda_0\colon B_0 \to B$ is compact.

Then, the set

$$\mathcal{W} = \{v \in L^{p_0}(0, T; B_0) \mid (\Lambda_1 \Lambda_0 v)' \in L^{p_1}(0, T; V)\}$$

is a real Banach space with respect to the norm

$$\|v\|_{\mathcal{W}} = \|v\|_{L^{p_0}(0,T;B_0)} + \|(\Lambda_1 \Lambda_0 v)'\|_{L^{p_1}(0,T;B_1)}$$

and the mapping $v \mapsto \Lambda_0 v$ embeds $\mathcal{W}$ into $L^{p_0}(0, T; B)$ algebraically and topologically, i.e., it is a continuous linear injection and $\Lambda_0 \mathcal{W}$ is dense in $L^{p_0}(0, T; B)$. Moreover, this embedding is compact.

Bibliographical note. For background material for Section 1.1, refer to [Adams], [Agmon], [Brézis1], [Brézis2], [Dieudo], [Lions], [LioMag], [Schwa], [Yosida].

1.2   Monotone operators, convex functions, and subdifferentials

Let X be a real Banach space with the dual X^*, the dual pairing $\langle \cdot, \cdot \rangle$, and the associated norms $\| \cdot \|_X$ and $\| \cdot \|_{X^*}$. By a multivalued operator $A \colon D(A) \subset X \to X^*$ we mean a mapping that assigns to each $x \in D(A)$ a set $Ax \subset X^*$. The graph of A is defined by

$$G(A) := \big\{ (x, y) \in X \times X^* \mid x \in D(A),\ y \in Ax \big\}.$$

Obviously, for every subset of $X \times X^*$, there exists a unique multivalued operator A such that $G(A)$ coincides with that subset. So, every multivalued operator A can be identified with $G(A)$ and we shall write $(x, y) \in A$ instead of $x \in D(A)$ and $y \in Ax$. We also write briefly $A \subset X \times X^*$ instead of $A \colon D(A) \subset X \to X^*$. The *range* of a multivalued operator $A \colon D(A) \subset X \to X^*$ is defined by

$$R(A) := \bigcup_{x \in D(A)} Ax.$$

If Ax is a singleton then we shall often identify Ax with its unique element. Define

$$A^{-1} := \{ (y, x) \mid (x, y) \in A \}.$$

Obviously, A^{-1} is a multivalued operator, here viewed as a subset of $X^* \times X$, with $D(A^{-1}) = R(A)$ and $R(A^{-1}) = D(A)$. If $A \colon D(A) \subset X \to X^*$, $B \colon D(B) \subset X \to X^*$ are multivalued operators, and $\lambda \in \mathbb{R}$, we define, as usual,

$$A + B = \{ (x, y + z) \mid (x, y) \in A \text{ and } (x, z) \in B \},$$
$$\lambda A = \{ (x, \lambda y) \mid (x, y) \in A \}.$$

Obviously, $D(A + B) = D(A) \cap D(B)$ and $D(\lambda A) = D(A)$. Recall that $A \colon D(A) \subset X \to X^*$ (possibly multivalued) is said to be *monotone* if

$$\langle x_1 - x_2, y_1 - y_2 \rangle \geq 0 \text{ for all } (x_1, y_1), (x_2, y_2) \in A. \tag{1.2.1}$$

A is called *strictly monotone* if it satisfies (1.2.1) with ">" instead of "$\geq$" for $x_1 \neq x_2$. If the following stronger inequality holds

$$\langle x_1 - x_2, y_1 - y_2 \rangle \geq a \| x_1 - x_2 \|_X^2 \text{ for all } (x_1, y_1), (x_2, y_2) \in A, \tag{1.2.2}$$

for some fixed $a > 0$, then A is called *strongly monotone*. Actually, this means that $A - aF$ is monotone, where $F \subset X \times X^*$ is the duality operator given by

$$Fx = \big\{ x^* \in X^* \mid \langle x, x^* \rangle = \| x \|_X^2 = \| x^* \|_{X^*}^2 \big\}. \tag{1.2.3}$$

An operator A is said to be ω-*monotone*, $\omega > 0$, if $A + \omega F$ is monotone. For example, if A is single-valued and Lipschitz continuous, with the Lipschitz constant ω, then A is ω-monotone.

If A is single-valued then (1.2.1) can be written as

$$\langle x_1 - x_2, Ax_1 - Ax_2 \rangle \geq 0 \text{ for all } x_1, x_2 \in D(A). \qquad (1.2.4)$$

For the sake of simplicity, we shall sometimes use (1.2.4) instead of (1.2.1) even for multivalued A. In the case $X = \mathbb{R}^N$ we shall often use the word "mapping" instead of "operator".

Now, we recall the following important concept: a monotone operator $A \colon D(A) \subset X \to X^*$ is called *maximal monotone*, if A has no proper monotone extension (in other words, A, viewed as a subset of $X \times X^*$, cannot be extended to any $A' \subset X \times X^*$, $A' \neq A$, such that the corresponding multivalued operator A' is monotone).

In what follows we restrict ourselves to the case when X is a real Hilbert space and redenote it by H in order to remind the reader of the fact that we shall work in a Hilbert framework. We identify H with its dual. Then the duality mapping is the identity mapping I in H.

THEOREM 1.2.1

(R.T. Rockafellar). If $A \colon D(A) \subset H \to H$ is monotone, then A is locally bounded at every point $x_0 \in \text{Int}\, D(A)$ (i.e., there exists a ball $B(x_0, r) \subset D(A)$ such that the set $\{y \in Ax \mid x \in B(x_0, r)\}$ is bounded).

A characterization of the concept of maximal monotone operator is given by the following classical result:

THEOREM 1.2.2

(G. Minty). Let $A \colon D(A) \subset H \to H$ be a monotone operator. It is maximal monotone if and only if $R(I + A) = H$. In this case $R(I + \lambda A) = H$ for all $\lambda > 0$.

THEOREM 1.2.3

Let $A \colon D(A) \subset H \to H$ be a maximal monotone operator. Then:

(a) A^{-1} is maximal monotone;

(b) For every $x \in D(A)$, the set Ax is convex and closed;

(c) A is demiclosed, i.e., if (x_n) converges strongly toward x, (y_n) converges weakly toward y, and $(x_n, y_n) \in A$ for all $n = 1, 2, \ldots$, then $(x, y) \in A$ (hence, in particular, A is closed).

(d) If (x_n) and (y_n) converge weakly toward x and y, respectively, (x_n, y_n)
$\in A$ for all $n = 1, 2, \ldots$, and

$$\liminf_{n \to \infty} (x_n, y_n)_H \leq (x, y)_H,$$

then $(x, y) \in A$.

The proof of Theorem 1.2.3 relies on elementary arguments.

Now, for A maximal monotone and $\lambda > 0$, we define the operators

$$J_\lambda = (I + \lambda A)^{-1}, \quad A_\lambda = \frac{1}{\lambda}(I - J_\lambda),$$

which are called the *resolvent* and the *Yosida approximation* of A, respectively. It is easily seen that (see Theorem 1.2.2) $D(J_\lambda) = D(A_\lambda) = H$ and that J_λ, A_λ are single-valued, for every $\lambda > 0$. Other well known properties of J_λ and A_λ are collected in the next result.

THEOREM 1.2.4

If $A: D(A) \subset H \to H$ is a maximal monotone operator, then for every $\lambda > 0$, we have

(P1) J_λ is nonexpansive (i.e., Lipschitz continuous with the Lipschitz constant $= 1$);

(P2) $A_\lambda x \in A J_\lambda x$ for all $x \in H$;

(P3) A_λ is monotone and Lipschitz continuous, with the Lipschitz constant $1/\lambda$;

(P4) $\|A_\lambda x\|_H \leq \|A^0 x\|_H$ for all $x \in D(A)$;

(P5) $\lim_{\lambda \to 0+} \|A_\lambda x - A^0 x\|_H = 0$ for all $x \in D(A)$;

(P6) $\overline{D(A)}$ is a convex set (hence, $\overline{R(A)} = \overline{D(A^{-1})}$ is convex, too);

(P7) $\lim_{\lambda \to 0+} \|J_\lambda x - \mathrm{Pr}_{\overline{D(A)}} x\|_H = 0$ for all $x \in H$,

where $\mathrm{Pr}_{\overline{D(A)}} x$ denotes the projection of x on $\overline{D(A)}$.

We have denoted by A^0 the so-called *minimal section* of A, which is defined by

$$A^0 x = \mathrm{Pr}_{Ax}\, 0 \quad \text{for all} \quad x \in D(A),$$

i.e., $A^0 x$ is the element of minimal norm of Ax.

PROPOSITION 1.2.1

If $A: D(A) \subset H \to H$ is a maximal monotone operator and, in addition, A is strictly monotone, then A_λ is strictly monotone too, for each $\lambda > 0$.

PROOF   Fix $\lambda > 0$. Let $x, y \in H$ be such that

$$(A_\lambda x - A_\lambda y, x - y)_H = 0,$$

By the definition of the Yosida approximation, we have

$$\lambda \|A_\lambda x - A_\lambda y\|_H^2 + (A_\lambda x - A_\lambda y, J_\lambda x - J_\lambda y)_H = 0. \qquad (1.2.5)$$

Since A is strictly monotone, it follows from (1.2.5) and Theorem 1.2.4, (P2), that $J_\lambda x = J_\lambda y$ and $A_\lambda x = A_\lambda y$. Therefore, $x = J_\lambda x + \lambda A_\lambda x = y$.
∎

PROPOSITION 1.2.2

Let $A: D(A) = \mathbb{R} \to \mathbb{R}$ be a single-valued maximal monotone mapping. Then, the following implications are valid for each $\lambda > 0$:

(1) If $\lim_{x \to \infty} Ax = \infty$, then $\lim_{x \to \infty} A_\lambda x = \infty$;

(2) If $\lim_{x \to -\infty} Ax = -\infty$, then $\lim_{x \to -\infty} A_\lambda x = -\infty$.

PROOF   We shall prove only the first implication, because the second one can be derived similarly. Let $\lambda > 0$ be fixed and $\lim_{x \to \infty} Ax = \infty$. Since A_λ is monotone, the counter assumption is that $\lim_{x \to \infty} A_\lambda x < \infty$. Therefore, there exists a constant C such that

$$x - J_\lambda x \leq C \quad \text{for all } x > 0.$$

So, $\lim_{x \to \infty} J_\lambda x = \infty$ and this implies that

$$\lim_{x \to \infty} A_\lambda x = \lim_{x \to \infty} A J_\lambda x = \lim_{x \to \infty} Ax = \infty,$$

which contradicts our counter assumption.   ∎

Now, recall that a single-valued operator $A: D(A) = H \to H$ is said to be *hemicontinuous* if for every $x, y \in H$

$$\lim_{t \to 0} A(x + ty) = Ax, \quad \text{weakly in } H.$$

THEOREM 1.2.5

(G. Minty). If $A: D(A) = H \to H$ is single-valued, monotone, and hemicontinuous, then A is maximal monotone.

THEOREM 1.2.6
Let $A\colon D(A) \subset H \to H$ be maximal monotone and coercive with respect to some $x_0 \in H$, i.e.,

$$\frac{(x - x_0, y)_H}{\|x\|_H} \to \infty, \quad as \ \ \|x\|_H \to \infty \ and \ (x, y) \in A.$$

Then A is surjective, i.e., $R(A) = H$.

Obviously, if A is strongly monotone, then A is coercive with respect to every $x_0 \in D(A)$.

THEOREM 1.2.7
(H. Attouch). If $A\colon D(A) \subset H \to H$ and $B\colon D(B) \subset H \to H$ are two maximal monotone operators and $0 \in \text{Int}\,\big(D(A) - D(B)\big)$, then $A + B$ is maximal monotone, too.

We have denoted above by $D(A) - D(B)$ the algebraic difference of the two sets, i.e., $D(A) - D(B) = \{x - y \mid x \in D(A), \ y \in D(B)\}$.

REMARK 1.2.1　The last result is a generalization of the well known perturbation theorem by R.T. Rockafellar, which says that: if A, B are both maximal monotone and

$$\big(\text{Int}\, D(A)\big) \cap D(B) \neq \emptyset, \tag{1.2.6}$$

then $A + B$ is maximal monotone, too.　█

Indeed, (1.2.6) can be expressed as

$$0 \in D(B) - \text{Int}\, D(A)$$

and this implies $0 \in \text{Int}\,\big(D(A) - D(B)\big)$. Thus (1.2.6) is stronger than Attouch's condition.

A very important class of monotone operators is that of subdifferentials. Before introducing the concept of subdifferential, let us recall that a function $\psi\colon H \to (-\infty, \infty]$ is said to be *proper* if $\psi \not\equiv +\infty$ (i.e., ψ takes at least one finite value). A function $\psi\colon H \to (-\infty, \infty]$ is called *convex* if

$$\psi\big(tx + (1 - t)y\big) \leq t\psi(x) + (1 - t)\psi(y) \tag{1.2.7}$$
$$\text{for all } t \in (0, 1) \text{ and } x, y \in H.$$

In (1.2.7) we use the classical conventions concerning the computations involving ∞. Clearly, if $\psi\colon H \to (-\infty, \infty]$ is a convex function, then its *effective domain*

$$D(\psi) = \{x \in H \mid \psi(x) < \infty\}$$

is a convex set. We also recall that a function $\psi\colon H \to (-\infty, \infty]$ is said to be *lower semicontinuous* at $x_0 \in H$ if

$$\psi(x_0) \leq \liminf_{x \to x_0} \psi(x).$$

Clearly, a convex function $\psi\colon H \to (-\infty, \infty]$ is lower semicontinuous at $x_0 \in H$ if and only if

$$\psi(x_0) = \liminf_{x \to x_0} \psi(x). \tag{1.2.8}$$

It is easily seen that ψ is lower semicontinuous on H (i.e., lower semicontinuous at every $x_0 \in H$) if and only if the level set $\{x \in H \mid \psi(x) \leq \lambda\}$ is closed, for each $\lambda \in \mathbb{R}$. On the other hand, we recall that every convex set is closed if and only if it is weakly closed (cf. Mazur's theorem). Therefore, a convex function ψ is lower semicontinuous on H if and only if it is weakly lower semicontinuous on H (i.e., (1.2.8) holds with $x \to x_0$ weakly in H, for every $x_0 \in H$).

THEOREM 1.2.8

If $\psi\colon H \to (-\infty, \infty]$ is proper, convex, and lower semicontinuous on H, then ψ is bounded from below by an affine function, i.e., there exists a point $(y_0, a) \in H \times \mathbb{R}$ such that

$$\psi(x) \geq (y_0, x)_H + a \quad \text{for all } x \in H. \tag{1.2.9}$$

THEOREM 1.2.9

If $\psi\colon H \to (-\infty, \infty]$ is proper, convex, lower semicontinuous on H, and

$$\lim_{\|x\|_H \to \infty} \psi(x) = \infty, \tag{1.2.10}$$

then there exists $x^* \in D(\psi)$ such that

$$\psi(x^*) = \inf\{\psi(x) \mid x \in H\}.$$

THEOREM 1.2.10

If $\psi\colon H \to (-\infty, \infty]$ is proper, convex, and lower semicontinuous on H such that the interior of $D(\psi)$ is nonempty, then ψ is continuous on the interior of $D(\psi)$.

THEOREM 1.2.11

Let $\psi\colon H \to (-\infty, \infty]$ be proper, convex, and lower semicontinuous on H. Then its conjugate ψ^* is also proper, convex, and lower semicontinuous on

H, where ψ^* is defined by

$$\psi^*(x^*) = \sup\left\{(x, x^*)_H - \psi(x) \mid x \in H\right\}.$$

Let $\psi : H \to (-\infty, \infty]$ be a proper convex function. Its *subdifferential at* $x \in D(\psi)$ is defined by

$$\partial\psi\, x = \left\{y \in H \mid \psi(x) + (y, v - x)_H \le \psi(v) \ \text{for all } v \in H\right\}.$$

The operator $\partial\psi \subset H \times H$ is called the *subdifferential* of ψ. Clearly, its domain is included in $D(\psi)$, i.e., $D(\partial\psi) \subset D(\psi)$.

THEOREM 1.2.12

If $\psi\colon H \to (-\infty, \infty]$ is a proper convex lower semicontinuous function, then $\partial\psi$ is a maximal monotone operator and, furthermore, $\overline{D(\partial\psi)} = \overline{D(\psi)}$, $\operatorname{Int} D(\partial\psi) = \operatorname{Int} D(\psi)$, and $(\partial\psi)^{-1} = \partial\psi^*$, where ψ^* is the conjugate of ψ.

Let us recall that the directional derivative of the function $F\colon H \to (-\infty, \infty]$ at point $u \in H$ to the direction $v \in H$ is

$$F'(u; v) = \lim_{\epsilon \to 0^+} \frac{F(u + \epsilon v) - F(u)}{\epsilon}$$

if this limit exists. If $F'(u; \cdot)$ is a linear mapping from H into itself, then F is said to be *Gâteaux-differentiable at* $u \in H$ and the unique point $F'(u) \in H$, given by the Riesz theorem and

$$(F'(u), v)_H = F'(u; v) \ \text{for all } v \in H,$$

is called the *Gâteaux differential of F at u*. Now, we can present a theorem, which relates the notions of subdifferential and Gâteaux differential (see [EkeTem, p. 23]).

THEOREM 1.2.13

Let $\psi\colon H \to (-\infty, \infty]$ be a convex function. If ψ is Gâteaux-differentiable at a point $u \in H$, then it is subdifferentiable at $u \in H$ and $\partial\psi u = \{\psi'(u)\}$. Conversely, if ψ is finite and continuous and has only one subgradient at a point $u \in H$, then ψ is Gâteaux-differentiable at u and $\partial\psi u = \{\psi'(u)\}$.

REMARK 1.2.2   If $\psi\colon H \to (-\infty, \infty]$ is proper and convex, then the operator $A = \partial\psi$ is *cyclically monotone*, i.e., for every $n \in \mathbb{N}^*$ we have

$$(x_0 - x_1, x_0^*)_H + (x_1 - x_2, x_1^*)_H + \ldots + (x_{n-1} - x_n, x_{n-1}^*)_H +$$
$$+ (x_n - x_0, x_n^*)_H \ge 0$$

for all $(x_i, x_i^*) \in A$, $i = 0, 1, \ldots, n$. ∎

An operator $A: D(A) \subset H \to H$ is called *maximal cyclically monotone* if A cannot be properly extended to another cyclically monotone operator. Obviously, if $\psi: H \to (-\infty, \infty]$ is a proper convex lower semicontinuous function, then $A = \partial\psi$ is maximal cyclically monotone. The converse implication is also true:

THEOREM 1.2.14

If $A: D(A) \subset H \to H$ is a maximally cyclically monotone operator, then there exists a proper convex lower semicontinuous function $\psi: H \to (-\infty, \infty]$, uniquely determined up to an additive constant, such that $A = \partial\psi$.

In the special case $H = \mathbb{R}$, we have:

THEOREM 1.2.15

For every maximal monotone mapping $\beta: D(\beta) \subset \mathbb{R} \to \mathbb{R}$, there exists a proper convex lower semicontinuous function $j: \mathbb{R} \to (-\infty, \infty]$, uniquely determined up to an additive constant, such that $\beta = \partial j$. More precisely, such a function j is given by

$$j(x) = \begin{cases} \int_{x_0}^x \beta^0(s)\, ds & \text{if } x \in \overline{D(\beta)}, \\ +\infty & \text{otherwise,} \end{cases}$$

where x_0 is a fixed point in $D(\beta)$ and β^0 denotes the minimal section of β.

We continue with the following result that is probably known:

PROPOSITION 1.2.3

Let $j: \mathbb{R} \to (-\infty, \infty]$ be a proper convex function such that $D(j)$ is not a singleton. Then, j is strictly convex (i.e., j satisfies (1.2.7) with "$<$" instead of "$\leq$" for $x \neq y$) if and only if $\beta = \partial j$ is a strictly monotone mapping.

PROOF　Suppose that j is strictly convex. Let $\xi_1, \xi_2 \in D(\beta)$ be such that $\beta(\xi_1) \cap \beta(\xi_2) \neq \emptyset$. We have to show that $\xi_1 = \xi_2$. Assume, by contradiction, that $\xi_1 \neq \xi_2$, say $\xi_1 < \xi_2$. Let $w \in \beta(\xi_1) \cap \beta(\xi_2)$ and $t \in (0, 1)$. We set $\xi_t = t\xi_1 + (1 - t)\xi_2$. Then

$$j(\xi_t) < tj(\xi_1) + (1 - t)j(\xi_2). \tag{1.2.11}$$

On the other hand,

$$tj(\xi_1) + (1-t)j(\xi_2) - j(\xi_t) = t\big(j(\xi_1) - j(\xi_t)\big) +$$
$$+(1-t)\big(j(\xi_2) - j(\xi_t)\big) \le tw(\xi_1 - \xi_t) + (1-t)w(\xi_2 - \xi_t) = 0,$$

and therefore

$$tj(\xi_1) + (1-t)j(\xi_2) \le j(\xi_t). \tag{1.2.12}$$

But (1.2.12) contradicts (1.2.11), hence $\xi_1 = \xi_2$.

Now, in order to prove the converse implication, suppose that β is strictly monotone, but j is not strictly convex. So there exist $\xi_1, \xi_2 \in D(j)$, $\xi_1 < \xi_2$, and $t \in (0,1)$ such that

$$j\big(t\xi_1 + (1-t)\xi_2\big) = tj(\xi_1) + (1-t)j(\xi_2). \tag{1.2.13}$$

Actually, (1.2.13) implies that j is an affine function on the interval $[\xi_1, \xi_2]$, because j is convex. More precisely, (1.2.13) holds for all $t \in [0,1]$, i.e.,

$$j(\xi) = \big(j(\xi_1) - j(\xi_2)\big)\frac{\xi_2 - t}{\xi_2 - \xi_1} + j(\xi_2) \ \text{ for all } \xi \in [\xi_1, \xi_2].$$

Therefore,

$$\beta(\xi) = \frac{j(\xi_2) - j(\xi_1)}{\xi_2 - \xi_1} \ \text{ for all } \xi \in (\xi_1, \xi_2),$$

which contradicts the strict monotonicity of β. $\blacksquare$

DEFINITION 1.2.1   *Let $\lambda > 0$ and $\psi\colon H \to (-\infty, \infty]$ be convex. The function $\psi_\lambda\colon H \to \mathbb{R}$,*

$$\psi_\lambda(x) = \inf\left\{\frac{1}{2\lambda}\|x - \xi\|_H^2 + \psi(x) \ \Big| \ \xi \in H\right\},$$

is called the Moreau-Yosida regularization of ψ.

THEOREM 1.2.16

(H. Brézis & J.J. Moreau). Let $\psi\colon H \to (-\infty, \infty]$ be a proper convex lower semicontinuous function, whose subdifferential is denoted by A. Then:

(Q1) The Moreau-Yosida regularization $\psi_\lambda\colon H \to \mathbb{R}$ is convex, Fréchet differentiable on H, and $\partial\psi_\lambda = A_\lambda$ for all $\lambda > 0$;

(Q2) $\psi_\lambda(x) = \frac{1}{2\lambda}\|x - J_\lambda x\|_H^2 + \psi(J_\lambda x)$ for all $x \in H$ and $\lambda > 0$, where $J_\lambda = (I + \lambda A)^{-1}$;

(Q3) $\psi(J_\lambda x) \le \psi_\lambda(x) \le \psi(x)$ for all $x \in H$ and $\lambda > 0$;

(Q4) $\lim_{\lambda \to 0+} \psi_\lambda(x) = \psi(x)$ for all $x \in H$.

Up to now we have presented the convex functions and their subdifferentials on a real Hilbert space. However, the theory has been extended to locally convex separated vector spaces (see [EkeTem, Ch. 1]). We shall need the following chain rule in Chapter 10.

THEOREM 1.2.17
Let X and Y be real locally convex spaces with duals X^* and Y^*, respectively. Let $\Lambda\colon X \to Y$ be a linear continuous mapping, whose adjoint is $\Lambda^*\colon Y^* \to X^*$, and let $\Phi\colon Y \to (-\infty, \infty]$ be a proper convex lower semicontinuous function. Then the composed function

$$\phi \circ \Lambda\colon X \to (-\infty, \infty], \ (\phi \circ \Lambda)(x) = \phi(\Lambda x),$$

is a proper convex lower-semicontinuous function. If, in addition, there exists a $p \in Y$, where ϕ is finite and continuous, then

$$\partial(\phi \circ \Lambda)(x) = \Lambda^* \partial\phi(\Lambda x) \ \text{ for all } x \in X.$$

The following chain rule for convex functions will also be useful. For the case where $\phi_t = 0$, see [Brézis1, p. 73]. We denote by ϕ_t the partial derivative of ϕ with respect to t.

THEOREM 1.2.18
([Hokk1, p. 119]). Let $T > 0$ be fixed. Let $\phi\colon [0,T] \times H \to (-\infty, \infty]$, $g\colon \mathbb{R} \to \mathbb{R}$, $u \in H^1(0,T;H)$, and $v \in L^2(0,T;H)$ satisfy:

  (i) $\phi(t, \cdot)$ is a proper, convex, and lower semicontinuous function for all $t \in [0,T]$;

  (ii) $\big(u(t), v(t)\big) \in \partial\phi(t, \cdot)$ for a.a. $t \in (0,T)$;

  (iii) the functions $\phi(\cdot, z)\colon [0,T] \to \mathbb{R}$ are differentiable for all $z \in R(u)$;

  (iv) $\phi(\cdot, u), g(u) \in L^1(0,T)$;

  (v) $|\phi_t(t, z)| \leq g(z)$ for all $z \in R(u)$.
Then $\phi(\cdot, x) \in W^{1,1}(0,T)$ and

$$\frac{d}{dt}\phi\big(t, u(t)\big) = \phi_t\big(t, u(t)\big) + \big(v(t), u'(t)\big)_H \ \text{ for a.a. } t \in (0,T).$$

We complete this section with a discussion on the convex integrands and integral functions. For further details, see [BarPr, pp. 116-120]. Let $\Omega \subset \mathbb{R}^n$, $n \in \mathbb{N}^*$, be an open set, $p \in [1, \infty)$, and let p' be its conjugate exponent, i.e., $(p')^{-1} + p^{-1} = 1$. A function $g\colon \Omega \times \mathbb{R}^m \to (-\infty, \infty]$, $m \in \mathbb{N}^*$, is called a *normal convex integrand on $\Omega \times \mathbb{R}^m$* if the following two conditions are satisfied:

(i) For a.a. $x \in \Omega$, $g(x, \cdot): \mathbb{R}^m \to (-\infty, \infty]$ is a proper convex lower semicontinuous function;

(ii) The function g is measurable with respect to the σ-field generated by products of Lebesgue sets in Ω and Borel sets in $\mathbb{R}^m$.

Clearly, if g is a normal convex integrand and $y: \Omega \to \mathbb{R}^m$ is measurable, then $x \mapsto g(x, y(x))$ is Lebesgue measurable. Condition (ii) is a generalization of the classical Caratheodory condition (i.e., $g(\cdot, y)$ measurable and $g(x, \cdot)$ continuous for a.a x, y).

We need a couple of conditions more:

(iii) There exist functions $\alpha \in L^{p'}(\Omega; \mathbb{R}^m)$ and $\beta \in L^1(\Omega; \mathbb{R}^m)$ such that

$$g(x, z) \geq \big(z, \alpha(x)\big)_{\mathbb{R}^m} + \beta(x) \quad \text{for all } (x, z) \in \Omega \times \mathbb{R}^m;$$

(iv) There exists at least one function $y_0 \in L^p(\Omega; \mathbb{R}^m)$ such that $g(\cdot, y_0) \in L^1(\Omega)$.

The conditions (iii) and (iv) are satisfied if $g(x, \cdot)$ is independent of x.

THEOREM 1.2.19
Assume (i)-(iv) and define $G: L^p(\Omega; \mathbb{R}^m) \to (-\infty, \infty]$,

$$G(y) = \begin{cases} \int_\Omega g(x, y(x))\, dx & \text{if } g(\cdot, y) \in L^1(\Omega), \\ +\infty & \text{otherwise.} \end{cases}$$

Then G *is a proper convex lower semicontinuous function and at every* $y \in L^p(\Omega; \mathbb{R}^m)$

$$\partial G y = \big\{ w \in L^{p'}(\Omega; \mathbb{R}^m) \, \big| \, w(x) \in \partial g(x, \cdot) y(x) \text{ for a.a. } x \in \Omega \big\}.$$

Bibliographical note. For background material concerning the topics discussed in this section we refer the reader to [Barbu1], [BarPr], [Brézis1], [EkeTem], and [Moro1].

1.3 Some elements of spectral theory

Let H be a real separable Hilbert space with the inner product $(\cdot, \cdot)_H$, which induces the norm $\| \cdot \|_H$; $\|u\|_H^2 = (u, u)_H$. We assume that a linear operator $B: D(B) \subset H \to H$ satisfies the conditions (B.1)-(B.4) below. We shall reintroduce the *energetic extension* B_E of the operator B; see [Zeidler] for details. Let us recall some basic concepts needed in this theory. We begin by stating our hypotheses on the linear operator B.

(B.1) The operator B is *symmetric*, i.e., $D(B)$ is a dense subset of H and

$$(Bu, v)_H = (u, Bv)_H \text{ for all } u, v \in D(B). \qquad (1.3.1)$$

(B.2) The operator B is *strongly monotone*, i.e., there exists a constant $c > 0$ such that

$$(Bu, u)_H \geq c\|u\|_H^2 \text{ for all } u \in D(B). \qquad (1.3.2)$$

(B.3) The domain $D(B)$ of B is an *infinite dimensional subspace* of H.

On the domain $D(B)$ of B we define an inner product $(\cdot, \cdot)_{H_E}$ by

$$(u, v)_{H_E} = (Bu, v)_H \text{ for all } u, v \in D(B). \qquad (1.3.3)$$

It is called the *energetic inner product*. Moreover, it induces a norm on $D(B)$, which is denoted by $\| \cdot \|_{H_E}$ and is said to be the *energetic norm*. We call the *energetic space* of B the set of all vectors of H that are limits in H of sequences (u_n) of elements of $D(B)$ such that (u_n) is a Cauchy sequence with respect to the energetic norm $\| \cdot \|_{H_E}$. The energetic space of B, denoted by H_E, is a Hilbert space, if the energetic inner product and norm are extended by

$$(u, v)_{H_E} = \lim_{n \to \infty} (u_n, v_n)_{H_E}, \quad \|u\|_{H_E}^2 = (u, u)_{H_E}, \qquad (1.3.4)$$

where (u_n) and (v_n) are sequences in $D(B)$, corresponding to u and v, respectively. Indeed, H_E is obtained by completing $D(B)$ with respect to the energetic norm. Using the strong monotonicity of B we see that H_E is embedded continuously into H by the identity mapping $H_E \mapsto H$; more precisely

$$\|u\|_H \leq c^{-1/2}\|u\|_{H_E} \text{ for all } u \in H_E. \qquad (1.3.5)$$

Now we can state our last assumption on B.

(B.4) The embedding $H_E \subset H$ is *compact*, i.e., the identity mapping $H_E \mapsto H$ is compact.

Hence H is embedded continuously into H_E^*, the dual of H_E, by the linear mapping $j: H \mapsto H_E^*$, which is given by

$$j(h)(v) = (h, v)_H \text{ for all } v \in H_E, \ h \in H. \qquad (1.3.6)$$

Indeed, if $j(h)$ is identified with h, then H becomes a subspace of H_E^* and we can write

$$H_E \subset H \subset H_E^* \text{ and } h(v) = (h, v)_H \text{ for all } v \in H_E, \ h \in H. \qquad (1.3.7)$$

The duality mapping B_E from H_E into H_E^* is given by

$$B_E u(v) = (u, v)_{H_E} \quad \text{for all } u, v \in H_E. \tag{1.3.8}$$

It is an extension of B; we call it the *energetic extension* of B.

We recall that the linear operator $A\colon D(A) \subset H \to H$, given by

$$D(A) = \{u \in H_E \mid B_E u \in H\}, \quad Au = B_E u, \tag{1.3.9}$$

is called the *Friedrichs extension* of B. According to [Zeidler, p. 280], A is self-adjoint and strongly monotone, hence maximal monotone (see [Haraux, p. 48]). Clearly,

$$D(B) \subset D(A) \subset D(B_E) = H_E \subset H. \tag{1.3.10}$$

Therefore A is the maximal monotone extension of B in H.

We need a result from spectral theory.

THEOREM 1.3.1

Assume (B.1)-(B.4). Then there exist eigenvalues $\lambda_n > 0$ and eigenvectors $e_n \in D(A)$ of A, $n \in \mathbb{N}^*$, which satisfy:

   (i) The set $\{e_n \mid n \in \mathbb{N}^*\}$ is a complete orthonormal basis of H_E;

   (ii) The set $\{\sqrt{\lambda_n} e_n \mid n \in \mathbb{N}^*\}$ is a complete orthonormal basis of H;

   (iii) The set $\{\lambda_n e_n \mid n \in \mathbb{N}^*\}$ is a complete orthonormal basis of H_E^*;

   (iv) The sequence (λ_n) is increasing and $\lim_{n \to \infty} \lambda_n = \infty$.

PROOF   Let $f \in H$. By the Riesz Theorem, the problem

$$(u, v)_{H_E} = (f, v)_H \quad \text{for all } v \in H_E \tag{1.3.11}$$

has a unique solution u_f. Thus we have a mapping $P\colon H \mapsto H_E, Pf = u_f$, and its restriction to H_E, $Q\colon H_E \mapsto H_E, Qf = Pf$. Clearly, Q is symmetric. Let (f_n) be a bounded sequence in H_E. Since H_E is embedded compactly into H, there exists a subsequence (f_{n_j}) converging toward some $f \in H$ in H. By (1.3.11),

$$\|Qf_{n_j} - Pf\|_{H_E} \le \|f_{n_j} - f\|_H \to 0, \quad \text{as } j \to \infty.$$

Hence Q is compact. For any nonzero $f \in H_E$,

$$\|Qf\|_{H_E} = \sup_{g \in H_E,\, g \ne 0} \frac{(Qf, g)_{H_E}}{\|g\|_{H_E}} \ge \frac{(Qf, f)_{H_E}}{\|f\|_{H_E}} = \frac{\|f\|_H^2}{\|f\|_{H_E}} > 0,$$

whence the kernel of Q is $\{0\}$ and its possible eigenvalues are positive. Since H is infinite dimensional, we obtain from the Hilbert-Schmidt Theorem, e.g., [Zeidler, p. 232], that there exists eigenvectors $e_1, e_2, \ldots$ of Q and corresponding eigenvalues $\mu_1, \mu_2, \ldots$ of Q such that $\{e_1, e_2, \ldots\}$ is a complete orthonormal basis of H_E and (μ_n) is a decreasing sequence, converging toward zero. Let $v \in H_E$. Then

$$\mu_n(Ae_n, v)_H = \mu_n(Be_n, v)_H = \mu_n(e_n, v)_{H_E} = (Qe_n, v)_{H_E} = (e_n, v)_H.$$

Thus e_n is an eigenvector of A, corresponding to the eigenvalue $\lambda_n = 1/\mu_n$. We have proved (i) and (iv).

The set $S = \{\sqrt{\lambda_1}e_1, \sqrt{\lambda_2}e_2, \ldots\}$ is orthonormal in H, since

$$(\lambda_n e_n, e_m)_H = (\lambda_n Qe_n, e_m)_{H_E} = (e_n, e_m)_{H_E}.$$

For the completeness of S it suffices to prove that

$$(g, \sqrt{\lambda_n}e_n)_H = 0 \ \text{ for all } n \in \mathbb{N}^* \tag{1.3.12}$$

implies $g = 0$; see, e.g., [Zeidler, pp. 202, 222]. Let $g \in H$ satisfy (1.3.12). By (1.3.11), $(Pg, e_n)_{H_E} = 0$ for all $n \in \mathbb{N}^*$. Since $\{e_1, e_2, \ldots\}$ is a complete orthonormal basis of H_E, $Pg = 0$. Hence $(g, v)_H = 0$ for all $v \in H_E$. Since H_E is dense in H, $g = 0$.

It remains to prove (iii). Using (1.3.11)

$$\|e_n + e_m\|_{H_E^*} = \sup_{v \in H_E, v \neq 0} \frac{(e_n + e_m, v)_H}{\|v\|_{H_E}} = \sup_{v \in H_E, v \neq 0} \frac{(Qe_n + Qe_m, v)_{H_E}}{\|v\|_{H_E}} =$$

$$= \sup_{v \in H_E, v \neq 0} \frac{(\mu_n e_n + \mu_m e_m, v)_{H_E}}{\|v\|_{H_E}} = \sqrt{\mu_n^2 + 2\mu_n\mu_m(e_n, e_m)_{H_E} + \mu_m^2}.$$

By the parallelogram identity,

$$2(e_n, e_m)_{H_E^*} = \|e_n + e_m\|_{H_E^*}^2 - \|e_n\|_{H_E^*}^2 - \|e_m\|_{H_E^*}^2 = \frac{2(e_n, e_m)_{H_E}}{\lambda_n \lambda_m}.$$

Thus $S' = \{\lambda_1 e_1, \lambda_2 e_2, \ldots\}$ is orthonormal in H_E^*. Let $g \in H_E^*$ be such that

$$(g, \lambda_n e_n)_{H_E^*} = 0 \ \text{ for all } n \in \mathbb{N}^*.$$

Let $y \in H$. So $y = \sum_n \alpha_n e_n$ for some coefficients α_n. Thus $(g, y)_{H_E^*} = 0$. Since H is dense in H_E^*, $g = 0$.

Theorem 1.3.1 is proved. ∎

For our regularity considerations we shall need some subspaces of H_E, which we define by powers of the operator A.

THEOREM 1.3.2

Assume (B.1)-(B.4) and let $\lambda_1, \lambda_2, \ldots$ and $e_1, e_2, \ldots$ be as in Theorem 1.3.1, and $\gamma \geq 0$. Define A^γ by

$$A^\gamma u = \sum_{n=1}^\infty \lambda_n^\gamma (u, \sqrt{\lambda_n} e_n)_H \sqrt{\lambda_n} e_n \qquad (1.3.13)$$

and $u \in D(A^\gamma)$ whenever this series converges. Then A^γ is selfadjoint and strongly monotone.

PROOF   Since H is separable, A^γ is selfadjoint, by [Zeidler, p. 294]. Since $\{\lambda_n\}$ is increasing,

$$(A^\gamma u, u)_H = \sum_{k=1}^\infty \lambda_k^\gamma \left(u, \sqrt{\lambda_k} e_k\right)_H^2 \geq \sum_{k=1}^\infty \lambda_1^\gamma \left(u, \sqrt{\lambda_k} e_k\right)_H^2 = \lambda_1^\gamma \|u\|_H^2.$$

Thus A^γ is strongly monotone.   ∎

Now we can define for $k \in \mathbb{N}^*$, the Hilbert spaces $(V_k, (\cdot, \cdot)_k)$, by

$$V_k = D(A^{k/2}) \text{ and } (u, v)_k = \left(A^{k/2} u, A^{k/2} u\right)_H, \qquad (1.3.14)$$

where $A^{k/2}$ is the square root of A^k. Let us denote $V_0 = H$, $V_{-k} = V_k^*$, and identify $H^* = H$.

THEOREM 1.3.3

Assume (B.1)-(B.4) and let $k \in \mathbb{N}^*$. Then $\lambda_1, \lambda_2, \ldots$ and $e_1, e_2, \ldots$ of Theorem 1.3.1 satisfy:

(i) The set $\{\lambda_n^{(1-k)/2} e_n \mid n \in \mathbb{N}^*\}$ is a complete orthonormal basis of V_k;

(ii) The set $\{\lambda_n^{(1+k)/2} e_n \mid n \in \mathbb{N}^*\}$ is a complete orthonormal basis of V_k^*;

(iii) In the following inclusion chain every embedding is continuous and compact:

$$\ldots \subset V_{k+1} \subset V_k \subset \ldots \subset V_2 = D(A) \subset V_1 =$$
$$= H_E \subset H = H^* \subset V_1^* \subset \ldots \subset V_k^* \subset V_{k+1}^* \subset \ldots.$$

PROOF   Clearly, $V_1 = H_E$ and $(\cdot, \cdot)_1 = (\cdot, \cdot)_{H_E}$; see [Zeidler, p. 296]. One can easily see that (i) is satisfied. Moreover, for each $y_0 \in V_k$ and

$k \in \mathbb{N}^*$,

$$\|y_0\|_k^2 = \sum_{n=1}^{\infty} \left(y_0, \lambda_n^{(1-k)/2} e_n\right)_k^2 = \qquad (1.3.15)$$

$$= \sum_{n=1}^{\infty} \left(y_0, A^k \lambda_n^{(1-k)/2} e_n\right)_H^2 = \sum_{n=1}^{\infty} \lambda_n^{k+1} (y_0, e_n)_H^2.$$

Since (λ_n) is an increasing sequence of positive numbers, (1.3.15) implies that V_{k+1} is embedded continuously into V_k. Since V_1 is embedded compactly into H, then also $V_{k+1} \subset V_k$ and $V_k^* \subset V_{k+1}^*$ compactly, so that (iii) holds. Moreover, $V_k \subset H$ compactly. The duality mapping $J_k \colon V_k \mapsto V_k^*$ is given by

$$J_k(u)(v) = (u, v)_k \text{ for all } u, v \in V_k. \qquad (1.3.16)$$

Clearly, $J_1 = B_E$. We easily see that

$$J_k\left(\lambda_n^{(1-k)/2} e_n\right) = \lambda_n^{(1+k)/2} e_n \text{ for all } n, k \in \mathbb{N}^*.$$

Hence (ii) is also satisfied. ∎

1.4 Linear evolution equations and semigroups

We are interested in linear evolution equations of the type

$$u'(t) + Bu(t) = f(t) \text{ for a.a. } t > 0, \qquad (1.4.1)$$

where $B \colon D(B) \subset X \to X$ is a linear unbounded operator, $f \colon \mathbb{R}_+ \to X$, and X is a Banach space. In this section we briefly recall some definitions and results from the theory of semigroups of bounded linear operators. For a more complete discussion we refer to [Pazy] and [Yosida]. The linear semigroup theory will be the main tool in Chapter 8.

Let X be a Banach space. A one parameter family of bounded linear operators $S(t) \colon X \to X$, $t \geq 0$, is said to be a *semigroup of linear bounded operators on X*, if

(i) $S(0) = I$, the identity operator on X;

(ii) $S(t + s) = S(t)S(s)$ for all $t, s \geq 0$.

The set $\{S(t) \colon X \to X \mid t \geq 0\}$ is said to be a *C_0-semigroup*, if, in addition,

(iii) $\lim_{t \to 0+} S(t)x = x$ for all $x \in X$.

The linear operator A, given by

$$D(A) = \left\{ x \in X \mid \lim_{t \to 0+} \frac{S(t)x - x}{t} \text{ exists in } X \right\}, \quad Ax = \lim_{t \to 0+} \frac{S(t)x - x}{t},$$

is called the *infinitesimal generator* of the semigroup $\{S(t): X \to X \mid t \geq 0\}$.

THEOREM 1.4.1

Let $\{S(t): X \to X \mid t \geq 0\}$ be a C_0-semigroup. Then, there exist constants $M \geq 1$ and $\omega \geq 0$ such that

$$\|S(t)\|_{L(X;X)} \leq M e^{\omega t} \text{ for all } t \geq 0.$$

The *resolvent set* $\rho(A)$ of a linear operator A is given by

$$\rho(A) = \{\lambda \in \mathbb{C} \mid (\lambda I + A)^{-1}: X \to X \text{ is defined and bounded}\}.$$

The *resolvent* of A is the operator

$$R(\lambda: A) = (\lambda I + A)^{-1}: X \to X, \ \lambda \in \rho(A).$$

The next two theorems tell, how the resolvent is continuous and how it behaves, as λ is an increasing real number.

THEOREM 1.4.2

A linear operator A is the infinitesimal generator of a C_0-semigroup, if and only if both (i) and (ii) below are satisfied:

(i) $D(A)$ is dense in X and A is closed;

(ii) The constants M and ω of Theorem 1.4.1 satisfy $(\omega, \infty) \subset \rho(A)$ and

$$\|R(\lambda: A)^n\|_{L(X;X)} \leq \frac{M}{(\lambda - \omega)^n} \text{ for all } \lambda > \omega, \ n \in \mathbb{N}^*.$$

THEOREM 1.4.3

Let A satisfy (i) and (ii) of Theorem 1.4.2. Then

$$\lim_{\lambda \to \infty} \lambda R(\lambda: A)x = x \text{ for all } x \in X.$$

Let $f \in L^1(0, T; X)$, $x \in X$, A satisfy (i) and (ii) of Theorem 1.4.2, and consider the Cauchy problem

$$u'(t) + Au(t) = f(t), t \in (0, T), \ u(0) = x. \qquad (1.4.2)$$

Let $\{S(t)\colon X \to X \mid t \geq 0\}$ be the C_0-semigroup generated by $-A$. The function $u \in C\big([0,T];X\big)$, given by

$$u(t) = S(t)x + \int_0^t S(t-s)f(s)\,ds, \tag{1.4.3}$$

is called the *mild solution* of (1.4.2). A function $u \in W^{1,1}(0,T;X)$ is called a *strong solution* of (1.4.2) if $u(t) \in D(A)$ and $f(t) - u'(t) = Au(t)$ for a.a. $t \in (0,T)$.

THEOREM 1.4.4

Let $-A$ be the infinitesimal generator of a C_0-semigroup $\{S(t)\colon X \to X \mid t \geq 0\}$, $f \in C^1\big([0,T];X\big)$, and $x \in D(A)$. Then (1.4.2) has a unique (classical) solution $u \in C^1\big([0,T);X\big)$. If $f \in W^{1,1}(0,T;X)$, then the mild solution of (1.4.2) is also its unique strong solution and, for a.a. $t \in (0,T)$,

$$u'(t) = -S(t)Ax + S(t)f(0) + \int_0^t S(t-s)f'(s)\,ds. \tag{1.4.4}$$

1.5 Nonlinear evolution equations

Throughout this section H is a real Hilbert space, whose scalar product and norm are again denoted by $(\cdot,\cdot)_H$ and $\|\cdot\|_H$, respectively ($\|x\|_H^2 = (x,x)_H, x \in H$). Consider in H the following Cauchy problem

$$u'(t) + Au(t) \ni f(t), \ 0 < t < T, \tag{1.5.1}$$
$$u(0) = u_0, \tag{1.5.2}$$

where $A\colon D(A) \subset H \to H$ is a nonlinear operator (possibly multivalued), and $f \in L^1(0,T;H)$.

DEFINITION 1.5.1 *A function $u \in C\big([0,T];H\big)$ is called a strong solution of the Cauchy problem (1.5.1)-(1.5.2) if*

(a) u is absolutely continuous on every compact subinterval of $(0,T)$;

(b) $u(t) \in D(A)$ for a.a. $t \in (0,T)$;

(c) $u(0) = u_0$ and u satisfies (1.5.1) for a.a. $t \in (0,T)$.

DEFINITION 1.5.2 *A function $u \in C\big([0,T];H\big)$ is said to be a weak solution of (1.5.1)-(1.5.2) if there exist $u_n \in W^{1,\infty}(0,T;H)$ and $f_n \in$*

$L^1(0,T;H)$, $n \in \mathbb{N}^*$, *such that*

$$u'_n(t) + Au_n(t) \ni f_n(t) \ \textit{for a.a.} \ t \in (0,T), \ n \in \mathbb{N}^*; \tag{1.5.3}$$

$$u_n \to u \ \textit{in} \ C\big([0,T];H\big), \ \textit{as} \ n \to \infty; \tag{1.5.4}$$

$$u(0) = u_0 \ \textit{and} \ f_n \to f \ \textit{in} \ L^1(0,T;H), \ \textit{as} \ n \to \infty. \tag{1.5.5}$$

THEOREM 1.5.1

If $A\colon D(A) \subset H \to H$ is a maximal monotone operator, $u_0 \in D(A)$ and $f \in W^{1,1}(0,T;H)$, then the Cauchy problem (1.5.1)-(1.5.2) has a unique strong solution $u \in W^{1,\infty}(0,T;H)$. Moreover, $u(t) \in D(A)$, for all $t \in [0,T]$, u is differentiable from the right at every $t \in [0,T)$, and

$$\frac{d^+u}{dt}(t) = \big(f(t) - Au(t)\big)^0 \ \textit{for all} \ t \in [0,T), \tag{1.5.6}$$

$$\left\|\frac{d^+u}{dt}(t)\right\|_H \leq \left\|\big(f(0) - Au_0\big)^0\right\|_H + \int_0^t \|f'(s)\|_H \, ds \tag{1.5.7}$$

for all $t \in [0,T)$, where $\big(f(t) - Au(t)\big)^0$ denotes the element of minimal norm of the convex and closed set $f(t) - Au(t)$. If u_1, u_2 are the strong solutions corresponding to $(u_0, f) := (u_{01}, f_1), (u_{02}, f_2) \in D(A) \times W^{1,1}(0,T;H)$ then, for all $t \in [0,T]$,

$$\|u_1(t) - u_2(t)\|_H \leq \|u_{01} - u_{02}\|_H + \int_0^t \|f_1(s) - f_2(s)\|_H \, ds. \tag{1.5.8}$$

REMARK 1.5.1   Theorem 1.5.1 is still valid if $A + \omega I$ is maximal monotone for some $\omega > 0$ (and this allows Lipschitzian perturbations), with the exception of the estimates (1.5.7) and (1.5.8), which are slightly modified. ∎

We shall later need the following ordinary Gronwall's inequality [Brézis1, p. 156]:

LEMMA 1.5.1

Let $a, b, c \in \mathbb{R}$ with $a < b$ and $c \geq 0$, $g \in L^1(a,b)$ with $g \geq 0$ a.e. on (a,b), and $h \in C[a,b]$ such that

$$h(t) \leq c + \int_a^t g(s)h(s) \, ds \ \textit{for all} \ t \in [a,b].$$

Then

$$h(t) \leq c \exp \int_a^t g(s) \, ds \ \textit{for all} \ t \in [a,b].$$

An ingredient in the proof of Theorem 1.5.1 which will also be used later, is the following variant of Gronwall's inequality [Brézis1, p. 157]:

LEMMA 1.5.2
Let $a, b, c \in \mathbb{R}$ with $a < b$, $g \in L^1(a,b)$ with $g \geq 0$ a.e. on (a,b), and $h \in C[a,b]$ such that

$$\frac{1}{2}h(t)^2 \leq \frac{1}{2}c^2 + \int_a^t g(s)h(s)\,ds \ \text{ for all } t \in [a,b].$$

Then

$$|h(t)| \leq |c| + \int_0^t g(s)\,ds \ \text{ for all } t \in [a,b].$$

This lemma is still valid if c is a real function such that $t \to |c(t)|$ is nondecreasing on $[a,b]$.

The basic idea in proving Theorem 1.5.1 is to start with an approximating equation, obtained by replacing A in (1.5.1) by its Yosida approximation, which is Lipschitz continuous. The solution of this regularized problem is guaranteed by the following lemma ([Brézis1, p. 10]), which will also be useful later on:

LEMMA 1.5.3
Let $C \subset H$ be a nonempty closed convex set, $u_0 \in C$, $T, L > 0$, and let the mappings $J(t) \colon C \to C$ satisfy:

(i) $\|J(t)x - J(t)y\|_H \leq L\|x - y\|_H$ for all $x, y \in C$, $t \in [0,T]$;

(ii) $t \mapsto J(t)x$ is integrable for all $x \in C$.

Then there exists a unique $u \in W^{1,1}(0,T;H)$ such that $u(0) = u_0$ and

$$u'(t) + u(t) - J(t)u(t) = 0 \ \text{for a.a. } t \in (0,T).$$

THEOREM 1.5.2
If $A \colon D(A) \subset H \to H$ is maximal monotone, $u_0 \in \overline{D(A)}$, and $f \in L^1(0,T;H)$, then the problem (1.5.1)-(1.5.2) has a unique weak solution $u \in C([0,T];H)$. If u_1, u_2 are the weak solutions corresponding to $(u_0, f) := (u_{01}, f_1)$, $(u_{02}, f_2) \in \overline{D(A)} \times L^1(0,T;H)$, then u_1, u_2 still satisfy (1.5.8).

The last theorem above has an immediate proof, based on a density argument. Indeed, we can approximate (u_0, f) by $(u_{0n}, f_n) \in D(A) \times W^{1,1}(0,T;H)$, for which Theorem 1.5.1 guarantees the existence of strong solutions for problem (1.5.1)-(1.5.2) with $(u_0, f) := (u_{0n}, f_n)$. Then, we can use (1.5.8) to conclude the proof.

THEOREM 1.5.3

(H. Brézis). If A is the subdifferential of a proper convex lower semicontinuous function $\psi: H \to (-\infty, \infty]$, $u_0 \in \overline{D(A)}$, and $f \in L^2(0, T; H)$, then the problem (1.5.1)-(1.5.2) has a unique strong solution u such that $t \mapsto t^{\frac{1}{2}} u'(t)$ belongs to $L^2(0, T; H)$, $t \mapsto \psi\big(u(t)\big)$ is integrable on $[0, T]$ and absolutely continuous on $[\delta, T]$, for all $\delta \in (0, T)$. If, in addition, $u_0 \in D(\psi)$, then $u' \in L^2(0, T; H)$, $t \mapsto \psi\big(u(t)\big)$ is absolutely continuous on $[0, T]$, and

$$\psi\big(u(t)\big) \le \psi(u_0) + \frac{1}{2} \int_0^T \|f(s)\|_H^2 \, ds \ \text{ for all } t \in [0, T]. \tag{1.5.9}$$

DEFINITION 1.5.3   *Let C be a nonempty closed subset of H. A continuous semigroup of contractions on C is a family of operators $S(t): C \to C$, $t \ge 0$, satisfying:*

(A1) $S(0)x = x$ for all $x \in C$;

(A2) $S(t + s)x = S(t)S(s)x$ for all $x \in C$, $t, s \ge 0$;

(A3) for every $x \in C$, the mapping $t \mapsto S(t)x$ is continuous on $[0, \infty)$;

(A4) $\|S(t)x - S(t)y\|_H \le \|x - y\|_H$ for all $x, y \in C$, $t \ge 0$.

The infinitesimal generator of a semigroup $\{S(t): C \to C \mid t \ge 0\}$, say G, is given by

$$Gx = \lim_{h \to 0+} \frac{1}{h}\big(S(h)x - x\big) \tag{1.5.10}$$

with $D(G)$ consisting of all $x \in C$ for which the limit in (1.5.10) exists. We shall also say that the operator G generates the semigroup $\{S(t): C \to C \mid t \ge 0\}$.

REMARK 1.5.2   Let $A: D(A) \subset H \to H$ be a maximal monotone operator. From Theorem 1.5.1 we know that for every $x \in D(A)$ there exists a unique strong solution $u(t)$, $t \ge 0$, of the Cauchy problem

$$u'(t) + Au(t) \ni 0, \ t > 0, \ u(0) = x. \tag{1.5.11}$$

We set $S(t)x := u(t)$, $t \ge 0$. Then it is easily seen that $S(t)$ is a contraction on $D(A)$ (see (1.5.8)) and so $S(t)$ can be extended as a contraction on $\overline{D(A)}$, for each $t \ge 0$. Moreover, it is obvious that the family $\{S(t): \overline{D(A)} \to \overline{D(A)}, t \ge 0\}$ is a continuous semigroup of contractions and its infinitesimal generator is $-A^0$, where A^0 denotes the minimal section of A (see (1.5.6)). We shall say that this semigroup is generated by $-A$. Obviously, if $x \in \overline{D(A)}$, then $u(t) = S(t)x$ is the weak solution of (1.5.11) (more precisely, it is a weak solution of (1.5.11) on $[0, T]$, for each $T > 0$). ∎

Now, we are going to recall some facts concerning the long-time behavior of the solution of (1.5.1) considered on $[0, \infty)$.

THEOREM 1.5.4
Let $A: D(A) \subset H \to H$ be a maximal monotone operator and let the family $\{S(t): \overline{D(A)} \to \overline{D(A)} \mid t \geq 0\}$ be the semigroup generated by $-A$. If $S(t)x$ converges strongly, as $t \to \infty$, for every $x \in D(A)$, then the weak solution $u(t)$ of (1.5.1)-(1.5.2) converges strongly, as $t \to \infty$, for every $u_0 \in \overline{D(A)}$, $f \in L^1(0, \infty; H)$, and the limit of $u(t)$, if it exists, is an element of $F := A^{-1}0$.

REMARK 1.5.3   The above result reduces the study of asymptotic behavior of solutions $u(t)$ in the case $f \in L^1(0, \infty; H)$ to the asymptotic behavior of $S(t)x$, $x \in D(A)$. On the other hand, the condition $F \neq \emptyset$ is a necessary one for such an asymptotic behavior.   ∎

THEOREM 1.5.5
(R.E. Bruck). Let A be the subdifferential of a proper convex lower semi-continuous function $\psi: H \to (-\infty, \infty]$ such that $F := A^{-1}0$ is nonempty (or, equivalently, ψ has at least one minimum point). Then, for every $x \in \overline{D(A)}$, $S(t)x$ converges weakly to a point of F, as $t \to \infty$.

THEOREM 1.5.6
(C.M. Dafermos and M. Slemrod). Let $A: D(A) \subset H \to H$ be a maximal monotone operator and let $S(t): \overline{D(A)} \to \overline{D(A)}$, $t \geq 0$, be the semigroup generated by $-A$. Assume that for some $x \in \overline{D(A)}$ the ω-limit set $\omega(x)$ of A is nonempty, where

$$\omega(x) := \{p \in \overline{D(A)} \mid \text{ there exists a sequence } (t_n) \text{ such that}$$
$$\lim_{n \to \infty} t_n = \infty \text{ and } \lim_{n \to \infty} \|S(t_n)x - p\|_H = 0\}.$$

Then, we have:

(a) For every $t \geq 0$, $S(t)$ is an isometric homeomorphism on $\omega(x)$;

(b) If $a \in F : A^{-1}0$, then $\omega(x)$ lies on a sphere $\{y \in H \mid \|y - a\|_H = r\}$, with $r \leq \|x - a\|_H$;

(c) If $\omega(x)$ is compact, then there exists a $y_0 \in \omega(x)$ such that

$$\lim_{t \to \infty} \|S(t)x - S(t)y_0\|_H = 0;$$

(d) If $x \in D(A)$, then $\omega(x) \subset D(A)$.

The remainder of this section is dedicated to recalling some existence results for nonautonomous evolution equations.

THEOREM 1.5.7

(T. Kato, [Kato]). Let $A(t): D \subset H \to H$, $t \in [0,T]$, be a family of single-valued maximal monotone operators (with $D\big(A(t)\big) = D$ independent of t) satisfying the following condition

$$\|A(t)x - A(s)x\|_H \leq L|t - s|\big(1 + \|x\|_H + \|A(s)x\|_H\big) \qquad (1.5.12)$$

for all $x \in D$, $s,t \in [0,T]$, where L is a positive constant. Then, for every $u_0 \in D$, there exists a unique function $u \in W^{1,1}(0,T;H)$ such that $u(0) = u_0$ and

$$u(t) \in D \ \text{ for all } t \in [0,T], \qquad (1.5.13)$$

$$u'(t) + A(t)u(t) = 0 \ \text{ for a.a. } t \in (0,T). \qquad (1.5.14)$$

THEOREM 1.5.8

(H. Attouch and A. Damlamian, [AttDam]). Let $A(t) = \partial\psi(t,\cdot)$, $t \in [0,T]$, where $\psi(t,\cdot): H \to (-\infty,\infty]$ are all proper, convex, and lower semicontinuous. Assume further that there exist some positive constants C_1, C_2 and a nondecreasing function $\gamma: [0,T] \to \mathbb{R}$ such that

$$\psi(t,x) \leq \psi(s,x) + \big(\gamma(t) - \gamma(s)\big)\big(\psi(s,x) + C_1\|x\|_H^2 + C_2\big) \qquad (1.5.15)$$

for all $x \in H$, $0 \leq s \leq t \leq T$. Then, for all $u_0 \in D(\psi(0,\cdot))$ and $f \in L^2(0,T;H)$, there exists a unique function $u \in W^{1,2}(0,T;H)$ such that $u(0) = u_0$ and

$$u'(t) + A(t)u(t) \ni f(t) \ \text{ for a.a. } t \in (0,T). \qquad (1.5.16)$$

Moreover, there exists a function $h \in L^1(0,T)$ such that

$$\psi\big(t,u(t)\big) \leq \psi\big(s,u(s)\big) + \int_s^t h(\sigma)\,d\sigma \text{ for all } 0 \leq s \leq t \leq T. \qquad (1.5.17)$$

THEOREM 1.5.9

(D. Tătaru, [Tătaru]). Let $A(t): D\big(A(t)\big) \subset H \to H$, $t \in [0,T]$, be a family of maximal monotone operators satisfying the following condition

$$-\big(x - y, A(t)x - A(s)y\big)_H \leq M\|x - y\|_H^2 + \qquad (1.5.18)$$

$$+|t - s| \cdot \big|g(t) - g(s)\big|\big(1 + \|x\|_H^2 + \|y\|_H^2 + \|A(t)x\|_H^2 + \|A(s)y\|_H^2\big)$$

for all $t,s \in [0,T]$, $x \in D\big(A(t)\big)$, and $y \in D\big(A(s)\big)$, where M is a positive constant and g is a function of bounded variation on $[0,T]$. Then, for each

$u_0 \in D\big(A(0)\big)$, *there exists a unique function* $u \in W^{1,\infty}(0,T;H)$ *such that* $u(0) = u_0$ *and*

$$u'(t) + A(t)u(t) \ni 0 \ \textit{for a.a.} \ t \in (0,T). \qquad (1.5.19)$$

REMARK 1.5.4 An easy computation shows that (1.5.12) is stronger than (1.5.18) and so Theorem 1.5.7 can be derived from Theorem 1.5.9. However, for some applications it is easier to apply Theorem 1.5.9. Notice, that Theorem 1.5.7 still holds with $L = L(\|x\|_H)$, where $L(\cdot)$ is a nondecreasing function (actually, this is Kato's original assumption). On the other hand, Theorem 1.5.9 holds under more general conditions so that Kato's original result can again be derived as a special case (see [Tătaru]). ∎

REMARK 1.5.5 The concepts of strong solution and weak solution can be extended to time-dependent equations. Actually, the last three results give the existence and uniqueness of strong solutions for the corresponding time-dependent equations. Then, the existence and uniqueness of weak solutions follow by a simple density argument (involving the monotonicity of $A(t)$). ∎

Bibliographical note. This section is based on the books [Brézis1] and [Moro1] with the exception of the last three theorems for which we have indicated specific references.

References

[Adams] R.A. Adams, *Sobolev Spaces*, Academic Press, New York, 1975.

[Agmon] S. Agmon, *Lectures on Elliptic Boundary Value Problems*, Van Nostrand, New York, 1965.

[AttDam] H. Attouch & A. Damlamian, Strong solutions for parabolic variational inequalities, *Nonlin. Anal.*, 2 (1978), no. 3, 329-353.

[Barbu1] V. Barbu, *Nonlinear Semigroups and Differential Equations in Banach Spaces*, Noordhoff, Leyden, 1976.

[BarPr] V. Barbu & Th. Precupanu, *Convexity and Optimization in Ba-*

nach Spaces, Editura Academiei and Reidel, Bucharest and Dordrecht, 1986.

[Brézis1] H. Brézis, *Opérateurs maximaux monotones et semi-groupes de contractions dans les espaces de Hilbert*, North-Holland, Amsterdam, 1973.

[Brézis2] H. Brézis, *Analyse fonctionelle*, Masson, Paris, 1983.

[Dieudo] J..Dieudonné, *Eléments d'analyse. Tome I. Eléments de l'analyse moderne*, Dunod, Gauthier-Villars, Paris, 1969.

[EkeTem] I. Ekeland & R. Temam, *Analyse convexe et problèmes variationnels*, Dunod, Gauthier-Villars, Paris, 1974.

[Haraux] A. Haraux, *Nonlinear Evolution Equations. Global Behavior of Solutions*, Lecture Notes in Math. 841, Springer-Verlag, Berlin, 1981.

[Hokk1] V.-M. Hokkanen, An implicit non-linear time dependent equation has a solution, *J. Math. Anal. Appl.*, 161 (1991), no. 1, 117–141.

[Kato] T. Kato, Nonlinear semigroups and evolution equations, *J. Math. Soc. Japan*, 19 (4) (1967), 508–520.

[Lions] J.-L. Lions, *Quelques méthodes de résolution des problèmes aux limites non-linéaires*, Dunod, Gauthier-Villars, Paris, 1969.

[LioMag] J.-L. Lions & E. Magenes, *Problèmes aux limites non homogènes et applications*, Vol.1, Dunod, Paris, 1968.

[Moro1] Gh. Moroşanu, *Nonlinear Evolution Equations and Applications*, Editura Academiei and Reidel, Bucharest and Dordrecht, 1988.

[Pazy] A. Pazy, *Semigroups of Linear Operators and Applications to Partial Differential Equations*, Springer-Verlag, Berlin, 1983.

[Schwa] L. Schwartz, *Théorie des distributions*, Herman, Paris, 1967.

[Tătaru] D. Tătaru, Time dependent m-accretive operators generating differentiable evolutions, *Diff. Int. Eq.*, 4 (1991), 137–150.

[Yosida] K. Yosida, *Functional Analysis, fifth ed.*, Springer-Verlag, Berlin, 1978.

[Zeidler] E. Zeidler, *Applied Functional Analysis*, Appl. Math. Sci. 108, Springer-Verlag, 1995.

Chapter 2

Elliptic boundary value problems

In this chapter we investigate different types of elliptic boundary value problems, mainly one-dimensional problems.

The first section is dedicated to a nonlinear, one-dimensional, nondegenerate, elliptic boundary value problem. Both the variational case and the nonvariational one are investigated. Some consistent results are obtained by choosing an appropriate functional framework and by using some elements of convex analysis and of monotone operator theory. In the second section we investigate existence and uniqueness for one-dimensional, doubly nonlinear and multivalued, degenerate, second order boundary value problems of the form (2.2.12)-(2.2.13) below. Actually, we begin with the classical problem of capillarity, which is a two-dimensional model. This model can be viewed as a particular case of a more general problem considered in an N-dimensional domain Ω (see problem (2.2.4)-(2.2.5) below). If Ω is the unit sphere and u is radially symmetric, we can use spherical coordinates and so we obtain the degenerate one-dimensional problem (2.2.9)-(2.2.10). Then, a generalization of this problem, more precisely problem (2.2.12)-(2.2.13), is investigated in detail. The main result, Theorem 2.2.1, has very general assumptions, some of them being even minimal.

2.1 Nondegenerate elliptic boundary value problems

In this section we deal with a class of one-dimensional, nonlinear, nondegenerate elliptic boundary value problems. The results presented here will later be used for investigating some parabolic problems. The fact that the equations taken into consideration are nondegenerate allows us to associate very general boundary conditions that include as particular cases different types of classical boundary conditions.

The results of this section are nontrivial generalizations of those in

[Moro1, pp. 233-235], [MorPe]. The main novelty here is the nonlinearity of the differential operator governing the equation. For the sake of simplicity, we restrict our investigation to the case of second order differential equations. Our results have been inspired by the recent paper [Lin2]. Although our assumptions are more general, the existence results presented here are stronger than those of [Lin2]. This fact is a consequence of the choice of an appropriate functional framework.

Consider the boundary value problem

$$-\frac{d}{dr}G\big(r, u'(r)\big) + K\big(r, u(r)\big) = f(r), \ 0 < r < 1, \qquad (2.1.1)$$

$$\Big(G\big(0, u'(0)\big), -G\big(1, u'(1)\big)\Big) \in \beta\big(u(0), u(1)\big), \qquad (2.1.2)$$

under the following assumptions:

(I.1) The function $G: [0,1] \times \mathbb{R} \to \mathbb{R}$, $(r, \xi) \mapsto G(r, \xi)$ is continuously differentiable and

$$\frac{\partial G}{\partial \xi}(r, \xi) \geq k_0 \text{ for all } (r, \xi) \in [0,1] \times \mathbb{R}, \qquad (2.1.3)$$

where k_0 is a fixed positive constant;

(I.2) The function $K: [0,1] \times \mathbb{R} \to \mathbb{R}$ is continuous and

$$\big(K(r, \xi_1) - K(r, \xi_2)\big)(\xi_1 - \xi_2) \geq k_1(\xi_1 - \xi_2)^2 \qquad (2.1.4)$$

for all $\xi_1, \xi_2 \in \mathbb{R}$, $r \in [0, 1]$; here k_1 is some positive constant;

(I.3) The mapping $\beta \subset \mathbb{R}^2 \times \mathbb{R}^2$ is maximal monotone (possibly multivalued);

(I.4) The function $f: (0, 1) \to \mathbb{R}$ belongs to $\in L^2(0, 1)$.

DEFINITION 2.1.1   *A solution of problem (2.1.1)-(2.1.2) is a function* $u \in C^1[0, 1]$ *with* $G\big(\cdot, u'(\cdot)\big) \in H^1(0, 1)$, u *satisfies equation (2.1.1) for a.a.* $r \in (0, 1)$, *as well as condition (2.1.2).*

In order to illustrate the generality of problem (2.1.1)-(2.1.2) we are going to consider some examples.

Example 2.1.1
Assume that G, K, f satisfy (I.1), (I.2), (I.4) and β is the subdifferential of the function $j: \mathbb{R}^2 \to (-\infty, \infty]$,

$$j(x_1, x_2) = \begin{cases} 0 & \text{if } x_2 = a, \ x_2 = b, \\ \infty & \text{otherwise,} \end{cases}$$

where a, b are some fixed real numbers. In this case we have existence of solution by Theorem 2.1.1 below. Boundary condition (2.1.2) reads:

$$u(0) = a, \ u(1) = b, \tag{2.1.5}$$

i.e., we have bilocal conditions of Dirichlet type. $\quad\Box$

Example 2.1.2

Assume that $G(r, \xi) = \xi$, K and f satisfy (I.2) and (I.4), while β is the subdifferential of the function $j_1 \colon \mathbb{R}^2 \to (-\infty, \infty]$,

$$j(x_1, x_2) = \begin{cases} 0 & \text{if } x_1 = x_2, \\ \infty & \text{otherwise.} \end{cases}$$

In this case, we have a periodic problem

$$\begin{cases} u''(r) = K\big(r, u(r)\big) + f(r), \ 0 < r < 1, \\ u(0) = u(1), \ u'(0) = u'(1). \end{cases}$$

$\quad\Box$

Example 2.1.3

Assume that $G(r, \xi) = \xi$, K and f satisfy (I.2) and (I.4), while β is the subdifferential of the function $j_2 \colon \mathbb{R}^2 \to (-\infty, \infty]$,

$$j_2(x_1, x_2) = \begin{cases} -bx_2 & \text{if } x_1 = a, \\ \infty & \text{otherwise.} \end{cases}$$

In this case, we have the following bilocal problem

$$\begin{cases} u''(r) = K\big(r, u(r)\big) + f(r), \ 0 < r < 1, \\ u(0) = a, \ u'(0) = b. \end{cases}$$

$\quad\Box$

Other choices of G, K, and β lead to other classical boundary value problems, as the reader can easily see.

In order to solve problem (2.1.1)-(2.1.2), we shall regard this problem as an equation in the space $H = L^2(0, 1)$, endowed with the usual scalar product and the associated norm. Consider the operator $T \colon D(T) \subset H \to H$, defined by

$$D(T) = \left\{ u \in H^2(0, 1) \ \middle| \ \begin{pmatrix} G\big(0, u'(0)\big) \\ -G\big(1, u'(1)\big) \end{pmatrix} \in \beta \begin{pmatrix} u(0) \\ u(1) \end{pmatrix} \right\}, \tag{2.1.6}$$

$$(Tu)(r) = -\frac{d}{dr} G\big(r, u'(r)\big) \ \text{ for a.a. } r \in (0, 1). \tag{2.1.7}$$

PROPOSITION 2.1.1

If assumptions (I.1) and (I.3) hold, then the operator T defined above is maximal monotone, with $D(T)$ dense in $H = L^2(0,1)$.

Before proving Proposition 2.1.1, we present an auxiliary result.

LEMMA 2.1.1

([HokMo]). Let $\gamma \subset \mathbb{R}^2 \times \mathbb{R}^2$ be a maximal monotone operator and let $\tilde{F}: D(\tilde{F}) \subset H \to H$ be defined by

$$D(\tilde{F}) = \left\{ u \in H^2(0,1) \mid \big(u'(0), u'(1)\big) \in \gamma\big(u(0), -u(1)\big) \right\}, \quad (2.1.8)$$

$$(\tilde{F}u)(r) = -u''(r) \ \text{ for a.a. } r \in (0,1). \tag{2.1.9}$$

Then, the operator $\tilde{F}$ is maximal monotone in H.

PROOF   Obviously, $\tilde{F}$ is monotone. Let us show that $R(I + \tilde{F}) = H$, i.e., for all $y \in H$, there exists $u \in H^2(0,1)$, which satisfies

$$-u''(r) + u(r) = y(r) \text{ for a.a. } r \in (0,1), \tag{2.1.10}$$

$$\big(u'(0), u'(1)\big) \in \gamma\big(u(0), -u(1)\big). \tag{2.1.11}$$

The general solution of (2.1.10) is given by

$$u(r) = c_1 \mathrm{e}^r + c_2 \mathrm{e}^{-r} + u_1(r), \tag{2.1.12}$$

where $u_1 \in H^2(0,1)$ is a particular solution of (2.1.10). Clearly, u satisfies (2.1.11) whenever $\gamma z + \hat{\gamma} z = z_1$, where $z_1 \in \mathbb{R}^2$ depends only on u_1 and

$$\hat{\gamma} = \frac{1}{\mathrm{e}^2 - 1} \begin{pmatrix} 1 + \mathrm{e}^2 & 2\mathrm{e} \\ 2\mathrm{e} & 1 + \mathrm{e}^2 \end{pmatrix}, \quad z = \begin{pmatrix} 1 & 1 \\ -\mathrm{e} & -\mathrm{e}^{-1} \end{pmatrix} \begin{pmatrix} c_1 \\ c_2 \end{pmatrix}.$$

As $\hat{\gamma}$ is a strongly positive matrix, it follows that $\gamma + \hat{\gamma}$ is a surjective mapping. Therefore, the equation $\gamma z + \hat{\gamma} z = z_1$ has a solution $z \in \mathbb{R}^2$. ∎

PROOF of Proposition 2.1.1. By (I.1) we can see that $G(0, \cdot)$ and $G(1, \cdot)$ are surjective mappings. Therefore, there exist $a, b, c, d \in \mathbb{R}$ such that $\big(G(0,c), -G(1,d)\big) \in \beta(a,b)$. It follows that $D(T)$ is a nonempty set, as it contains $\hat{v}$,

$$\hat{v}(r) = (2a - 2b + c + d)r^3 + (-3a + 3b - 2c - d)r^2 + cr + a. \tag{2.1.13}$$

Moreover, we have

$$\{\hat{v} + \phi \mid \phi \in C_0^\infty(0,1)\} \subset D(T).$$

Therefore, $D(T)$ is a dense set in H. It is also easy to check that T is monotone.

Let $y \in H$ and consider the problem

$$u(r) - w'(r) = y(r) \text{ for a.a. } r \in (0,1), \tag{2.1.14}$$

$$w(r) = G(r, u'(r)) \text{ for a.a. } r \in (0,1), \tag{2.1.15}$$

$$\big(w(0), -w(1)\big) \in \beta\big(u(0), u(1)\big). \tag{2.1.16}$$

Without any loss of generality, one can assume that $(0,0) \in D(\beta)$ and $(0,0) \in \beta(0,0)$. Otherwise, if $a = (a_1, a_2) \in D(\beta)$ and $b = (b_1, b_2) \in \beta a$, one replaces $\beta, G, u,$ and w by

$$\tilde{\beta}(\xi) = \beta(\xi + a) - b,$$
$$\tilde{G}(r, \xi) = G(r, \xi - a_1 + a_2) - (1-r)b_1 + rb_2,$$
$$\tilde{u}(r) = u(r) - a_1(1-r) - a_2 r,$$
$$\tilde{w}(r) = w(r) - b_1(1-r) + b_2 r.$$

Now, let $y_n \in C_0^\infty(0,1)$, $n \in \mathbb{N}^*$, be such that $y_n \to y$ in H, as $n \to \infty$. Consider the following problems

$$-w_n''(r) + G(r, \cdot)^{-1} w_n(r) + \frac{1}{n} w_n(r) = y_n'(r), \ 0 < r < 1, \tag{2.1.17}$$

$$\big(w_n'(0), w_n'(1)\big) \in \beta^{-1}\big(w_n(0), -w_n(1)\big). \tag{2.1.18}$$

By (I.1), $G(r, \cdot)$ is a surjective mapping for any $r \in [0,1]$ and the mapping $(r, \xi) \mapsto G(r, \cdot)^{-1}\xi$ belongs to $C^1([0,1] \times \mathbb{R})$. Moreover, the operator $P \colon H \to H$, defined by

$$(Pu)(r) = G(r, \cdot)^{-1} u(r) \text{ for a.a. } r \in (0,1),$$

is monotone and Lipschitzian. It follows that $\tilde{F} + P$ is maximal monotone (see Lemma 2.1.1). So, for each $n \in \mathbb{N}^*$, there exists a unique $w_n \in H^2(0,1)$, which satisfies (2.1.17)-(2.1.18). Denote

$$u_n(r) = w_n'(0) + \int_0^r G(\sigma, \cdot)^{-1} w_n(\sigma)\, d\sigma, \tag{2.1.19}$$

$$z_n(r) = u_n(r) + \frac{1}{n} \int_0^r w_n(\sigma)\, d\sigma. \tag{2.1.20}$$

Obviously, we have (see (2.1.17)-(2.1.18))

$$-w_n'(r) + z_n(r) = y_n(r) \text{ for all } r \in [0,1], \tag{2.1.21}$$

$$\big(w_n(0), -w_n(1)\big) \in \beta\big(z_n(0), z_n(1)\big). \tag{2.1.22}$$

Multiplying (2.1.21) by z_n and integrating by parts, one obtains

$$\int_0^1 w_n(r) z_n'(r)\, dr - w_n(1) z_n(1) + w_n(0) z_n(0) + \|z_n\|_H^2 = (z_n, y_n)_H.$$

By (2.1.20), (2.1.22), and (I.1)

$$\frac{1}{2}\|z_n\|_H^2 + k_0 \|u_n'\|_H^2 + \frac{1}{n}\|w_n\|_H^2 \leq \frac{1}{2}\|y_n\|_H^2 \leq C, \tag{2.1.23}$$

where C is some positive constant. Therefore (see (2.1.21)),

$$\|u_n\|_{H^1(0,1)} + \|w_n'\|_H \leq C_1 \tag{2.1.24}$$

for some positive constant C_1. We are now going to prove that (w_n) is a bounded sequence in $H^1(0,1)$. Indeed, if this is not the case, then $\big(w_n(r)\big)$ is unbounded for each $r \in [0,1]$ (see (2.1.24)). Then, by Fatou's lemma

$$\liminf_{n\to\infty} \int_0^1 u_n'(r)^2\, dr \geq \int_0^1 \liminf_{n\to\infty} \left(G(r,\cdot)^{-1} w_n(r) \right)^2 dr = \infty,$$

which contradicts (2.1.23). Therefore, for some constant $C_2 > 0$,

$$\|u_n\|_{H^1(0,1)} + \|w_n\|_{H^1(0,1)} \leq C_2. \tag{2.1.25}$$

So, one can take the limit in (2.1.21) and (2.1.22), thus obtaining the existence of $u, w \in H^1(0,1)$, which satisfy (2.1.14)-(2.1.16).

Actually, $u \in H^2(0,1)$, since $u'(r) = G(r,\cdot)^{-1} w(r)$ for all $r \in [0,1]$. Proposition 2.1.1 is now completely proved.

We continue with a perturbation result, which can not be derived as a consequence of any classical perturbation theorem:

PROPOSITION 2.1.2
If (I.1), (I.2) with $k_1 = 0$ and (I.3) hold, then the operator $A\colon D(A) = D(T) \subset H \to H$, given by

$$Au = Tu + K(\cdot, u), \tag{2.1.26}$$

is maximal monotone.

PROOF   The monotonicity of A is obvious. In order to show its maximality, we fix an arbitrary $y \in H$ and consider the equation

$$u_\lambda + Tu_\lambda + K_\lambda(\cdot, u_\lambda) = y, \tag{2.1.27}$$

where $K_\lambda(r, \cdot)$ is the Yosida approximation of $K(r, \cdot)$ and $\lambda > 0$. Obviously, the operator $u \mapsto K_\lambda(\cdot, u)$ is monotone and Lipschitzian in H for any $\lambda > 0$.

So, the operator $u \mapsto Tu + K_\lambda(\cdot, u)$ is maximal monotone, with domain $D(T)$. Hence (2.1.27) has a unique solution $u_\lambda \in D(T)$ for every $\lambda > 0$. Without any loss of generality, one can assume that $(0,0) \in D(\beta)$ and $(0,0) \in \beta(0,0)$. By multiplying (2.1.27) by u_λ and using the same argument as in the proof of Proposition 2.1.1, one infers that the set $\{K_\lambda(\cdot, u_\lambda) \mid \lambda > 0\}$ is bounded in $C[0,1]$. Moreover, $\{w_\lambda \mid \lambda > 0\}$ is bounded in $H^1(0,1)$, where

$$w_\lambda(r) = G\big(r, u_\lambda'(r)\big) \ \text{ for all } r \in (0,1).$$

So, on a subsequence, (u_λ) converges in $C[0,1]$ to a function $u \in H^1(0,1)$, as $\lambda \to 0^+$. Actually, $u \in H^2(0,1)$ and u satisfies the equation $u + Au = y$, which we can see by passing to limit in (2.1.27). $\blacksquare$

Let us now formulate the main result of this section:

THEOREM 2.1.1
If (I.1)-(I.4) hold, then problem (2.1.1)-(2.1.2) has a unique solution $u \in H^2(0,1)$.

PROOF Problem (2.1.1)-(2.1.2) can be expressed as an equation in H, namely $Au = f$, where A is a maximal monotone operator (cf. Proposition 2.1.2). Moreover, (2.1.4) implies that A is strongly monotone and hence $R(A) = H$. $\blacksquare$

The variational case

If β is the subdifferential of a proper, convex and lower semicontinuous function $j \colon \mathbb{R}^2 \to (-\infty, \infty]$, then the solution of problem (2.1.1)-(2.1.2) is a minimum point of some appropriate convex function. Indeed, let us consider the function $\Psi \colon H \to (-\infty, \infty]$, $H = L^2(0,1)$, defined by

$$\Psi(v) = \begin{cases} \int_0^1 \Big(g\big(r, v'(r)\big) + k\big(r, v(r)\big) - f(r)v(r) \Big) \, dr + j\big(v(0), v(1)\big) \\ \qquad \text{if } v \in H^1(0,1), \ g(\cdot, v') \in L^1(0,1), \ \big(v(0), v(1)\big) \in D(j), \\ \infty \ \text{ otherwise,} \end{cases}$$

where

$$g(r, \xi) = \int_0^\xi G(r, s) \, ds \ \text{ and } \ k(r, \xi) = \int_0^\xi K(r, s) \, ds.$$

PROPOSITION 2.1.3
([AiMoP]). If (I.1), (I.2) with $k_1 = 0$, and (I.4) are fulfilled, and $\beta = \partial j$, where $j \colon \mathbb{R}^2 \to (-\infty, \infty]$ is a proper, convex, and lower semicontinuous

function, then the function Ψ defined above is also proper, convex, and lower semicontinuous.

PROOF   If $(a, b) \in D(j)$, then the function

$$\hat{v}\colon [0, 1] \to \mathbb{R}, \ \hat{v}(r) = a(1 - r) + br,$$

belongs to the set $D(\Psi)$, that is $D(\Psi) \neq \emptyset$. As the convexity of the function Ψ is obvious, it remains to show that Ψ is lower semicontinuous. To this end, it suffices to prove that for any $\lambda \in \mathbb{R}$ the level set $\{v \in H \mid \Psi(v) \leq \lambda\}$ is closed in H. So, for some fixed λ we consider a sequence (v_n) with $\Psi(v_n) \leq \lambda$ and that converges to some function v in H. By (2.1.3) it is easily seen that

$$g(r, \xi) \geq \frac{k_0}{3}\xi^2 - c_1 \ \text{ for all } (r, \xi) \in [0, 1] \times \mathbb{R}, \tag{2.1.28}$$

where c_1 is some positive constant.

Now, using (2.1.28), the boundedness of the sequence $\left(\|v_n\|_H\right)$ and the fact that j is bounded from below by an affine function, one can see that (v_n) is bounded in $H^1(0, 1)$. Therefore, $v_n \to v$ in $C[0, 1]$ and $v_n' \to v'$ weakly in H, as $n \to \infty$, on a subsequence. It follows that

$$\liminf_{n \to \infty} \Psi(v_n) \geq \Psi(v). \tag{2.1.29}$$

Here, we have used the fact that the function $w \mapsto \int_0^1 g\big(r, w(r)\big)\, dr$ is lower semicontinuous on H (hence, equivalently, it is also weakly lower semicontinuous on H). This can easily be proved with the help of inequality (2.1.28). By (2.1.29) one can deduce that $\Psi(v) \leq \lambda$, i.e., the level set $\{v \in H \mid \Psi(v) \leq \lambda\}$ is closed, as asserted. ∎

REMARK 2.1.1   Assume that the assumptions of the preceding proposition are fulfilled. It is easily seen that

$$-f + Au \in \partial \Psi u \ \text{ for all } u \in D(A).$$

As A is maximal monotone, the operator $-f + A$ coincides with $\partial \Psi$, which implies that every solution of problem (2.1.1)-(2.1.2) (i.e., of the equation $Au = f$) is a minimizer of the function Ψ. ∎

2.2   Degenerate elliptic boundary value problems

In this section we shall study a class of degenerate, nonlinear, second order boundary value problems. In order to illustrate the applicability of such problems, we begin with a classical model from the theory of capillarity, which is less discussed in literature but very important. As we shall see, this problem leads to some interesting one-dimensional or N-dimensional generalizations, which can also be associated with other applications.

The classical model of capillarity

Denote by $u = u(x,y)$ the height of the liquid surface in a vertical tube with respect to the reference plane $u = 0$ (see the figure below). Denote by Ω the domain occupied by the tube in the plane $u = 0$. It is well known (see [Finn], [GilTru, pp. 262-263], [LanLif, § 61]) that the equilibrium shape of the liquid surface in the tube, in the case when the gravity field is uniform and the surface tension is constant, is described by the following classical *equation of capillarity*

$$\operatorname{div} \frac{\nabla u(x,y)}{\sqrt{1 + \|\nabla u(x,y)\|^2_{\mathbb{R}^2}}} = k_2 u(x,y) \text{ for all } (x,y) \in \Omega, \qquad (2.2.1)$$

where k_2 is a positive constant depending on the liquid. We have denoted by ∇u the gradient of the function $u = u(x,y)$, i.e., the two-dimensional vector with the components $\partial u/\partial x$, $\partial u/\partial y$, while $\|\cdot\|_{\mathbb{R}^2}$ stands for the Euclidean norm of the space $\mathbb{R}^2$. We associate with (2.2.1) the following natural boundary condition

$$\left(1 + \|\nabla u(x,y)\|^2_{\mathbb{R}^2}\right)^{-\frac{1}{2}} \frac{\partial u}{\partial n} = \cos\gamma, \qquad (2.2.2)$$

where γ represents the *contact angle*, i.e., the angle between the liquid surface and the lateral surface of the tube, while n is the corresponding outward normal to $\partial\Omega$ (which is assumed to be sufficiently smooth).

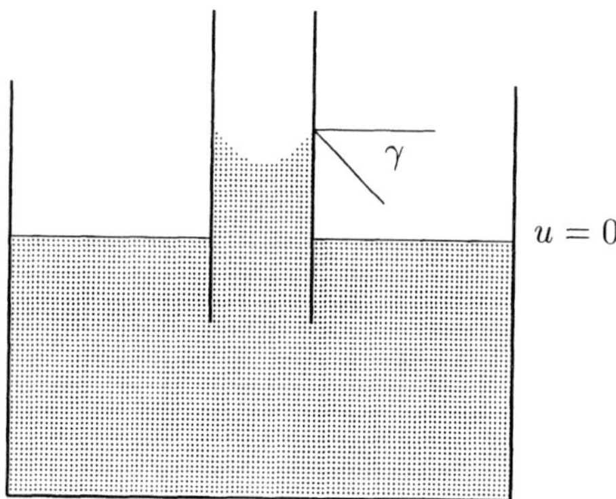

The derivation of (2.2.1) can be done by combining some physical and geometric considerations (see, e.g., [CoMo1]). Also, it is easily seen that problem (2.2.1)-(2.2.2) can be viewed as a particular case of a more general problem. Indeed, let us consider the functional

$$J(u) = \int_\Omega \left(\frac{1}{2}\Phi\big(\|\nabla u(x)\|^2\big) + h\big(u(x)\big) \right) dx - \int_{\partial\Omega} \phi(x)u(x)\, d\sigma, \qquad (2.2.3)$$

where Ω is a domain of $\mathbb{R}^N$, $N \in \mathbb{N}$, $n \geq 2$, with a sufficiently smooth boundary $\partial\Omega$. Without introducing the precise assumptions concerning the functions ϕ, h, and ϕ, one can easily check by a formal computation that every critical point of the functional J (i.e., every function u at which the Gâteaux derivative of J vanishes) is a solution of the following boundary value problem

$$\mathrm{div}\,\Phi'\left(\|\nabla u(x)\|^2_{\mathbb{R}^2}\right)\nabla u(x) = h'\big(u(x)\big), \ \ x \in \Omega, \qquad (2.2.4)$$

$$\Phi'\left(\|\nabla u(x)\|^2_{\mathbb{R}^2}\right)\frac{\partial u}{\partial n}(x) = \phi(x), \ \ x \in \partial\Omega. \qquad (2.2.5)$$

This problem contains as a particular case problem (2.2.1)-(2.2.2). Indeed, to see that, it suffices to take

$$N = 2, \ \ \Phi(\xi) = 2\big(\sqrt{1+\xi} - 1\big), \ \ \phi = \cos\gamma, \ \text{and } h(\xi) = \frac{k_2}{2}\xi^2.$$

From this discussion we can deduce that the capillarity problem can be put in a variational form. Of course, the functional (2.2.3) is very general and so is the boundary value problem (2.2.4)-(2.2.5); hence, many practical problems are particular cases of it. Let us notice that establishing the existence of critical points of the functional J is a difficult task. Even in the particular case of the capillarity problem, we can see the functional J is coercive at most on the space $W^{1,1}(\Omega)$, because of the presence of the term

$$\int_\Omega \sqrt{1 + \|\nabla u(x)\|^2_{\mathbb{R}^2}}\, dx.$$

As the space $W^{1,1}(\Omega)$ is not reflexive, the existence of critical points is not at all obvious. This remark underlines the difficulty of the general problem (2.2.4)-(2.2.5).

We shall consider in what follows the particular case in which Ω is the unit sphere of $\mathbb{R}^N$, i.e.,

$$\Omega = B(0,1) = \{x = (x_1, x_2, \ldots, x_N) \in \mathbb{R}^N \mid x_1^2 + x_2^2 + \ldots + x_N^2 < 1\}$$

and u depends only on

$$\|x\|_{\mathbb{R}^N} = \sqrt{x_1^2 + \ldots + x_N^2},$$

which means that u is a radially symmetric function. Consequently, we have to admit that the function ϕ of (2.2.5) is a constant function.

In the case of the capillarity problem (2.2.1)-(2.2.2) the assumption that $\Omega = B(0,1)$ means that the tube is circular. Then, the assumption that u is radially symmetric is superfluous. Indeed, it is known that the capillarity problem has at most one solution [Finn] and, on the other hand, one can prove the existence of a radially symmetric solution if $\Omega = B(0,1)$, as it is shown below. Actually, from a physical viewpoint, it is natural to suppose that u is a radially symmetric function in the case of a circular tube, because the liquid surface in the tube is a rotational surface.

Now, let us return to the general case $N \in \mathbb{N}$, $N \geq 2$, and suppose that $\Omega = B(0,1)$ and $u = u(r)$, $r = \|x\|_{\mathbb{R}^N}$. Naturally, we shall use for problem (2.2.4)-(2.2.5) the spherical coordinates $r, \theta_1, \ldots, \theta_{N-1}$:

$$\begin{cases} x_1 = r \cos \theta_1, \\ x_2 = r \sin \theta_1 \cos \theta_2, \\ x_3 = r \sin \theta_1 \sin \theta_2 \cos \theta_3, \\ \vdots \\ x_{N-1} = r \sin \theta_1 \sin \theta_2 \ldots \sin \theta_{N-2} \cos \theta_{N-1}, \\ x_N = r \sin \theta_1 \sin \theta_2 \ldots \sin \theta_{N-2} \sin \theta_{N-1}, \end{cases}$$

where $0 \leq r \leq 1$, $0 \leq \theta_1 \leq \pi, \ldots$, $0 \leq \theta_{N-2} \leq \pi$, $0 \leq \theta_{N-1} \leq 2\pi$. By straightforward computations, we get the following one-dimensional problem

$$\frac{d}{dr}\left(r^{N-1} u'(r) \Phi'\big(u'(r)^2\big)\right) = r^{N-1} h'\big(u(r)\big) \quad \text{for all } r \in (0,1), \quad (2.2.6)$$

$$u'(1)\phi'\big(u'(1)^2\big) = C. \quad (2.2.7)$$

Also, by a usual computation (see, e.g., [Fichte, p. 366]), it is easily seen that the functional J can be written as

$$J(u) = \frac{2\pi^{N/2}}{\Gamma(N/2)} \left(\int_0^1 r^{N-1} \left(\frac{1}{2}\Phi'\big(u'(r)^2\big) + h\big(u(r)\big)\right) dr - Cu(1)\right), \quad (2.2.8)$$

where Γ is the classical Euler function

$$\Gamma(p) = \int_0^\infty t^{p-1} \mathrm{e}^{-t}\, dt, \quad p > 0.$$

If one denotes

$$g(\xi) = \frac{1}{2}\Phi(\xi^2), \quad G(\xi) = g'(\xi) = \xi\Phi'(\xi^2), \quad K(\xi) = h'(\xi),$$

then the problem (2.2.6)-(2.2.7) can be written in the form

$$\frac{d}{dr}\left(r^{N-1} G\big(u'(r)\big)\right) = r^{N-1} K\big(u(r)\big), \quad r \in (0,1), \quad (2.2.9)$$

$$G\big(u'(1)\big) = C. \quad (2.2.10)$$

In addition, the functional J becomes

$$J(u) = \frac{2\pi^{N/2}}{\Gamma(N/2)} \left(\int_0^1 r^{N-1} \Big(g\big(u'(r)\big) + h\big(u(r)\big) \Big) \, dr - Cu(1) \right). \quad (2.2.11)$$

Notice that (2.2.9) is degenerate because the coefficient $p(r) = r^{N-1}$ vanishes for $r = 0$. We also observe that problem (2.2.9)-(2.2.10) has no condition at $r = 0$ as a consequence of its degenerate character. In what follows, we formulate and study a problem even more general than problem (2.2.9)-(2.2.10). More precisely, we are going to investigate the following doubly nonlinear, multivalued, possibly degenerate, second order boundary value problem

$$0 \in -\frac{d}{dr} \Big(p(r)G\big(u'(r)\big) \Big) + q(r)K\big(u(r)\big), \; r \in (0,1), \quad (2.2.12)$$

$$0 \in p(r)G\big(u'(r)\big)\Big|_{r=0^+}, \; C \in p(1)G\big(u'(1)\big). \quad (2.2.13)$$

The precise meaning of this problem will be explained later, but clearly it is a natural extension of the classical problem in which we have equalities instead of inclusions.

Let us now list our assumptions.

(A.1) The mapping $G: D(G) \subset \mathbb{R} \to \mathbb{R}$ is maximal monotone (possibly multivalued) and strictly monotone. Moreover, $(0,0) \in G$.

(A.2) The function p belongs to $C\big((0,1]\big)$, $p(r) > 0$ for all $r \in (0,1]$; $q \in L^1(0,1)$, $q(r) > 0$ for a.a. $r \in (0,1)$; for every Lipschitz continuous and nondecreasing function $z: C[0,1] \to \mathbb{R}_+ = [0,\infty)$, the mapping

$$r \mapsto \frac{1}{p(r)} \int_0^r q(s)z(s) \, ds$$

is also nondecreasing in $(0,1]$; and finally we have

$$\lim_{r \to 0^+} \frac{1}{p(r)} \int_0^r q(s) \, ds = 0. \quad (2.2.14)$$

(Remark that the limit in (2.2.14) always exists, because of the previous condition in which we take $z(r) = 1$, but we require this limit to be zero.)

(A.3) The real constant C satisfies $C/p(1) \in R(G)$, where $R(G)$ denotes the range of G.

(A.4) The mapping $K: D(K) \subset \mathbb{R} \to \mathbb{R}$ is maximal monotone (possibly multivalued) and $(0,0) \in K$; if $C > 0$, there exists $\gamma \in D(K)$, $\gamma > \beta := G^{-1}\big(C/p(1)\big)$, such that

$$\frac{CC_1}{p(1)} < \sup K(\gamma - \beta), \quad (2.2.15)$$

and, respectively, if $C < 0$, there exists $\gamma \in D(K)$, $\gamma < \beta$, such that

$$\frac{CC_1}{p(1)} > \inf K(\gamma - \beta), \tag{2.2.16}$$

where

$$C_1 := \frac{p(1)}{\int_0^1 q(s)\, ds}.$$

We continue with some remarks concerning our assumptions.

REMARK 2.2.1 By (A.2),

$$C_1 = \inf_{0 < r \le 1} \frac{p(r)}{\int_0^r q(s)\, ds}.$$

∎

REMARK 2.2.2 If $\gamma - \beta \in \operatorname{Int} D(K)$ it follows by Rockafellar's theorem (see Theorem 1.2.1) that $K(\gamma - \beta)$ is a bounded set. In fact, it is a bounded closed interval of real numbers (possibly a singleton), because K is a maximal monotone mapping. Therefore, $\sup K(\gamma - \beta)$ and $\inf K(\gamma - \beta)$ are finite numbers. It is also possible that $\beta = 0$ and γ be the right or left end point of the interval $D(K)$ and in this case the right hand side of (2.2.15) (respectively, (2.2.16)) is ∞ (respectively, $-\infty$). ∎

REMARK 2.2.3 As G and K are assumed to be nonlinear and multivalued, it is natural to say that problem (2.2.12)-(2.2.13) is *doubly nonlinear and multivalued.* ∎

REMARK 2.2.4 Our assumptions allow the function p to vanish at $r = 0$ (more precisely, $\lim_{r \to 0+} p(r) = 0$) or, even more, to have a singularity at $r = 0$ (for example, $p(r) = r^a$, $q(r) = r^b$ with $a, b \in \mathbb{R}$, $b + 1 > \max\{0, a\}$). That is why we call our boundary value problem (2.2.12)-(2.2.13) *possibly degenerate* (according, e.g., to the terminology of [Mikhl, Ch. 7] for linear elliptic partial differential equations). ∎

In what follows we shall suppose that assumptions (A.1)-(A.4) hold if not otherwise stated. In order to make clear the meaning of problem (2.2.12)-(2.2.13), let us give some notions of solution and discuss them by means of some appropriate examples.

DEFINITION 2.2.1   *By a solution of problem (2.2.12)-(2.2.13) we mean a function $u \in C^1[0,1]$ such that*

$$u(r) \in D(K), \ u'(r) \in D(G) \ \ \text{for all } r \in [0,1], \qquad (2.2.17)$$

$$u'(1) = \beta := G^{-1}\left(\frac{C}{p(1)}\right), \qquad (2.2.18)$$

and there exists a function $v \in AC[0,1]$ satisfying

$$v(r) \in p(r)G\big(u'(r)\big) \ \ \text{for all } r \in (0,1], \qquad (2.2.19)$$

$$v'(r) \in q(r)K\big(u(r)\big) \ \ \text{for a.a. } r \in (0,1), \qquad (2.2.20)$$

$$v(0) = 0. \qquad (2.2.21)$$

We have denoted as usual by $C^1[0,1]$ the space of all real valued continuously differentiable functions on $[0,1]$ and by $AC[0,1]$ the space of all absolutely continuous functions $[0,1] \to \mathbb{R}$.

We may also consider the following alternate concept of solution:

DEFINITION 2.2.2   *The function $u \in C^1[0,1]$ is a solution of problem (2.2.12)-(2.2.13) if u satisfies the conditions of Definition 2.2.1 except for (2.2.18), which is replaced by*

$$v(1) = C. \qquad (2.2.22)$$

Obviously, if u is a solution in the sense of the last definition, then it is also a solution in the sense of Definition 2.2.1. In general, the converse is not true, as Examples 2.2.1 and 2.2.2 below show. Hence, the second definition is stronger than the first one. As one can observe immediately, in the second case we have uniqueness, at least up to an additive constant, while in the first case this may not happen (see also Examples 2.2.1 and 2.2.2 below). That is a consequence of the fact that G is multivalued. Of course, if $CG^{-1}\big(C/p(1)\big) = C/p(1)$ then the two notions of solution are identical. This is the case for any C satisfying (A.3), if G is, in addition, a single-valued mapping.

Example 2.2.1

Let $p(r) = q(r) = r$, $C \geq 0$, and let $G, K \subset \mathbb{R} \times \mathbb{R}$ be defined by

$$K(\xi) = \xi, \ G(\xi) = \begin{cases} \xi & \text{if } \xi < 0, \\ [0,1] & \text{if } \xi = 0, \\ \xi + 1 & \text{if } \xi > 0. \end{cases}$$

It is easy to see that our assumptions (A.1)-(A.4) are all satisfied. Let u be a solution of the boundary value problem (2.2.12)-(2.2.13) in the sense of Definition 2.2.1 Then, for all $r \in [0, 1]$,

$$u'(r) = G^{-1}\left(\frac{1}{r}\int_0^r su(s)\,ds\right) = \tag{2.2.23}$$

$$= G^{-1}\left(\frac{1}{r}\int_0^r s\left(u(0) + \int_0^s u'(\sigma)\,d\sigma\right)ds\right).$$

From this equation we can see that if $u(0) > 0$ then $u' \geq 0$ on $[0, 1]$ and hence (see again (2.2.23)) u' is nondecreasing in $[0, 1]$. Similarly, if $u(0) < 0$, then u' is nonpositive and nonincreasing in $[0, 1]$. If $u(0) = 0$, then Gronwall's inequality applied to (2.2.23) shows that u is identically zero. On the other hand, (2.2.23) implies that $u'(0) = 0$. If $C \in [0, 1]$ then (2.2.18) reads $u'(0) = 0$ and hence u' is identically zero, because of its monotonicity. It is then easy to see that, for $C \in [0, 1]$, the constant functions $u(r) = C_1$, $C_1 \in [0, 2]$ are solutions in the sense of Definition 2.2.1. Hence, we have existence, without uniqueness. On the other hand, for each $C \in [0, 1]$, our boundary value problem admits the unique solution $u(r) = 2C$, in the sense of Definition 2.2.2.

For $C > 1$, the conditions (2.2.18) and (2.2.22) coincide and therefore the two concepts of solution are identical. In this case, Theorem 2.2.1 below guarantees existence and uniqueness. $\quad\square$

Example 2.2.2

Let $p(r) = q(r) = 1$, $C \geq 0$ and let $G, K \subset \mathbb{R} \times \mathbb{R}$ be given by

$$K(\xi) = \xi, \ G(\xi) = \begin{cases} \xi & \text{if } \xi < 1, \\ [1, 2] & \text{if } \xi = 1, \\ \xi + 1 & \text{if } \xi > 1. \end{cases}$$

In this case, it is easy to prove the uniqueness of the solution of our boundary value problem, in the sense of Definition 2.2.2. (Anyway, we shall reconsider this issue in the general framework of our assumptions).

Now, let $u \in C^1[0, 1]$ be a solution in the sense of Definition 2.2.1. Then, we can write the identity

$$v(r)u(r) = \int_0^r \left(v(s)u'(s) + u(s)^2\right)ds \ \text{ for all } r \in [0, 1].$$

This implies that

$$\{r \in (0, 1] \mid u(r) = 0\} = \{r \in (0, 1] \mid u'(r) = 0\}, \tag{2.2.24}$$

and this set is either the empty set or an interval of the form $(0, \delta]$. If $C = 0$, then clearly $u = 0$. Now, suppose that $C > 0$. Then $u'(1) > 0$

and hence, according to the above remark concerning the form of the set in (2.2.24), $u' \geq 0$ in $[0, 1]$. From the obvious equation

$$u'(r) = G^{-1}\left(\int_0^r u(s)\,ds\right) \quad \text{for all } r \in [0, 1], \tag{2.2.25}$$

we can deduce that $u(0) \geq 0$, and hence $u \geq 0$ in $[0, 1]$. Looking again at (2.2.25), we then deduce that u' is nondecreasing in $[0, 1]$. Now, if $0 < C < 1$, the set $U = \{r \in [0, 1] \mid u'(r) < 1\}$ coincides with $[0, 1]$. Therefore, u satisfies the problem

$$u'' = u \quad \text{a.e. in } (0, 1), \quad u'(0) = 0, \quad u'(1) = C,$$

which has the unique solution:

$$u(r) = \frac{C}{e - e^{-1}}\left(e^r + e^{-r}\right) \quad \text{for all } r \in [0, 1]. \tag{2.2.26}$$

In fact, in this case the two concepts of solution coincide.

Now, for every $C \in [1, 2]$, we have the same boundary conditions

$$u'(0) = 0, \quad u'(1) = 1.$$

As u' is nondecreasing, the interval $[0, 1]$ can be decomposed into two subintervals, as follows:

$$\{r \in [0, 1] \mid u'(r) < 1\} = [0, r_0), \quad \{r \in [0, 1] \mid u'(r) = 1\} = [r_0, 1],$$

where $r_0 \in (0, 1]$. An elementary computation reveals to us that u is given by

$$u(r) = \begin{cases} \dfrac{\cosh r}{\sinh r_0} & \text{if } 0 \leq r \leq r_0, \\ r - r_0 + \coth r_0 & \text{if } r_0 < r \leq 1 \end{cases} \tag{2.2.27}$$

for all $r_0 \in (0, 1]$ verifying the inequality

$$(1 - r_0)\coth r_0 + \frac{(1 - r_0)^2}{2} \leq 1. \tag{2.2.28}$$

So, we may conclude that for every $C \in [1, 2]$, our boundary value problem has an infinite number of solutions in the sense of Definition 2.2.1 (the same solutions for every $C \in [1, 2]$).

Now, we may ask ourselves, what about the solutions in the sense of Definition 2.2.2 for $C \in [1, 2]$.

First, for $C = 1$, the (unique) solution in the sense of Definition 2.2.2 is given by (2.2.26) with $C = 1$. Let us now take $C \in (1, 2]$ and denote by u_C the corresponding solution in the sense of Definition 2.2.2, assuming that it does exist. Then clearly there exists a number $r_0 \in (0, 1)$ such that u_C coincides with u given by (2.2.27). An easy computation, involving all the conditions of Definition 2.2.2, shows that r_0 should necessarily satisfy the following condition

$$(1 - r_0)\coth r_0 + \frac{1}{2}(1 - r_0)^2 = C - 1. \qquad (2.2.29)$$

But (2.2.29) has a unique solution and hence, for every $C \in [1,2]$, our boundary value problem has a unique solution in the sense of Definition 2.2.2.

Finally, for $C > 2$ the two notions of solution coincide again, because (2.2.18) and (2.2.22) are identical: $u'(1) = C - 1$. Therefore, in this case there exists a unique solution, given by Theorem 2.2.1 below. In fact, we can find the explicit solution in this case, namely

$$u(r) = \frac{C - 1}{e - e^{-1}}\left(e^{r} - e^{-r}\right), \; 0 \le r \le 1. \qquad (2.2.30)$$

Observe that we have the same solution u_C for $C = 1$ and $C = 2$. On the other hand, we can see form (2.2.29) that r_0 depends continuously on $C \in [1,2]$. Therefore, taking into account (2.2.26), (2.2.27), and (2.2.30), we can deduce that u_C depends continuously on C. □

We recommend that the reader also examines the same example but with $p(r) = q(r) = r$. This is a multivalued and degenerate problem and similar aspects can be observed. Of course, in this case the boundary condition at $r = 0$ is automatically satisfied, so it is superfluous.

The main result of this section is the following theorem.

THEOREM 2.2.1

[MoAZ] If assumptions (A.1)-(A.4) are satisfied, then problem (2.2.12)-(2.2.13) has a solution in the sense of Definition 2.2.2, which is unique up to an additive constant. If, in addition, K is strictly increasing, then the solution in the sense of Definition 2.2.2 is unique.

Before proving this result, let us discuss our assumptions, by using several adequate examples. First of all, we remark that the strict monotonicity of G is essential. Otherwise, problem (2.2.12)-(2.2.13) may have no solutions, even if all other assumptions of Theorem 2.2.1 hold. Here is an example in this sense.

Example 2.2.3

Let $G: \mathbb{R} \to \mathbb{R}$ be the single valued (but not strictly monotone) function defined by

$$G(\xi) = \begin{cases} \xi & \text{if } \xi < 1, \\ 1 & \text{if } 1 \le \xi \le 2, \\ \xi - 1 & \text{if } \xi > 2. \end{cases}$$

Consider the following boundary value problem (which satisfies (A.1)-(A.4) except for the strict monotonicity of G)

$$\frac{d}{dr}\Big(rG\big(u'(r)\big)\Big) = ru(r) \text{ for } r \in (0,1), \ G\big(u'(1)\big) = 2. \qquad (2.2.31)$$

Observe that in this case the first condition of (2.2.13) is superfluous whereas the second one coincides with $u'(1) = 3$, that is, the two concepts of solution are now identical. Let us suppose that (2.2.31) has a solution u. From the obvious equation

$$G\big(u'(r)\big) = \frac{1}{r}\int_0^r su(s)\,ds \text{ for all } r \in [0,1],$$

we can see that $u'(0) = 0$. As $u'(1) = 3$, the set of values of u' contains the interval $[0,3]$, because u' has the Darboux property. On the other hand, in the open set

$$U = \{r \in (0,1) \mid 1 < u'(r) < 2\}$$

the function u satisfies the equation $1 = ru(r)$ and this implies that $U = \emptyset$. The contradiction we have arrived at shows that (2.2.31) has no solution.
$\square$

As the next two examples show, there are, however, special situations when existence of solution is possible without the strict monotonicity of G.

Example 2.2.4
Take $C \geq 0$, $p(r) = q(r) = 1$, $K(\xi) = \xi$, and $G\colon \mathbb{R} \to \mathbb{R}$,

$$G(\xi) = \begin{cases} 0 & \text{if } \xi \leq 1, \\ \xi - 1 & \text{if } \xi > 1. \end{cases}$$

It is easy to see that the solution of the corresponding boundary value problem in the sense of Definition 2.2.2 is unique for any $C \geq 0$ (if it exists). For $C = 0$ this is identically zero. In this case, the condition (2.2.18) of Definition 2.2.1 should be reformulated. (In fact, even in the previous example, (2.2.18) does not make sense in that form if $C = 1$.) If $C > 0$, then (2.2.18) and (2.2.22) are identical: $u'(1) = C + 1$. On the other hand, we have

$$\{r \in [0,1] \mid u'(r) < 1\} \subset \{r \in [0,1] \mid u(r) = 0\}$$

and consequently, as u' has the Darboux property and $u'(1) > 1$, we necessarily have $u' \geq 1$ in $[0,1]$. Therefore, for $C > 0$ our boundary value problem becomes

$$\begin{cases} u''(r) = u(r), \ r \in (0,1), \\ u'(0) = 1, \ u'(1) = C + 1, \\ u'(r) \geq 1 \ \text{for all } r \in [0,1]. \end{cases} \tag{2.2.32}$$

But (2.2.32) has the unique solution

$$u(r) = C_1(e^r + e^{-r}), \ C_1 = \frac{C + 1 - e}{e - e^{-1}}.$$

$\square$

Example 2.2.5

Take $0 \leq C \leq 1$, $p(r) = q(r) = r$, $K(\xi) = \xi$ and let $G \subset \mathbb{R} \times \mathbb{R}$ be the multi-valued Heaviside function,

$$G(\xi) = \begin{cases} 0 & \text{if } \xi < 0, \\ [0,1] & \text{if } \xi = 0, \\ 1 & \text{if } \xi > 0. \end{cases}$$

It is easily seen that for each $C \in [0,1]$ this particular case of problem (2.2.12)-(2.2.13) has the unique solution $u(r) = 2C$, in the sense of Definition 2.2.2. On the other hand, for every $C \in [0,1]$, the constant functions $u(r) = C_1$, $0 \leq C_1 \leq 2$ are solutions in the sense of Definition 2.2.1. $\square$

The reader is encouraged to consider the same example but with G replaced by the multivalued sign function.

The last two examples also show that our assumption (A.1) is not even minimal.

The Darboux property indicates to us that, in order to have existence, G and K should be assumed to be maximal monotone mappings. The next example will clarify this remark.

Example 2.2.6

Set $C = 3$, $p(r) = q(r) = 1$, $H(\xi) = \xi$, $G \colon \mathbb{R} \to \mathbb{R}$,

$$G(\xi) = \begin{cases} \xi & \text{if } \xi \leq 1, \\ \xi + 1 & \text{if } \xi > 1, \end{cases}$$

and $C = 3$. If the boundary value problem corresponding to (2.2.12)-(2.2.13) admits a solution $u \in C^1[0,1]$, it follows that $u'(0) = 0$ and $u'(1) = 2$. Hence the range of u' is an interval I that includes $[0,2]$. But $G(I)$ is not an interval and therefore u cannot satisfy the equation

$$G\big(u'(r)\big) = \int_0^r u(s)\,ds \ \text{for all } r \in [0,1].$$

This situation will no longer appear if G is replaced by the corresponding multivalued extension $\tilde{G} \subset \mathbb{R} \times \mathbb{R}$,

$$\tilde{G}(\xi) = \begin{cases} G(\xi) & \text{if } \xi \neq 1, \\ [1,2] & \text{if } \xi = 1, \end{cases}$$

which is a maximal monotone mapping. Similar arguments show that K must also be maximal monotone. In fact, as we shall see, it is enough to assume that G and K are restrictions of maximal monotone operators, such that their graphs are continuous curves in $\mathbb{R}^2$. $\quad\Box$

As regards the assumption (A.2), this is technical and perhaps it could be weakened. But in its current form it covers a wide class of applications.

In what follows, we shall construct two examples that show that condition (2.2.15) (or (2.2.16)) is not only essential but also minimal for existence.

Example 2.2.7

Take $p(r) = q(r) = 1$, $C = 1$, and $G, K : \mathbb{R} \to \mathbb{R}$,

$$G(\xi) = \xi^{2k+1}, \quad K(\xi) = \begin{cases} \xi & \text{if } \xi \leq a, \\ a & \text{if } \xi > a, \end{cases}$$

where $k \in \mathbb{N}$ and $a \in \mathbb{R}$.

As G is strictly monotone, the two notions of solution coincide. If $a > 1$, then (2.2.15) is satisfied and the existence for the boundary value problem corresponding to (2.2.12)-(2.2.13) is guaranteed by Theorem 2.2.1 above.

Now, we consider the case $0 < a < 1$ for which (2.2.15) is no longer valid. We suppose that our boundary value problem has a solution $u \in C^1[0,1]$. We multiply (2.2.12) by $u(r)$ and then integrate the result over $[0,r]$ to obtain:

$$u(r)u'(r)^{2k+1} = \int_0^r \left(u'(s)^{2k+2} + u(s)K\big(u(s)\big) \right) ds. \tag{2.2.33}$$

From (2.2.33) we can see that

$$\{r \in (0,1] \mid u(r) = 0\} = \{r \in (0,1] \mid u'(r) = 0\} \tag{2.2.34}$$

and this set is either an empty set or an interval of the form $(0,\delta]$. As $u'(1) = 1$ it follows that $u' \geq 0$ in $[0,1]$; so, u is nondecreasing in $[0,1]$. By (2.2.33) and (2.2.34) it follows that $u \geq 0$ in $[0,1]$ and this implies that u' is nondecreasing, because (see (2.2.12))

$$u'(r) = \left(\int_0^r K\big(u(s)\big) ds \right)^{\frac{1}{2k+1}}.$$

In particular,

$$0 = u'(0) \le u'(r) \le 1 \ \text{ for all } r \in [0,1]. \tag{2.2.35}$$

On the other hand, multiplying (2.2.12) by u' and then integrating over $[0,r]$, one gets

$$\frac{2k+1}{2k+2} u'(r)^{2k+2} = h\big(u(r)\big) + C_2, \tag{2.2.36}$$

where C_2 is some real number and

$$h(\xi) = \begin{cases} \frac{1}{2}\xi^2 & \text{if } \xi < a, \\ \frac{a}{2}(2\xi - a) & \text{if } \xi \ge a. \end{cases}$$

From (2.2.35) and the Mean Value Theorem it follows that there exists a point $\alpha \in (0,1)$ such that

$$\frac{2k+1}{2k+2} = h\big(u(1)\big) - h\big(u(0)\big) = K\big(u(\alpha)\big) u'(\alpha).$$

Therefore (see also (2.2.35)),

$$\frac{2k+1}{2k+2} \le a,$$

but this inequality is impossible for k large enough. Consequently, for such k our boundary value problem has no solution!

The limit case $a = 1$, for which we have equality in (2.2.15), remains open. ∎

Example 2.2.8

We invite the reader to consider the same data as in the previous example except for K, which is replaced by the following (strictly increasing) function

$$K : \mathbb{R} \to \mathbb{R}, \ \ K(\xi) = a \arctan \xi,$$

where $a \in (0, 2/\pi)$. Clearly, (2.2.15) is not satisfied, and repeating, step by step, the reasoning used in the previous example we can show that our boundary value problem (2.2.12)-(2.2.13) has no solution for large k. For $a > 2/\pi$, inequality (2.2.15) holds and Theorem 2.2.1 above says that our boundary value problem has a unique solution for any positive integer k (see Remark 2.2.7 below). In the limit case $a = 2/\pi$, the inequality (2.2.15) is still not satisfied. In this case, our boundary value problem has a unique solution. However, this is a limit case. ∎

The above two examples show very clearly that, even in the case in which our boundary value problem is nondegenerate, the contribution of the nonlinear K is very important for existence (by (2.2.15) or (2.2.16) K should be "big enough").

Let us finish this long but necessary discussion by presenting a very simple example (in fact, a counterexample) that shows that if in Theorem 2.2.1 K is not strictly increasing, then the solution of problem (2.2.12)-(2.2.13) may not be unique (of course, it is however unique up to an additive constant).

Example 2.2.9

Let $p(r) = q(r) = 1$, $C = 1$, and let $G, K \colon \mathbb{R} \to \mathbb{R}$ be given by

$$G(\xi) = \xi, \quad K(\xi) = \begin{cases} \xi & \text{if } \xi < 1, \\ 1 & \text{if } 1 \leq \xi \leq 2, \\ \xi - 1 & \text{if } \xi > 2. \end{cases}$$

As (A.1)-(A.4) are all satisfied, the existence is assured by Theorem 2.2.1. Moreover, it is easily seen that all the functions

$$u(r) = \frac{r^2}{2} + C_1, \quad 1 \leq C_1 \leq \frac{3}{2}, \tag{2.2.37}$$

are solutions of the corresponding boundary value problem:

$$u'' = K(u), \quad r \in (0, 1), \quad u'(0) = 0, \quad u'(1) = 1.$$

In fact, there are no other solutions of this boundary value problem. Indeed, by Theorem 2.2.1 above, we have uniqueness up to an additive constant and, on the other hand, u given by (2.2.37), with $C_1 < 1$ or $C_1 > \frac{3}{2}$, cannot satisfy (2.2.12). $\Box$

Now, we are going to prove Theorem 2.2.1. Notice that particular cases of this result can be found in [CoMo1], [CoMo2], [CoMo3], [Moro2], [Moro3], [Moro4], [MoZo1], [MoZo2].

PROOF of Theorem 2.2.1. The proof is divided into several steps.

Step 1. *Uniqueness.* Let $u_1, u_2 \in C^1[0, 1]$ be two solutions of problem (2.2.12)-(2.2.13) in the sense of Definition 2.2.2 and let $v_1, v_2 \in AC[0, 1]$ be the corresponding selections given by that definition. Using (2.2.19)-(2.2.21) we can easily deduce that

$$0 = \int_0^1 \Big(\big(v_1(r) - v_2(r)\big)\big(u_1'(r) - u_2'(r)\big) + \tag{2.2.38}$$

$$+ q(r)\big(u_1(r) - u_2(r)\big)\big(w_1(r) - w_2(r)\big) \Big) \, dr,$$

where

$$w_i(r) \in K\big(u_i(r)\big) \quad \text{for a.a. } r \in (0, 1), \quad i = 1, 2,$$

and
$$v_i'(r) = q(r)w_i(r) \text{ for a.a. } r \in (0,1), \ i = 1,2.$$
As K is nondecreasing and G is strictly increasing, (2.2.38) yields $u_1' = u_2'$. If, in addition, K is strictly increasing too, then (2.2.38) implies that $u_1 = u_2$.

Step 2. *Reduction to the case $C > 0$ and $\beta > 0$.* Clearly, for $C = 0$, the null function is a solution of problem (2.2.12)-(2.2.13). In what follows we shall discuss only the case $C > 0$, because for $C < 0$ we can use similar arguments. Furthermore, we shall assume that $\beta > 0$. The case $\beta = 0$ is a little bit different and will be analyzed separately.

Step 3. *An auxiliary boundary value problem.* We fix a $C > 0$ satisfying (A.3). Assuming that $\beta > 0$, we define $\tilde{G}, \tilde{K} \colon \mathbb{R} \to \mathbb{R}$ by

$$\tilde{G}(\xi) = \begin{cases} \xi & \text{if } \xi \leq 0, \\ G(\xi) & \text{if } 0 < \xi < \beta, \\ [G^0(\xi), C/p(1)] & \text{if } \xi = \beta, \\ \xi - \beta + C/p(1) & \text{if } \xi \geq \beta, \end{cases}$$

$$\tilde{K}(\xi) = \begin{cases} \xi & \text{if } \xi \leq 0, \\ K(\xi) & \text{if } 0 < \xi < \gamma, \\ \xi - \gamma + K^0(\gamma) & \text{if } \xi \geq \gamma, \end{cases}$$

where β and γ are the constants appearing in (A.4) and

$$G^0(\beta) := \inf G(\beta), \quad K^0(\gamma) := \inf K(\gamma).$$

As $G(\beta)$ and $K(\gamma)$ are closed intervals, we have $G^0(\beta) \in G(\beta)$ and $K^0(\gamma) \in K(\gamma)$. Clearly, $\tilde{G}$ and $\tilde{K}$ are maximal monotone mappings. By replacing G, K in the original boundary value problem (2.2.12)-(2.2.13) by $\tilde{G}, \tilde{K}$ we obtain the problem

$$0 \in -\frac{d}{dr}\Big(p(r)\tilde{G}\big(u'(r)\big)\Big) + q(r)\tilde{K}\big(u(r)\big), \ r \in (0,1), \qquad (2.2.39)$$

$$0 \in p(r)\tilde{G}\big(u'(r)\big)\Big|_{r=0+}, \quad C \in p(1)\tilde{G}\big(u'(1)\big). \qquad (2.2.40)$$

Step 4. *A Cauchy problem associated to a regularized equation.* As $\tilde{G}$ is strictly monotone and maximal monotone, the operator $\tilde{F} = \tilde{G}^{-1}$ is single valued and maximal monotone too. For the time being we assume, in addition to (A.1)-(A.4), that

$$\tilde{F}, \tilde{K} \colon \mathbb{R} \to \mathbb{R} \text{ are Lipschitz continuous.} \qquad (2.2.41)$$

We are going to solve the problem

$$u'(r) = \tilde{F}\Big(\frac{1}{p(r)} \int_0^r q(r)\tilde{K}\big(u(s)\big)\,ds\Big) \text{ for all } r \in [0,1], \qquad (2.2.42)$$

$$u'(1) = \beta. \qquad (2.2.43)$$

First we consider the Cauchy problem consisting of (2.2.42) and the initial condition

$$u(0) = u_0. \qquad (2.2.44)$$

Denoting $y = u'$, this Cauchy problem can be rewritten as the following integral equation

$$y(r) = \tilde{F}\left(\frac{1}{p(r)} \int_0^r q(s)\tilde{K}\left(u_0 + \int_0^s y(\sigma)\,d\sigma\right) ds\right) \quad \text{for all } r \in [0,1]. \quad (2.2.45)$$

Due to (A.2), (2.2.45) makes sense at $r = 0$. Now, we state a lemma.

LEMMA 2.2.1

If (A.2) and (2.2.41) hold, then, for every $u_0 \in \mathbb{R}$, (2.2.45) has a unique solution $y = y(r_0, u_0) \in C^1[0,1]$.

PROOF   We apply the Banach Fixed Point Theorem to the operator $T: C[0,1] \to C[0,1]$, where $(Ty)(r)$ is the right hand side of (2.2.45). It suffices to observe that T is a contraction with respect to the metric

$$d(f,g) = \sup\left\{|f(t) - g(t)|\, e^{-2Lt} \mid t \in [0,1]\right\}$$

if L is a positive and sufficiently large constant.   ∎

     Step 5. *Prove that for regular $\tilde{G}$ and $\tilde{K}$, the problem (2.2.39)-(2.2.40) has a solution in the sense of Definition 2.2.1.* Suppose that (A.1)-(A.4) and (2.2.41) hold. We recall that $y = y(r, u_0)$ denotes the solution of the Cauchy problem (2.2.42)-(2.2.44). In what follows, the following equality

$$\{y(1, u_0) \mid u_0 \geq 0\} = [0, \infty) \qquad (2.2.46)$$

will be proved. In order to do this, we need some properties of $y(r, u_0)$. First, it is clear that

$$y(r, 0) = 0 \quad \text{for all } r \in [0,1]. \qquad (2.2.47)$$

Now, it is easily seen that

$$u_0 > 0 \text{ implies } y(r, u_0) \geq 0 \text{ for all } r \in [0,1], \qquad (2.2.48)$$

and

$$y(0, u_0) = 0 \quad \text{for all } u_0 \in \mathbb{R}. \qquad (2.2.49)$$

Indeed, if $u_0 > 0$, then $u_0 + \int_0^s y(t, u_0)\,dt \geq 0$ in some interval $[0, \delta]$ and hence, by (2.2.45), $y(r, u_0) \geq 0$ for $0 \leq r \leq \delta$. In fact, this interval can be extended to the right up to the interval $[0, \delta_{max}]$ in which $y(r, u_0) \geq 0$.

Moreover, $\delta_{max} = 1$ and so (2.2.48) is proved. As regards (2.2.49), this is a consequence of (2.2.14).

Now, using again (2.2.45) and Gronwall's inequality we can derive the Lipschitz continuity:

$$|y(r, u_0) - y(r, \hat{u}_0)| \leq K_0 |u_0 - \hat{u}_0| \text{ for all } u_0, \hat{u}_0 \in \mathbb{R}, \ r \in [0, 1], \quad (2.2.50)$$

where K_0 is some positive constant. On the other hand, since

$$y(1, u_0) \geq \tilde{F}\left(\frac{1}{p(1)} \tilde{K}(u_0) \int_0^1 q(s)\, ds\right),$$

we have

$$y(1, u_0) \to \infty, \text{ as } u_0 \to \infty. \quad (2.2.51)$$

From (2.2.47), (2.2.48), (2.2.50), and (2.2.51) one derives (2.2.46) as a consequence of the Darboux property. Clearly, (2.2.46) shows that there exists a $\hat{u}_0 \geq 0$ such that $y(1, \hat{u}_0) = \beta$ and hence the function

$$u(r) = \hat{u}_0 + \int_0^r y(s, \hat{u}_0)\, ds$$

is a solution of problem (2.2.42)-(2.2.43). In fact, $\hat{u}_0 > 0$, because $y(1, 0) = 0$ (see (2.2.47)).

Step 6. *Eliminating the assumption (2.2.41).* We admit only the assumptions (A.1)-(A.4) and replace the functions $\tilde{F}$, $\tilde{K}$ by their Yosida approximations $\tilde{F}_\lambda$, $\tilde{K}_\lambda$, i.e.,

$$\tilde{F}_\lambda = \frac{1}{\lambda}(I - J_\lambda) = \tilde{F} J_\lambda, \ J_\lambda = (I + \lambda \tilde{F})^{-1} \text{ for all } \lambda > 0.$$

It is well known that $\tilde{F}_\lambda$ and $\tilde{K}_\lambda$, $\lambda > 0$, are Lipschitz continuous. Therefore, according to Lemma 2.2.1 and Step 5, for each $\lambda > 0$, there exists a solution y_λ of (2.2.45) with $(\tilde{F}_\lambda, \tilde{K}_\lambda)$ instead of $(\tilde{F}, \tilde{K})$ and satisfying $y_\lambda(1) = \beta$. Notice that the conditions of Step 5 are all satisfied by the mappings $\tilde{F}_\lambda$ and $\tilde{K}_\lambda$, including the fact that $\tilde{F}_\lambda$ is linear for sufficiently large arguments and for $0 < \lambda \leq \lambda_0$, where $\lambda_0 > 0$ (this restriction for λ is not essential, because we intend to pass to the limit as $\lambda \to 0^+$).

Indeed, an elementary computation shows us that for such ξ and λ we have

$$\tilde{F}_\lambda(\xi) = \frac{1}{1 + \lambda}\left(\xi + \beta - \frac{C}{p(1)}\right). \quad (2.2.52)$$

Similarly, one gets

$$\tilde{K}_\lambda(\xi) = \frac{1}{1 + \lambda}\left(\xi - \gamma + K^0(\gamma)\right) \quad (2.2.53)$$

for ξ large enough and $\lambda \in (0, \lambda_0]$.

So, for each $\lambda \in (0, \lambda_0]$, there exists a unique solution y_λ of (2.2.45) with $\tilde{F}_\lambda$, $\tilde{K}_\lambda$ instead of $\tilde{F}$, $\tilde{K}$, satisfying the condition $y_\lambda(1) = \beta$. Actually, for each $\lambda \in (0, \lambda_0]$, there exists a $u_{0\lambda} > 0$ such that the function $u_\lambda \in C^1[0, 1]$ defined by

$$u_\lambda(r) = u_{0\lambda} + \int_0^r y_\lambda(s)\, ds \qquad (2.2.54)$$

satisfies the problem

$$u_\lambda'(r) = \tilde{F}_\lambda\left(\frac{1}{p(r)} \int_0^r q(s)\tilde{K}_\lambda\big(u_\lambda(s)\big)\, ds\right), \qquad (2.2.55)$$

$$u_\lambda'(1) = \beta. \qquad (2.2.56)$$

As $y_\lambda \geq 0$ (see (2.2.48)) it follows by (2.2.54) that u_λ is nondecreasing. Now, (A.2) comes again into play, showing that u_λ' is also nondecreasing (see (2.2.55). In particular, we have that

$$0 = u_\lambda'(0) \leq u_\lambda'(r) \leq \beta \ \text{ for all } \lambda > 0, \ r \in [0, 1]. \qquad (2.2.57)$$

Now, we are going to prove that, for some $\lambda_0 > 0$ fixed, the set

$$\{u_\lambda \mid 0 < \lambda \leq \lambda_0\} \ \text{ is bounded in } C[0, 1]. \qquad (2.2.58)$$

To this purpose it suffices to show that the set $\{u_{0\lambda} \mid 0 < \lambda \leq \lambda_0\}$ is bounded (cf. (2.2.54) and (2.2.57)). Indeed, we have that

$$\beta = u_\lambda'(1) \geq \tilde{F}_\lambda\left(\frac{1}{p(1)}\tilde{K}_\lambda(u_{0\lambda}) \int_0^1 q(s)\, ds\right) \geq 0. \qquad (2.2.59)$$

By (2.2.52), (2.2.53), and (2.2.59), we get the boundedness of the set $\{u_{0\lambda} \mid 0 < \lambda \leq \lambda_0\}$, as claimed. From (2.2.57) and (2.2.58) it follows by virtue of the Arzelà-Ascoli Theorem that there exists a function $u \in C[0, 1]$ such that, on a subsequence,

$$u_\lambda \to u \ \text{ in } C[0, 1], \ \text{ as } \lambda \to 0^+. \qquad (2.2.60)$$

We are now going to justify the passage to the limit in (2.2.55)-(2.2.56). The resolvent of $\tilde{K}$, say $J_\lambda^{\tilde{K}}$, is nonexpansive and $J_\lambda^{\tilde{K}}(0) = 0$. Hence

$$|J_\lambda^{\tilde{K}} u_\lambda(r)| \leq |u_\lambda(r)| \leq \text{Const. for all } \lambda \in (0, \lambda_0]. \qquad (2.2.61)$$

Obviously, $\tilde{K}$ is bounded on bounded sets and this implies, by virtue of (2.2.61), that

$$|\tilde{K}_\lambda(u_\lambda(r))| \leq C_1 \ \text{ for all } \lambda \in (0, \lambda_0], \ r \in [0, 1]. \qquad (2.2.62)$$

Therefore,

$$|J_\lambda^{\tilde{K}} u_\lambda(r) - u(r)| \le |J_\lambda^{\tilde{K}} u_\lambda(r) - u_\lambda(r)| + |u_\lambda(r) - u(r)| \le$$
$$\le C_1 \lambda + |u_\lambda(r) - u(r)|,$$

which implies (see (2.2.60)) that

$$J_\lambda^{\tilde{K}} u_\lambda \to u \text{ in } C[0,1], \text{ as } \lambda \to 0^+, \tag{2.2.63}$$

on the same subsequence as in (2.2.60). Using (2.2.62)-(2.2.63) and the fact that $\tilde{K}$ is closed (as a multivalued mapping) we can see that there exists a function $w \in L^\infty(0,1)$ such that

$$\tilde{K}_\lambda\big(u_\lambda(r)\big) \to w(r) \in \tilde{K}\big(u(r)\big), \text{ as } \lambda \to 0^+ \text{ for all } r \in [0,1]. \tag{2.2.64}$$

Consequently, by the Lebesgue Dominated Convergence Theorem (see also (2.2.62)), we have for each $r \in [0,1]$

$$\frac{1}{p(r)} \int_0^r q(s)\tilde{K}_\lambda\big(u_\lambda(s)\big)\,ds \to \frac{1}{p(r)} \int_0^r q(s)w(s)\,ds, \text{ as } \lambda \to 0^+. \tag{2.2.65}$$

In fact, (2.2.64)-(2.2.65) also hold with respect to the weak-star topology of $L^\infty(0,1)$. By a similar reasoning for $\tilde{F}_\lambda$ we can pass to the limit in (2.2.55)-(2.2.56) to conclude that u belongs to $C^1[0,1]$ and satisfies

$$u'(r) = \tilde{F}\Big(\frac{1}{p(r)} \int_0^r q(s)w(s)\,ds\Big) \text{ for all } r \in [0,1], \tag{2.2.66}$$
$$u'(1) = \beta, \tag{2.2.67}$$

i.e., u is a solution of problem (2.2.39)-(2.2.40) in the sense of Definition 2.2.1.

Step 7. *Existence for problem (2.2.12)-(2.2.13).* Consider a sequence (C_n) such that $C_n > C$ and $C_n \to C$, and denote by (P_n) our boundary value problem (2.2.12)-(2.2.13) with $(\tilde{G}, \tilde{K}, C_n)$ instead of (G, K, C). Set

$$\beta_n := \tilde{G}^{-1}\Big(\frac{C_n}{p(1)}\Big) = \beta + \frac{C_n - C}{p(1)}.$$

Taking into account the above reasoning, we can say that for each n problem (P_n) has a solution in the sense of Definition 2.2.1, say u_n. More precisely, for each n there exists a $w_n \in L^\infty(0,1)$ such that

$$w_n(r) \in \tilde{K}\big(u_n(r)\big) \text{ for all } r \in [0,1], \tag{2.2.68}$$
$$u_n'(r) = \tilde{F}\Big(\frac{1}{p(r)} \int_0^r q(s)w_n(s)\,ds\Big) \text{ for all } r \in [0,1], \tag{2.2.69}$$
$$u_n'(1) = \beta_n. \tag{2.2.70}$$

Moreover, u_n are solutions for (P_n) in the sense of Definition 2.2.2. Using again the above arguments we can obtain

$$0 \leq u_n'(r) \leq \beta_n \ \text{ for all } r \in [0,1], \qquad (2.2.71)$$

$$0 \leq u_n(0) \leq u_n(r) \ \text{ for all } r \in [0,1]. \qquad (2.2.72)$$

We are now going to prove that

$$u_n(0) \leq \gamma - \beta \ \text{ for } n \text{ sufficiently large.} \qquad (2.2.73)$$

Indeed, otherwise we would have

$$\frac{C_n C_1}{p(1)} = C_1 \tilde{G}\big(u_n'(1)\big) = \frac{C_1}{p(1)} \int_0^1 q(s) \tilde{K}\big(u_n(s)\big) \, ds \geq$$

$$\geq \tilde{K}\big(u_n(0)\big) \geq \sup \tilde{K}(\gamma - \beta) = \sup K(\gamma - \beta),$$

and this contradicts (2.2.15) for n great enough. Now, by (2.2.71)-(2.2.73), we find that

$$0 \leq u_n(r) = u_n(0) + \int_0^r u_n'(s) \, ds \leq \gamma + \beta_n - \beta \ \text{ for all } r \in [0,1]. \quad (2.2.74)$$

From (2.2.71) and (2.2.74) we deduce, by virtue of the Arzelà-Ascoli Theorem, that

$$u_n \to u \ \text{ in } C[0,1], \ \text{ as } n \to \infty, \qquad (2.2.75)$$

on a subsequence. We shall prove that u is a solution of problem (2.2.12)-(2.2.13) in the sense of Definition 2.2.2. First, we can pass to the limit in (2.2.69). Indeed, there exists a function $w \in L^\infty(0,1)$ with $w(s) \in \tilde{K}\big(u(s)\big)$ for a.a. $s \in (0,1)$ such that

$$u'(r) = \tilde{F}\left(\frac{1}{p(r)} \int_0^r q(s)w(s) \, ds\right) \ \text{ for all } r \in [0,1]. \qquad (2.2.76)$$

By (2.2.76) we can see that $u \in C^1[0,1]$. Moreover, according to (2.2.71) and (2.2.74), $u'(r) \in [0,\beta]$, $u(r) \in [0,\gamma]$ for all $r \in [0,1]$. Hence we can replace in (2.2.76) $(\tilde{F}, \tilde{K})$ by (F, K). On the other hand, let us denote

$$v_n(r) = \int_0^r q(s)w_n(s) \, ds, \ v(r) = \int_0^r q(s)w(s) \, ds.$$

Obviously, v_n are the functions associated to u_n in Definition 2.2.2. It is easy to see that

$$v_n \to v \ \text{ in } C[0,1], \ \text{ as } n \to \infty.$$

In particular, we have

$$v(1) = \lim_{n \to \infty} v_n(1) = \lim_{n \to \infty} C_n = C$$

and hence v satisfies Definition 2.2.2.

Step 8. *The case $C > 0$ and $\beta = 0$.* In this case we define $\tilde{G}$ as follows

$$\tilde{G}(\xi) = \begin{cases} \xi & \text{if } \xi \leq 0, \\ [0, C/p(1)] & \text{if } \xi = 0, \\ \xi + C/p(1) & \text{if } \xi > 0. \end{cases}$$

If $\gamma \in \text{Int } D(K)$, we define $\tilde{K}$ as above. If γ is such that $D(K) \cap [0, \infty) = [0, \gamma]$, we take $C_2 \in K(\gamma)$ such that $CC_1/p(1) < C_2$ and define $\tilde{K}$ as follows

$$\tilde{K}(\xi) = \begin{cases} \xi & \text{if } \xi \leq 0, \\ K(\xi) & \text{if } 0 < \xi < \gamma, \\ [K^0(\gamma), C_2] & \text{if } \xi = \gamma, \\ \xi + C_2 - \gamma & \text{if } \xi > \gamma. \end{cases}$$

With these slight modifications, the proof of existence can be done as above. In fact, in this case the solution is a constant function.

The proof of Theorem 2.2.1 is now complete.

REMARK 2.2.5   If $GG^{-1}(C/p(1)) = \{C/p(1)\}$, then we can take in Step 7 of the above proof $C_n = C$, because in this case the solution of problem (2.2.39)-(2.2.40) in the sense of Definition 2.2.1 is also a solution in the sense of Definition 2.2.2. The rest of the proof is unchanged.   ∎

REMARK 2.2.6   If u is the solution given by Theorem 2.2.1, then necessarily $u'(0) = 0$. In the case of the capillarity problem for circular tubes (see (2.2.9)-(2.2.10), with $N = 2$) this fact has a simple physical interpretation. In addition, as noticed before, u' is a nondecreasing function and so u is convex. This fact has also a simple explanation in the case of capillarity in circular tubes: the liquid surface in the tube is a convex one. ∎

REMARK 2.2.7   By revisiting Step 7 of the above proof, we can see that if in addition K is strictly increasing, then we can allow $\leq$ instead of $<$ in (2.2.15) for $C > 0$ (and, respectively, $\geq$ in (2.2.16) for $C < 0$).   ∎

REMARK 2.2.8   Theorem 2.2.1 gives the existence of at least one solution in the sense of Definition 2.2.2. So the weaker notion of solution in Definition 2.2.1 may seem to be just an artificial concept needed in the proof of Theorem 2.2.1. However, if (A.1)-(A.4) are not assumed, it may happen that only the weaker solution (i.e., the solution in the sense of Definition 2.2.1) does exist. The next example will illustrate this case.   ∎

Example 2.2.10

Let $p(r) = q(r) = r$, $C > 0$, $K \colon \mathbb{R} \to \mathbb{R}$, $K(\xi) = \xi$, and

$$G \subset \mathbb{R} \times \mathbb{R}, \quad G(\xi) = \begin{cases} 0 & \text{if } \xi < 1, \\ [0, \infty) & \text{if } \xi = 1, \\ \emptyset & \text{if } \xi > 1. \end{cases}$$

Obviously, (A.1)-(A.4) are all satisfied except for the strict monotonicity of G. On the other hand, if u is a solution of problem (2.2.12)-(2.2.13) in the sense of Definition 2.2.1, then we have

$$\{r \in [0, 1] \mid u'(r) < 1\} \subset \{r \in [0, 1] \mid u(r) = 0\}$$

and hence, due to the Darboux property, $u'(r) \geq 1$ for all $r \in [0, 1]$. Therefore, it is easily seen that the functions

$$u(r) = r + C_1, \quad C_1 \geq 0$$

are solutions of problem (2.2.12)-(2.2.13) in the sense of Definition 2.2.1. We wonder whether this problem has a solution in the sense of Definition 2.2.2. The existence of such a solution is equivalent to the existence of a function $v \in AC[0, 1]$ such that

$$v(r) \geq 0 \text{ for all } r \in (0, 1], \tag{2.2.77}$$
$$v'(r) = r(r + C_1) \text{ for a.a. } r \in (0, 1), \tag{2.2.78}$$
$$v(0) = 0, \ v(1) = C. \tag{2.2.79}$$

Clearly, for $0 < C < 1/3$, the system (2.2.77)-(2.2.79) has no solution; hence, problem (2.2.12)-(2.2.13) has no solution in the sense of Definition 2.2.2. For each $C \geq 1/3$, our boundary value problem (2.2.12)-(2.2.13) has a unique solution in the sense of Definition 2.2.2: $u(r) = r + 2(C - 1/3)$.

$\Box$

The variational interpretation of problems (2.2.12)-(2.2.13)

We assume again (A.1)-(A.4). It is well known that any maximal monotone operator from $\mathbb{R}$ into $\mathbb{R}$ is the subdifferential of some proper, convex and lower semicontinuous function, which is uniquely determined up to an additive constant. So, $G = \partial g$ and $K = \partial h$, where $g, h \colon \mathbb{R} \to (-\infty, \infty]$ are both proper, convex, and lower semicontinuous. More precisely, we know that $D(G)$ and $D(K)$ are intervals and g, h can be defined as follows (see Theorem 1.2.15)

$$g(\xi) = \begin{cases} \int_0^\xi G^0(s)\,ds & \text{if } \xi \in \overline{D(G)}, \\ \infty & \text{otherwise} \end{cases}$$

$$h(\xi) = \begin{cases} \int_0^\xi K^0(s)\,ds & \text{if } \xi \in \overline{D(K)}, \\ \infty & \text{otherwise.} \end{cases}$$

Of course, g is strictly convex since G is a strictly monotone mapping (see Proposition 1.2.3).

Now, let us define the functional $\Psi \colon W^{1,1}(0,1) \to (-\infty, \infty]$,

$$\Psi(v) = \int_0^1 \Big(p(r) g\big(v'(r)\big) + q(r) h\big(v(r)\big) \Big)\, dr - C v(1). \qquad (2.2.80)$$

It is only a simple exercise (involving the definition of the subdifferential) to see that the solution of problem (2.2.12)-(2.2.13) in the sense of Definition 2.2.2, given by Theorem 2.2.1 above, is a minimizer of the convex functional Ψ. Therefore, a solution of this boundary value problem in the sense of Definition 2.2.2 is a *variational solution*, while a solution in the sense of Definition 2.2.1 is not necessarily a variational one (in fact, it is a minimizer of Ψ given by (2.2.80) but possibly with another constant instead of C, which belongs to the interval $p(1)GG^{-1}\big(C/p(1)\big)$). This interpretation clarifies the meaning of the two solutions.

Obviously, from a physical point of view, the variational solutions are important. In the case of problem (2.2.9)-(2.2.10), the functional Ψ coincides with the functional J, up to a multiplicative constant,

$$\Psi(u) = \frac{\Gamma(N/2)}{2\pi^{N/2}} J(u), \quad p(r) = q(r) = r^{N-1}.$$

It is expected that the existence of a solution in the sense of Definition 2.2.2 (i.e., the existence of variational solution) for problem (2.2.12)-(2.2.13) can be derived as the minimizer of the functional Ψ. This is not a simple task, as noticed before. However, in the case of capillarity in circular tubes, we succeeded in using the variational approach to deduce existence [Moro3].

A similar variational interpretation can be done for the end point $r = 0$. In fact, the solution in the sense of Definition 2.2.2 is a variational solution with respect to both end points $r = 0$ and $r = 1$, as it appears as a minimizer of the functional Ψ. On the other hand, as seen in Remark 2.2.6, the solution of problem (2.2.12)-(2.2.13), in any of the two senses, satisfies $u'(0) = 0$. In general this is not equivalent with condition (2.2.21). Here is an example in this sense.

Example 2.2.11

[MoZo1]. Consider the equation

$$\frac{d}{dr}\big(r^{-1} u'(r)\big) = u(r), \quad r \in (0,1). \qquad (2.2.81)$$

According to Theorem 2.2.1, (2.2.81) with the boundary conditions

$$\lim_{r \to 0^+} r^{-1} u'(r) = 0, \quad u'(1) = C \qquad (2.2.82)$$

has a unique solution $u \in C^1[0,1]$. Now, let us associate with (2.2.81) the boundary conditions

$$u'(0) = 0, \ u'(1) = C. \tag{2.2.83}$$

The general solution of (2.2.81) is given by

$$u(r) = c_1 r I_{\frac{2}{3}}\left(\frac{2}{3}r^{\frac{3}{2}}\right) + c_2 r I_{-\frac{2}{3}}\left(\frac{2}{3}r^{\frac{3}{2}}\right), \tag{2.2.84}$$

where I_m represents the modified Bessel function of the first kind and of order m (see, e.g., [Cordu, p. 91] or [LomMa, p. 301]), while c_1 and c_2 are real constants. We notice that u given by (2.2.84) satisfies $u'(0) = 0$ for any constants c_1, c_2. Therefore, problem (2.2.81)-(2.2.83) has an infinite number of solutions.   ⬚

REMARK 2.2.9   In fact, we may consider, in the previous example, instead of (2.2.81) the following more general equation

$$\frac{d}{dr}\left(r^{-1}u'(r)\right) = r^b u(r), \ r \in (0,1), \tag{2.2.85}$$

where $b > -1$. Denote $a = \frac{2}{b+3}$. The reader can easily see that by means of the substitutions

$$x = ar^{\frac{1}{a}}, \ w = r^{-1}u$$

(2.2.85) can be written as the following modified Bessel equation

$$x^2 w''(x) + x w'(x) - (x^2 + a^2)w(x) = 0.$$

∎

Dependence on the data

Using again the technique from the proof of Theorem 2.2.1, we can easily obtain the following result of upper semicontinuity with respect to the parameter C.

THEOREM 2.2.2
Let (A.1)-(A.4) hold and let (C_n) be a sequence of real numbers converging toward a $C \in \mathbb{R}$. Then, there exist functions u_n, $n \in \mathbb{N}^*$, which are solutions in the sense of Definition 2.2.2 for problem (2.2.12)-(2.2.13) with C_n instead of C, such that a subsequence of (u_n) converges in $C^1[0,1]$ toward some function $u \in C^1[0,1]$, which is a solution of problem (2.2.12)-(2.2.13) in the sense of Definition 2.2.2.

It is expected that the same technique may be applied to prove a result of continuity with respect to p, q, G, K, and C. Also, in the case of uniqueness, some results of differentiability and sensitivity of the solutions with respect to some parameters are expected.

References

[AiMoP] S. Aizicovici, Gh. Moroşanu & N.H. Pavel, On a nonlinear second order differential operator, *Comm. Appl. Nonlinear Anal.*, 6 (1999), no. 4, 1–16.

[Cordu] A. Corduneanu, *Ecuaţii diferenţiale cu aplicaţii în electrotehnică*, Facla, Timişoara, 1981.

[CoMo1] A. Corduneanu & Gh. Moroşanu, Une équation intégro-differentielle de la théorie de la capillarité, *C.R. Acad. Sci.*, t. 319, Serie I (1994), 1171–1174.

[CoMo2] A. Corduneanu & Gh. Moroşanu, An integro-differential equation from the capillarity theory, *Libertas Math.*, 14 (1994), 115-123.

[CoMo3] A. Corduneanu & Gh. Moroşanu, A nonlinear integro-differential equation related to a problem from the capillarity theory, *Comm. Appl. Nonlin. Anal.* 3, (1996), 51–60.

[Fichte] G.M. Fichtenholz, *Differential and Integral Calculus, Vol. III (in Russian), third ed.*, Fizmatgiz, Moscow, 1963.

[Finn] R. Finn, *Equilibrium Capillary Surfaces*, Springer-Verlag, Berlin, 1988.

[GilTru] D. Gilbarg & N.S. Trudinger, *Elliptic Partial Differential Equations*, Springer-Verlag, Berlin, 1983.

[HokMo] V.-M. Hokkanen & Gh. Moroşanu, A class of nonlinear parabolic boundary value problems, *Math. Sci. Res. Hot-Line*, 3 (4) (1999), 1–22.

[LanLif] L.D. Landau & E.M. Lifschitz, *Fluid Mechanics (in Russian)*, Nauka, Moscow, 1988.

[Lin2] C.-Y. Lin, Quasilinear parabolic equations with nonlinear monotone boundary conditions, *Topol. Meth. Nonlin. Anal.*, 13 (1999), no. 2, 235–249.

[LomMa] D. Lomen & J. Mark, *Ordinary Differential Equations with Linear Algebra*, Prentice Hall, New Jersey, 1986.

[Mikhl] M.S. Mikhlin (Ed.), *Linear Equations of Mathematical Physics*, Holt, Rinehart and Winston, New York, 1967.

[Moro1] Gh. Moroşanu, *Nonlinear Evolution Equations and Applications*, Editura Academiei and Reidel, Bucharest and Dordrecht, 1988.

[Moro2] Gh. Moroşanu, On a problem arising in capillarity theory, *An. Şt. Univ. "Ovidius"*, Ser. Mat., 3 (1995), 127–137.

[Moro3] Gh. Moroşanu, A variational approach to a problem arising in capillarity theory, *J. Math Anal. Appl.*, 206 (1997), 442–447.

[Moro4] Gh. Moroşanu, On a second order boundary value problem related to capillary surfaces in *Analysis, Numerics and Applications of Differential and Integral Equations*, ed. by M. Bach et al., Pitman Res. Notes in Math. 379, Longman, London 1998, 149–152.

[MoAZ] Gh. Moroşanu, P. Anghel & D. Zofotă, A class of degenerate multivalued second order boundary value problems, *J. Math. Anal. Appl.*, 236 (1999), no. 1, 1–24.

[MorPe] Gh. Moroşanu & D. Petrovanu, Nonlinear monotone boundary conditions for parabolic equations, *Rend. Inst. Mat. Univ. Trieste*, 18 (1996), 136–155.

[MoZo1] Gh. Moroşanu & D. Zofotă, On a boundary value problem for a second order differential equation, *An. Şt. Univ. "Ovidius"*, 5 (1997), 89–94.

[MoZo2] Gh. Moroşanu & D. Zofotă, Un problème aux limites pour une équation différentielle non linéaire du deuxième ordre, *Bull. Cl. Sci. Acad. Roy. Belg.* (6), 10 (1999), no. 1-6, 79–88.

Chapter 3

Parabolic boundary value problems with algebraic boundary conditions

In this chapter we deal with doubly nonlinear equations of the form

$$u_t(r,t) - \frac{\partial}{\partial r} G\big(r, u_r(r,t)\big) + K\big(r, u(r,t)\big) =$$
$$= f(r,t), \quad (r,t) \in (0,1) \times (0,\infty), \tag{3.0.1}$$

to which we associate the boundary condition

$$\begin{pmatrix} -G\big(0, u_r(0,t)\big) \\ G\big(1, u_r(1,t)\big) \end{pmatrix} + \beta \begin{pmatrix} u(0,t) \\ u(1,t) \end{pmatrix} \ni s(t), \quad t > 0, \tag{3.0.2}$$

and the initial condition

$$u(r,0) = u_0(r), \quad r \in (0,1). \tag{3.0.3}$$

We have denoted $u_t = \partial u/\partial t$, $u_r = \partial u/\partial r$. The boundary condition (3.0.2) is called *algebraic* because it is an algebraic relation involving u and u_r at $(0,t)$ and $(1,t)$ as well as $s(t)$. In the next chapter we shall also meet *differential* boundary conditions, which involve u, u_r at $(0,t)$ and $(1,t)$ as well as some of their derivatives with respect to t.

We shall assume (I.1) and (I.3) of Chapter 2, Section 1. Instead of (I.2), a weaker assumption is introduced, namely (I.2'):

(I.2') The mapping $K \colon [0,1] \times \mathbb{R} \to \mathbb{R}$ is continuous and, in addition, it is nondecreasing with respect to its second variable.

Problem (3.0.1)-(3.0.3) is a very general mathematical model. In particular, (3.0.1) describes the heat propagation in a linear conductor and also diffusion phenomena. As we have already seen in Section 2.3, condition (3.0.2) includes classical boundary conditions, such as Dirichlet, Neumann, Robin-Steklov, periodic, etc. For the sake of simplicity, we consider here only the case where the equation is scalar and of second order.

The vectorial case is a model for electronic integrated circuits with negligible inductances [MMN1], [MMN2]. Notice that no essential difficulties appear in the vectorial case as compared to the scalar case. On the other hand, in the case where G is a linear function, problem (3.0.1)-(3.0.3) has been studied in [Moro5]. In this situation, a change of variable, of the form

$$\tilde{u}(r,t) = u(r,t) + \alpha(t)r^3 + \beta(t)r^2 + \gamma(t)r,$$

can transform the nonhomogeneous boundary condition (3.0.2) into a homogeneous boundary condition (i.e., a condition in which the right hand side $s(t)$ is zero). Notice, however, that by such change (3.0.1) becomes a nonautonomous equation, and this situation is covered by known results on time-dependent abstract evolution equations [Kato]. In the present case, G and β are nonlinear mappings and this does not allow the homogenization of condition (3.0.2). In what follows, we are going to study the homogeneous boundary conditions and the nonhomogeneous boundary conditions, separately. Finally, we notice that similar problems, with homogeneous boundary conditions in which nonlinear terms in u_r appear, have been studied by [GolLin], [Lin1], [LinFan].

3.1   Homogeneous boundary conditions

In this section, we deal with the situation where the right hand side $s(t)$ of (3.0.2) is a constant function. Actually, by modifying the mapping β, we can assume that $s(t) \equiv 0$, without any loss of generality. Let $H = L^2(0,1)$ with the usual scalar product and the associated Hilbertian norm. Define the operator $A : D(A) \subset H \to H$ by

$$D(A) = \left\{ v \in H^2(0,1) \,\middle|\, \begin{pmatrix} G\big(0,v'(0)\big) \\ -G\big(1,v'(1)\big) \end{pmatrix} \in \beta \begin{pmatrix} v(0) \\ v(1) \end{pmatrix} \right\}, \quad (3.1.1)$$

$$(Av)(r) = -\frac{d}{dr}G\big(r,v'(r)\big) + K\big(r,v(r)\big) \ \text{ for a.a. } r \in (0,1). \quad (3.1.2)$$

If assumptions (I.1), (I.2'), and (I.3) are satisfied, then the operator A is a maximal monotone in the space $H = L^2(0,1)$, with a dense domain $D(A)$ (see Propositions 2.1.1 and 2.1.2 above). If, in addition, β is cyclically monotone, i.e., β is the subdifferential of some proper, convex, and lower semicontinuous function $j : \mathbb{R}^2 \to (-\infty, \infty]$, then $A = \partial\Psi$, where $\Psi : H \to$

$(-\infty, \infty]$ is defined by (see Section 2.1 above)

$$\Psi(v) = \begin{cases} \int\limits_0^1 \Big(g\big(r, v'(r)\big) + k\big(r, v(r)\big)\Big)\, dr + j\big(v(0), v(1)\big) \\ \qquad \text{if } v \in H^1(0,1),\ g(\cdot, v') \in L^1(0,1) \\ \qquad\qquad \text{and } \big(v(0), v(1)\big) \in D(j), \\ \infty \quad \text{otherwise.} \end{cases} \tag{3.1.3}$$

Here

$$g(r, \xi) = \int_0^\xi G(r, \sigma)\, d\sigma \quad \text{and} \quad k(r, \xi) = \int_0^\xi K(r, \sigma)\, d\sigma.$$

By denoting $u(t) = u(\cdot, t)$ and $f(t) = f(\cdot, t)$, problem (3.0.1)-(3.0.3) can be expressed as a Cauchy problem in H, namely:

$$u'(t) + Au(t) = f(t),\ t > 0, \tag{3.1.4}$$
$$u(0) = u_0. \tag{3.1.5}$$

Therefore, we can apply known results of the theory of evolution equations (see Chapter 1, Section 1.5). In particular, we have:

THEOREM 3.1.1
Assume that (I.1) and (I.2') are satisfied and that β is the subdifferential of a proper, convex, and lower semicontinuous function $j: \mathbb{R}^2 \to (-\infty, \infty]$. Let $T > 0$ be fixed, $u_0 \in H$ and $f \in L^2(Q_T)$, $Q_T = (0,1) \times (0,T)$. Then, problem (3.1.4)-(3.1.5) has a unique strong solution $u: [0,T] \to H$ such that the mapping $(r,t) \mapsto \sqrt{t}\, u_t(r,t)$ belongs to $L^2(Q_T)$ (i.e., in particular, u satisfies (3.0.1) a.e. in Q_T as well as (3.0.2) with $s(t) \equiv 0$ a.e. in $(0,T)$). If, in addition, $u_0 \in D(\Psi)$ (see (3.1.3)), then $u \in H^1(Q_T)$.

PROOF The first part of the theorem is a direct consequence of Theorem 1.5.3. From the same result it follows that for $u_0 \in D(\Psi)$ we have $u_t \in L^2(Q_T)$. Therefore, in this case,

$$Au = f - u_t \in L^2(Q_T).$$

Let $\hat{v}$ be the function defined in the proof of Proposition 2.1.1 (see (2.1.13)). By our assumption (2.1.3) it follows that

$$k_0 \int_0^1 \big(u_r(r,t) - \hat{v}'(r)\big)^2 dr \leq \big(Au(t) - A\hat{v}, u(t) - \hat{v}\big)_H =$$
$$= \int_0^1 \big(f(r,t) - u_t(r,t) - A\hat{v}(r)\big)\big(u(r,t) - \hat{v}(r)\big)\, dr \tag{3.1.6}$$

for a.a. $t \in (0,T)$. This implies that $u_r \in L^2(Q_T)$, i.e., $u \in H^1(Q_T)$. ∎

THEOREM 3.1.2

Assume (I.1), (I.2'), and (I.3). Let $T > 0$ be fixed. If $u_0 \in D(A)$ and $f \in W^{1,1}\left(0, T; L^2(0,1)\right)$, then problem (3.0.1)-(3.0.3) has a unique solution

$$u \in W^{1,\infty}\left(0, T; L^2(0,1)\right) \cap L^\infty\left(0, T; H^2(0,1)\right) \cap W^{1,2}\left(0, T; H^1(0,1)\right).$$

PROOF   By Theorem 1.5.1, there exists a unique strong solution $u \in W^{1,\infty}(0,T;H)$, $H = L^2(0,1)$, of problem (3.1.4)-(3.1.5). It remains to show that u belongs to $L^\infty\left(0,T;H^2(0,1)\right) \cap W^{1,2}\left(0,T;H^1(0,1)\right)$. We have the standard estimate (see (1.5.7))

$$\left\|\frac{\partial^+ u}{\partial t}(\cdot, t)\right\|_H \leq \|f(0) - Au_0\|_H + \int_0^T \|f_\sigma(\cdot, \sigma)\|_H \, d\sigma \qquad (3.1.7)$$

for all $t \in [0, T)$. Actually, if we extend smoothly $f(t)$ to the right of $t = T$, then $u(t)$ can also be extended to the right of $t = T$ and so (3.1.7) is valid on the closed interval $[0, T]$.

By (3.1.6)-(3.1.7) we can deduce that

$$\sup\left\{\|u_r(\cdot, t)\|_H \mid t \in [0, T]\right\} < \infty. \qquad (3.1.8)$$

So, by the following obvious formula

$$u(r, t) = \int_0^1 \left(\sigma u_\sigma(\sigma, t) + u(\sigma, t)\right) d\sigma - \int_r^1 u_\sigma(\sigma, t) \, d\sigma, \qquad (3.1.9)$$

it follows that

$$\sup\left\{|u(r, t)| \mid r \in [0, 1], \; t \in [0, T]\right\} < \infty. \qquad (3.1.10)$$

Therefore,

$$\sup\left\{|K\left(r, u(r, t)\right)| \mid r \in [0, 1], \; t \in [0, T]\right\} < \infty. \qquad (3.1.11)$$

We are now going to show that $u_{rr} \in L^\infty(0, T; H)$. First, we have the equation

$$u(r, t) - \frac{\partial}{\partial r} G\left(r, u_r(r, t)\right) =$$

$$= u(r, t) - K\left(r, u(r, t)\right) + f(r, t) - \frac{\partial^+ u}{\partial t}(r, t). \qquad (3.1.12)$$

Taking into account (3.1.7), (3.1.10), and (3.1.11), we can see that the right hand side of (3.1.12), denoted by $q(r, t)$, satisfies

$$\sup\left\{\|q(\cdot, t)\|_H \mid t \in [0, T]\right\} < \infty.$$

Notice also that for each t (3.1.12) is similar to problem (2.1.14)-(2.1.16). So, if we argue in the same manner as in the proof of Proposition 2.1.1, we arrive at

$$\sup\left\{\|u(\cdot,t)\|_{H^1(0,1)} + \|w_r(\cdot,t)\|_H \mid t \in [0,T]\right\} < \infty, \qquad (3.1.13)$$

where the function w is defined by

$$w(r,t) = G\big(r, u_r(r,t)\big). \qquad (3.1.14)$$

Using now Fatou's lemma, we get that

$$\sup\left\{\|u(\cdot,t)\|_{H^1(0,1)} + \|w(\cdot,t)\|_{H^1(0,1)} \mid t \in [0,T]\right\} < \infty. \qquad (3.1.15)$$

By (3.1.14)-(3.1.15) and (I.1), it follows that $u_{rr} \in L^\infty(0,T;H)$, i.e., $u \in L^\infty(0,T;H^2(0,1))$.

Now, we are going to prove that $u \in W^{1,2}(0,T;H^1(0,1))$. To this end, we first write (3.0.1) for t and $t+\delta$ with $\delta \in (0,T)$, subtract the two equations, take the inner product of the resulting equation by $u(r,t+\delta) - u(r,t)$, and finally integrate over $[0,T-\delta]$. So, using our assumptions, we get

$$k_0 \int_0^{T-\delta} \|u_r(\cdot,t+\delta) - u_r(\cdot,t)\|_H^2 \, dt \le \frac{1}{2}\|u(\delta) - u_0\|_H^2 +$$

$$+ \int_0^{T-\delta} \|f(t+\delta) - f(t)\|_H \, \|u(t+\delta) - u(t)\|_H \, dt. \qquad (3.1.16)$$

On the other hand, there exist some positive constants C_1 and C_2 such that

$$\int_0^{T-\delta} \|f(t+\delta) - f(t)\|_H \, dt \le C_1\delta, \qquad (3.1.17)$$

$$\|u(t+\delta) - u(t)\|_H \le C_2\delta \quad \text{for all } t \in [0,T-\delta], \qquad (3.1.18)$$

since $f \in W^{1,1}(0,T;H)$ (see Theorem 1.1.2) and $u \in W^{1,\infty}(0,T;H)$. By (3.1.16)-(3.1.18) it follows that

$$\int_0^{T-\delta} \|u_r(\cdot,t+\delta) - u_r(\cdot,t)\|_H^2 \, dt \le C_3\delta^2, \qquad (3.1.19)$$

where $C_3 = (C_1 + \frac{1}{2}C_2)C_2/k_0$. But, the estimate (3.1.19) implies that $u_r \in W^{1,2}(0,T;H)$, i.e., $u \in W^{1,2}(0,T;H^1(0,1))$. $\blacksquare$

REMARK 3.1.1 If in Theorems 3.1.1 and 3.1.2 one assumes that the data are weaker, namely $u_0 \in L^2(0,1)$ and $f \in L^1(0,T;L^2(0,1))$, then one gets only the existence and uniqueness of a weak solution $u \in$

$C\big([0,T];L^2(0,1)\big)$ (see Theorem 1.5.2). However, this u also belongs to $L^2\big(0,T;H^1(0,1)\big)$.

Indeed, let (u_{0n}) and (f_n) be sequences in $D(A)$ and $W^{1,1}(0,T;H)$, respectively, such that $u_{0n} \to u_0$ in H and $f_n \to f$ in $L^1(0,T;H)$ as $n \to \infty$. Then the strong solutions u_n, corresponding to (u_{0n}, f_n), satisfy

$$\|u_n(t) - u_m(t)\|_H + \|u_{n,r} - u_{m,r}\|_{L^2(0,t;H)} \le$$
$$\le C_4\Big(\|u_{0n} - u_{0m}\|_H + \|f_n - f_m\|_{L^1(0,t;H)}\Big)$$

for all $t \in [0,T]$. This proves the assertion. ∎

In what follows, we are going to study the long time behavior of the solution of problem (3.0.1)-(3.0.3). Intuitively, for large t, the solution approaches a stationary solution of (3.0.1)-(3.0.2), provided such a stationary solution does exist. Indeed, we shall see that under appropriate conditions, this is true. The particular case when $G(r, \xi)$ is linear with respect to ξ has been studied in [Moro1, Ch. III, § 3]. We shall try to extend some results to the nonlinear case. To do that, let us start with the following result.

PROPOSITION 3.1.1

Assume (I.1), (I.2'), and (I.3). Denote by I the identity operator of $H = L^2(0,1)$. Then, for every $\lambda > 0$, the operator $(I + \lambda A)^{-1}$ maps bounded subsets of H into bounded subsets of $H^2(0,1)$.

PROOF   Let $\lambda > 0$ be fixed and let $Y \subset H$ be an arbitrary bounded set. Denote

$$u_p := (I + \lambda A)^{-1}p \quad \text{for all } p \in Y. \tag{3.1.20}$$

It is easily seen that the set $\{u_p \mid p \in Y\}$ is bounded in Y. On the other hand, (3.1.20) can equivalently be written as

$$u_p + \lambda A u_p = p \quad \text{for all } p \in Y. \tag{3.1.21}$$

Notice that (3.1.21) is similar to (3.1.12) (with a new parameter p, instead of t). We shall first show that the set $\{u_p \mid p \in Y\}$ is bounded in $H^1(0,1)$. Indeed, if $\hat{v}$ is the function defined in the proof of Proposition 2.1.1 (see (2.1.13)), then it is easily seen that

$$(p - \hat{v} - \lambda A\hat{v}, u_p - \hat{v})_H \ge \|u_p - \hat{v}\|_H^2 + \lambda k_0 \|u_p' - \hat{v}'\|_H^2,$$

which implies that

$$\|u_p - \hat{v}\|_H^2 + 2\lambda k_0 \|u_p' - \hat{v}'\|_H^2 \le \|p - \hat{v} - \lambda A\hat{v}\|^2.$$

Next, we can use an argument from the proof of Theorem 3.1.2 to get the boundedness of the set $\{u_p \mid p \in Y\}$ in $H^2(0,1)$. ∎

THEOREM 3.1.3

Assume (I.1), (I.2'), and (I.3). Denote $F := A^{-1}0$. Let $F \neq \emptyset$, $u_0 \in H$, $f \in L^1(\mathbb{R}_+; H)$, $\beta \subset \mathbb{R}^2 \times \mathbb{R}^2$ be a maximal cyclically monotone operator and let $u(t) = u(\cdot, t)$ be the weak solution of problem (3.0.1)-(3.0.3). Then, there exists a $\hat{p} \in F$ such that

$$u(t) \to \hat{p} \text{ in } H, \text{ as } t \to \infty. \tag{3.1.22}$$

If, in addition, $f \in W^{1,1}(\mathbb{R}_+; H)$, then $u(t)$ converges toward $\hat{p}$ in the weak topology of $H^2(0,1)$ and thus in the strong topology of $C^1[0,1]$, as $t \to \infty$.

PROOF   Let $f \in W^{1,1}(\mathbb{R}_+; H)$. We shall show that for any $\epsilon > 0$,

$$\{u(\cdot, t) \mid t \geq \epsilon\} \text{ is bounded in } H^2(0,1). \tag{3.1.23}$$

As u is a strong solution (see Theorem 3.1.1), there exists a $\delta \in (0, \epsilon)$ such that $u(\cdot, \delta) \in D(A)$. Then, $u(\cdot, t) \in D(A)$ for all $t \geq \epsilon$ and the following estimate holds (see (1.5.7)):

$$\left\| \frac{\partial^+ u}{\partial t}(\cdot, t) \right\|_H \leq \|f(\cdot, \epsilon) - Au(\cdot, \epsilon)\|_H +$$

$$+ \int_\epsilon^\infty \|f_\sigma(\cdot, \sigma)\|_H \, d\sigma < \infty \text{ for all } t \geq \epsilon. \tag{3.1.24}$$

On the other hand, as $F \neq \emptyset$, we have that the trajectory

$$\{u(\cdot, t) \mid t \geq 0\} \text{ is bounded in } H. \tag{3.1.25}$$

Indeed, it suffices to apply the estimate (1.5.8) (which is valid even for weak solutions) with $u_1(t) = u(\cdot, t)$, $u_2(t) = p \in F$, $f_1(t) = f(\cdot, t)$, and $f_2(t) = 0$. On the other hand, we have by (3.0.1) that

$$u(\cdot, t) = (I + A)^{-1}\left(f(\cdot, t) + u(\cdot, t) - \frac{\partial^+ u}{\partial t}(\cdot, t) \right) \text{ for all } t \geq \epsilon. \tag{3.1.26}$$

Now, by virtue of (3.1.24), (3.1.25) and (3.1.26), we obtain (3.1.23) (cf. Proposition 3.1.1). Clearly, it remains to show (3.1.22) for any $u_0 \in H$ and $f \in L^1(\mathbb{R}_+; H)$. Recall that $\overline{D(H)} = H$. According to Theorem 1.5.4 it suffices to prove (3.1.22), for $u_0 \in D(A)$ and $f(t) \equiv 0$. In this case, $u(\cdot, t) = S(t)u_0$, $t \geq 0$, where $\{S(t) \mid t \geq 0\}$ is the semigroup of contractions $H \to H$ generated by $-A$. Notice that the trajectory $\{S(t)u_0 \mid t \geq 0\}$ is precompact in H, for every $u_0 \in D(A)$ (see (3.1.23)). This combined with Theorem 1.5.5 concludes the proof. ∎

REMARK 3.1.2   Under the assumptions of Theorem 3.1.3, every two functions from $F = A^{-1}0$ differ only by an additive constant.

Indeed, if $p_1, p_2 \in F$, then

$$k_0 \|p_1' - p_2'\|_H \le (Ap_1 - Ap_2, p_1 - p_2)_H = 0.$$

Hence $p_1' = p_2'$, which proves the assertion.   ∎

REMARK 3.1.3   The condition $F \ne \emptyset$ is not always satisfied.

Indeed, let us consider the problem

$$p'' = 0, \ p'(0) = 0, \ p'(1) = 1.$$

Clearly, in this problem (I.1) and (I.2') are satisfied and β is a subdifferential, but it has no solution.   ∎

THEOREM 3.1.4

Assume (I.1),(I.2'), and (I.3). Let $F := A^{-1}0$ be a nonempty set, $\xi \mapsto K(r, \xi)$ strictly increasing for a.a. $r \in (0,1)$, $u_0 \in H = L^2(0,1)$, and $f \in L^1(\mathbb{R}_+ H)$. Then F contains only one element, say $\hat{p}$, and

$$u(\cdot, t) \to \hat{p} \text{ in } H, \text{ as } t \to \infty, \tag{3.1.27}$$

where $u(\cdot, t)$ is the corresponding weak solution of problem (3.0.1)-(3.0.3) or (3.1.4)-(3.1.5). If, in addition, $u_0 \in D(A)$ and $f \in W^{1,1}(\mathbb{R}_+; H)$, then $u(\cdot, t)$ converges toward $\hat{p}$ in the weak topology of $H^2(0,1)$, as $t \to \infty$.

PROOF   An argument already used in the proof of Theorem 3.1.3 shows that for $u_0 \in D(A)$ and $f \in W^{1,1}(\mathbb{R}_+; H)$, the corresponding trajectory is bounded in $H^2(0,1)$. To show that F is a singleton, let $p_1, p_2 \in F$. We have

$$0 = (Ap_1 - Ap_2, p_1 - p_2)_H \ge$$
$$\ge \int_0^1 \Big(K\big(r, p_1(r)\big) - K\big(r, p_2(r)\big) \Big) \big(p_1(r) - p_2(r)\big) \, dr \ge 0,$$

which implies that $p_1 = p_2$, since $K(r, \cdot)$ is strictly increasing for a.a. $r \in (0,1)$.

It remains to show (3.1.27). According to Theorem 1.5.4 it suffices to show (3.1.27) for every $u_0 \in D(A)$ and $f(t) \equiv 0$. Fix $u_0 \in D(A)$ and let p be a point of the ω-limit set $\omega(u_0)$. Obviously, $\omega(u_0)$ is nonempty, since the trajectory $\{S(t)u_0 \mid t \ge 0\}$ is bounded in $H^2(0,1)$ and so it is precompact in H. Moreover, according to Theorem 1.5.6, $S(t)p \in \omega(u_0)$, for all $t \ge 0$, and

$$\|S(t)p - \hat{p}\|_H^2 = \|p - \hat{p}\|_H^2 \text{ for all } t \ge 0. \tag{3.1.28}$$

By the same result, $p \in D(A)$ and so $S(t)p$ is a strong solution of (3.1.4) with $f(t) \equiv 0$. More precisely (see (1.5.6)),

$$\frac{d^+}{dt} S(t)p + AS(t)p = 0 \text{ for all } t \geq 0. \tag{3.1.29}$$

Now, by (3.1.28) and (3.1.29), it follows that

$$0 = \big(AS(t)p, S(t)p - \hat{p}\big)_H = \big(AS(t)p - A\hat{p}, S(t)p - \hat{p}\big)_H \text{ for all } t \geq 0.$$

Since $K(r, \cdot)$ are strictly increasing, this implies that $S(t)p = \hat{p}$ for all $t \geq 0$. Hence $p = \hat{p}$ and $\omega(u_0) = \{\hat{p}\}$. ∎

3.2   Nonhomogeneous boundary conditions

Next, we shall assume that the right hand side of (3.0.2) is a nonconstant function. As earlier indicated, (3.0.2) cannot be homogenized, because G and β are both nonlinear. The presence of the nonhomogeneous term $s(t)$ in (3.0.2) makes our problem more complex. Indeed, if we choose as in the case of homogeneous boundary conditions $H = L^2(0, 1)$ and we associate to our problem a Cauchy problem in H, then it easily turns out that this Cauchy problem is nonautonomous. Instead of the time independent operator A given by (3.1.1)-(3.1.2), we now have operators $A(t)$, $t \geq 0$, defined by

$$D\big(A(t)\big) = \Big\{ v \in H^2(0, 1) \,\Big|\, \Big(G\big(0, v'(0)\big), -G\big(1, v'(1)\big)\Big) +$$

$$+ s(t) \in \beta\big(v(0), v(1)\big) \Big\}, \tag{3.2.1}$$

$$\big(A(t)v\big)(r) = -\frac{d}{dr} G\big(r, v'(r)\big) + K\big(r, v(r)\big) - f(r, t) \tag{3.2.2}$$

for a.a. $r \in (0, 1)$. So, problem (3.0.1)-(3.0.3) can be expressed as the following Cauchy problem in H:

$$u'(t) + A(t)u(t) = 0, \ t > 0, \tag{3.2.3}$$

$$u(0) = u_0. \tag{3.2.4}$$

We have the following existence and uniqueness result:

THEOREM 3.2.1

Assume (I.1), (I.2'), and (I.3). Let $T > 0$ be fixed, $u_0 \in D(A)$, $s \in W^{1,2}(0, T; \mathbb{R}^2)$, and $f \in W^{1,1}(0, T; H)$. Then, problem (3.2.3)-(3.2.4) or (3.0.1)-(3.0.3) has a unique strong solution

$$u \in W^{1,\infty}(0, T; H) \cap L^\infty(0, T; H^2(0, 1)) \cap W^{1,2}(0, T; H^1(0, 1)).$$

PROOF   For the beginning, we assume that $f \in W^{1,\infty}(0,T;H)$ and $s \in W^{1,\infty}(0,T;\mathbb{R}^2)$. In this case, we can apply Theorem 1.5.9. Obviously, for each $t \in [0,T]$, the operator $A(t)$ is maximal monotone (actually, for any fixed $t \in [0,T]$, $A(t)$ has the same form as A defined by (3.1.1)-(3.1.2)). We are now going to show that the assumption (1.5.18) is fulfilled. Indeed, for $t_1, t_2 \in [0,T]$, $v_1 \in D\big(A(t_1)\big)$, and $v_2 \in D\big(A(t_2)\big)$, we have, by (3.2.1)-(3.2.3), integrating by parts, and using the monotonicity of $K(r,\cdot)$ and β,

$$-\big(v_1 - v_2, A(t_1)v_1 - A(t_2)v_2\big)_H =$$

$$= -\int_0^1 \Big(G\big(r,v_1'(r)\big) - G\big(r,v_2'(r)\big)\Big)\big(v_1'(r) - v_2'(r)\big)\,dr \,+$$

$$+\Big(G\big(r,v_1'(r)\big) - G\big(r,v_2'(r)\big)\Big)\big(v_1(r) - v_2(r)\big)\Big|_0^1 \,-$$

$$-\int_0^1 \Big(K\big(r,v_1(r)\big) - K\big(r,v_2(r)\big)\Big)\big(v_1(r) - v_2(r)\big)\,dr \,+$$

$$+\int_0^1 \big(f(r,t_1) - f(r,t_2)\big)\big(v_1(r) - v_2(r)\big)\,dr \,\leq$$

$$\leq -k_0\|v_1' - v_2'\|_H^2 + \big(f(t_1) - f(t_2), v_1 - v_2\big)_H \,+$$

$$+\Big(s(t_1) - s(t_2), \big(v_1(0), v_1(1)\big) - \big(v_2(0), v_2(1)\big)\Big)_{\mathbb{R}^2}.$$

Next we use the inequality

$$\|y\|_{C[0,1]} \leq \sqrt{2}\|y\|_{H^1(0,1)} \ \text{ for all } y \in H^1(0,1),$$

as well as the formulae,

$$f(t_1) = f(t_2) + \int_{t_1}^{t_2} f_t(\cdot,t)\,dt,$$

$$s(t_1) = s(t_2) + \int_{t_1}^{t_2} s_t(\cdot,t)\,dt,$$

which give us

$$-\big(v_1 - v_2, A(t_1)v_1 - A(t_2)v_2\big)_H \leq M\|v_1 - v_2\|_H^2 + L(t_1 - t_2)^2,$$

where

$$M = \frac{k_0 + 1}{2} \ \text{ and } \ L = \frac{1}{2}\|f_t\|_{L^\infty(0,T;H)}^2 + \frac{1}{k_0}\|s'\|_{L^\infty(0,T;\mathbb{R}^2)}^2.$$

Hence the condition (1.5.18) is satisfied ($g(t) = Lt$ is evidently of bounded variation). So, problem (3.2.3)-(3.2.4) has a unique strong solution $u \in W^{1,\infty}(0,T;H)$.

Now, suppose that $s \in W^{1,2}(0,T;\mathbb{R}^2)$ and $f \in W^{1,1}(0,T;H)$. Let (s_n) and (f_n) be sequences of elements of $W^{1,\infty}(0,T;\mathbb{R}^2)$ and $W^{1,\infty}(0,T;H)$, respectively, such that $s_n(0) = s(0)$ and

$$s_n \to s \text{ in } W^{1,2}(0,T;\mathbb{R}^2), \quad f_n \to f \text{ in } W^{1,1}(0,T;H), \text{ as } n \to \infty.$$

Denote by u_n the solution of problem (3.2.3)-(3.2.4), where (s,f) is replaced by (s_n, f_n). By a standard computation we get:

$$\frac{1}{2}\frac{d}{dt}\|u_n(t+h) - u_n(t)\|_H^2 + k_0\|u_{n,r}(\cdot, t+h) - u_{n,r}(\cdot, t)\|_H^2 \le$$

$$\le \|f_n(t+h) - f_n(t)\|_H \, \|u_n(t+h) - u_n(t)\|_H +$$

$$+\|s_n(t+h) - s_n(t)\|_{\mathbb{R}^2} \left\| \begin{pmatrix} u_n(0, t+h) - u_n(0,t) \\ u_n(1, t+h) - u_n(1,t) \end{pmatrix} \right\|_{\mathbb{R}^2} \quad (3.2.5)$$

for a.a. $0 < t < t + h < T$. On the other hand, as $H^1(0,1)$ is continuously embedded into $C[0,1]$, (3.2.5) implies that

$$\frac{1}{2}\|u_n(t+h) - u_n(t)\|_H^2 + \frac{k_0}{2}\int_0^t \|u_{n,r}(\cdot, \tau+h) - u_{n,r}(\cdot, \tau)\|_H^2 \, d\tau \le$$

$$\le \frac{1}{2}\|u_n(h) - u_0\|_H^2 + C_5 \int_0^t \|s_n(\tau+h) - s_n(\tau)\|_{\mathbb{R}^2}^2 \, d\tau +$$

$$+ \int_0^t \left(\|f_n(\tau+h) - f_n(\tau)\|_H + 2\|s_n(\tau+h) - s_n(\tau)\|_{\mathbb{R}^2} \right) \times$$

$$\times \|u_n(\tau+h) - u_n(\tau)\|_H \, d\tau \quad (3.2.6)$$

for a.a. $0 < t < t + h < T$, where C_5 is some positive constant. Similarly, we can see that

$$\frac{1}{2}\|u_n(h) - u_0\|_H^2 \le C_5 \int_0^h \|s_n(\tau) - s(0)\|_{\mathbb{R}^2}^2 \, d\tau +$$

$$+ \int_0^h \left(\|A(0)u_0\|_H + \|f_n(\tau) - f_n(0)\|_H + 2\|s_n(\tau) - s(0)\|_{\mathbb{R}^2} \right) \times$$

$$\times \|u_n(\tau) - u_0\|_H \, d\tau \text{ for all } h \in (0,T). \quad (3.2.7)$$

Notice also that

$$\|s_n(\tau) - s(0)\|_{\mathbb{R}^2}^2 = \left\| \int_0^\tau s_n'(\sigma) \, d\sigma \right\|_{\mathbb{R}^2}^2 \le \tau \int_0^T \|s_n'(\sigma)\|_{\mathbb{R}^2}^2 \le C_6 \tau \quad (3.2.8)$$

for all $\tau \in (0,T)$. Using (3.2.6)-(3.2.8), we can easily derive the following estimates (see Lemma 1.5.2):

$$\|u_n(h) - u_0\|_H \le C_7 h + \int_0^h \left(\|f_n(\tau) - f_n(0)\|_H + 2\|s_n(\tau) - s(0)\|_{\mathbb{R}^2} \right) d\tau,$$

$$\|u_n(t+h) - u_n(t)\|_H \leq$$

$$\leq \left(\|u_n(h) - u_0\|_H^2 + C_8 \int_0^t \|s_n(\tau+h) - s_n(\tau)\|_{\mathbb{R}^2}^2 \, d\tau\right)^{\frac{1}{2}} +$$

$$+ \int_0^t \left(\|f_n(\tau+h) - f_n(\tau)\|_H + 2\|s_n(\tau+h) - s(\tau)\|_{\mathbb{R}^2}\right) d\tau,$$

both for a.a. $0 < t < t + h < T$. Therefore, $(u_{n,t})$ is bounded in $L^\infty(0,T; H^1(0,1))$. Moreover, inequality (3.2.6) implies that $(u_{n,t})$ is also bounded in $L^2(0,T; H^1(0,1))$.

On the other hand, similar reasonings show that (u_n) is a Cauchy sequence in $C([0,T]; H) \cap L^2(0,T; H^1(0,1))$. Denote by u its limit. We are going to prove that u is a strong solution of problem (3.2.3)-(3.2.4).

Notice that

$$k_0\|u_{n,r}(\cdot,t) - u_0'\|_H^2 \leq \|f_n(t) - u_n'(t) - f(0) - A(0)u_0\|_H \, \|u_n(t) - u_0\| +$$

$$+2\|s_n(t) - s(0)\|_{\mathbb{R}^2}\left(\|u_n(t) - u_0\|_H + \|u_{n,r}(\cdot,t) - u_0'\|_H\right)$$

for a.a. $t \in (0,T)$, which implies that (u_n) is bounded in $L^\infty(0,T; H^1(0,1))$. Now, using the same argument as in the proof of Theorem 3.1.2 (see also the proof of Proposition 2.1.1) we can show that $(u_{n,rr})$ is bounded in $L^\infty(0,T; H)$. So, $u \in W^{1,\infty}(0,T; H) \cap L^\infty(0,T; H^2(0,1))$ and $u(0) = u_0$. Moreover, it is easily seen that (3.2.3) is satisfied for a.e. $t \in (0,T)$. Actually, we can pass to the limit in (3.0.1)-(3.0.2) using the information above. Finally, we have

$$\frac{k_0}{2} \int_0^{T-\delta} \left\|u_r(\cdot,t+\delta) - u_r(\cdot,t)\right\|_H^2 dt \leq \frac{1}{2}\|u(\delta) - u_0\|_H^2 +$$

$$+C_5 \int_0^{T-\delta} \left\|s(t+\delta) - s(t)\right\|_{\mathbb{R}^2}^2 dt + \int_0^{T-\delta}\left(\|f(t+\delta) - f(t)\|_H + \right.$$

$$\left. +4\|s(t+\delta) - s(t)\|_{\mathbb{R}^2}\right)\|u(t+\delta) - u(t)\|_H \, dt \leq C_9\delta^2$$

for all $\delta \in (0,T]$, since $u \in W^{1,\infty}(0,T; H)$, $f \in W^{1,1}(0,T; H)$, and $s \in W^{1,2}(0,T; \mathbb{R}^2)$ (see Theorem 1.1.2). Therefore, $u_r \in W^{1,2}(0,T; H)$ and this completes the proof of Theorem 3.2.1. ∎

REMARK 3.2.1   Of course, Theorem 3.1.2 can be obtained as a direct consequence of Theorem 3.2.1. However, we have preferred to discuss them separately for the convenience of the reader. Indeed, the last result involves some more complex estimates as well as a general existence result for time dependent equations. ∎

The case of a nonhomogeneous boundary condition, with β a subdifferential, can be treated in a specific manner to obtain other results concerning the existence of strong solutions. Actually, we shall reconsider this situation later, in an abstract framework.

THEOREM 3.2.2

Assume (I.1), (I.2'), and (I.3). Let $T > 0$ be fixed, $s \in L^2(0, T; \mathbb{R}^2)$, $f \in L^1(0, T; H)$, and $u_0 \in H$. Then, problem (3.2.3)-(3.2.4) or (3.0.1)-(3.0.2) has a unique weak solution $u \in C([0, T]; H) \cap L^2(0, T; H^1(0, 1))$.

PROOF Let $A_n(t)$, $t \in [0, T]$, be operators defined by (3.2.1)-(3.2.2) with (f_n, s_n) instead of (f, s). Consider the sequences (u_{0n}), (s_n), and (f_n) of elements of $D(A_n(0))$, $W^{1,2}(0, T; \mathbb{R}^2)$, and $W^{1,1}(0, T; H)$, respectively, such that, as $n \to \infty$,

$$u_{0n} \to u_0 \text{ in } H, \ s_n \to s \text{ in } L^2(0, T; \mathbb{R}^2), \text{ and } f_n \to f \text{ in } L^1(0, T; H).$$

Such sequences do exist: for example, we can choose s_n such that $s_n(0) = 0$ and $u_{0n} \in D(A)$ (see (3.1.1)), which is dense in H. Denote by u_n the strong solution of (3.2.3)-(3.2.4) in which (u_0, s, f) is replaced by (u_{0n}, f_n, s_n). By a standard calculation we get

$$\frac{1}{2}\|u_n(t) - u_m(t)\|_H^2 + \frac{k_0}{2}\int_0^t \|u_{n,r}(\cdot, \tau) - u_{m,r}(\cdot, \tau)\|_H^2 \, d\tau \leq$$

$$\leq \int_0^t \left(\|f_n(\tau) - f_m(\tau)\|_H + 2\|s_n(\tau) - s_m(\tau)\|_{\mathbb{R}^2} \right) \times$$

$$\times \|u_n(\tau) - u_m(\tau)\|_H \, d\tau +$$

$$+ \frac{1}{2}\|u_{0n} - u_{0m}\|_H^2 + C_5\|s_n - s_m\|_{L^2(0,T;\mathbb{R}^2)}^2 \tag{3.2.9}$$

for all $t \in [0, T]$. Dropping for the moment the second term on the left hand side of (3.2.9), one obtains

$$\|u_n(t) - u_m(t)\|_H \leq C_{10}\Big(\|u_{0n} - u_{0m}\|_H + \|s_n - s_m\|_{L^2(0,T;H)} +$$

$$+ \|s_n - s_m\|_{L^1(0,T;\mathbb{R}^2)} + \|f_n - f_m\|_{L^1(0,T;H)} \Big)$$

for all $t \in [0, T]$. This together with (3.2.9) yields

$$\|u_n(t) - u_m(t)\|_H + \|u_{n,r} - u_{m,r}\|_{L^2(0,T;H)} \leq$$

$$\leq C_{11}\Big(\|u_{0n} - u_{0m}\|_H + \|s_n - s_m\|_{L^2(0,T;\mathbb{R}^2)} +$$

$$+ \|s_n - s_m\|_{L^1(0,T;\mathbb{R}^2)} + \|f_n - f_m\|_{L^1(0,T;H)} \Big) \tag{3.2.10}$$

for all $t \in [0, T]$, where C_{11} is a positive constant, independent on n, m, and T. This concludes the proof. ∎

REMARK 3.2.2   Obviously, (3.2.10) extends to two arbitrary weak solutions, say u and $\tilde{u}$, corresponding to (u_0, f, s) and $(\tilde{u}_0, \tilde{f}, \tilde{s})$:

$$\|u(t) - \tilde{u}(t)\|_H + \|u_r - \tilde{u}_r\|_{L^2(0,T;H)} \leq$$

$$\leq C_{11} \Big(\|u_0 - \tilde{u}_0\|_H + \|s - \tilde{s}\|_{L^2(0,T;\mathbb{R}^2)} + \|s - \tilde{s}\|_{L^1(0,T;\mathbb{R}^2)} +$$

$$+ \|f - \tilde{f}\|_{L^1(0,T;H)} \Big) \text{ for all } t \in [0, T]. \tag{3.2.11}$$

We have kept the L^1-norm for later use of this estimate; the fact that C_{11} is independent of T will be essential.   ∎

The next result is concerned with the asymptotic behavior of solutions as $t \to \infty$. This is a difficult task for general nonautonomous evolution equations. Fortunately, our problem here contains only some nonhomogeneous terms, $f(t)$ and $s(t)$. So, we can easily extend our asymptotic results from the preceding section to the present case. The basic idea is to approximate $f(t)$ and $s(t)$ by their truncated functions.

THEOREM 3.2.3
Assume (I.1), (I.2'), and (I.3). Let $u_0 \in H$, $f \in L^1(\mathbb{R}_+; H)$, and $s \in L^2(\mathbb{R}_+; \mathbb{R}^2) \cap L^1(\mathbb{R}_+; \mathbb{R}^2)$. If the solutions of (3.0.1)-(3.0.2) with $f(t) \equiv 0$ and $s(t) \equiv 0$ converge in H, as $t \to \infty$, then the weak solution of (3.2.3)-(3.2.4) or (3.0.1)-(3.0.3) converges in H to a stationary solution, as $t \to \infty$.

PROOF   For each $n \in \mathbb{N}^*$, define the truncated functions

$$f_n(t) = \begin{cases} f(t) & \text{if } 0 < t < n, \\ 0 & \text{otherwise,} \end{cases} \quad \text{and} \quad s_n(t) = \begin{cases} s(t) & \text{if } 0 < t < n, \\ 0 & \text{otherwise.} \end{cases}$$

We denote by u_n the solution of problem (3.0.1)-(3.0.3) corresponding to (u_0, f_n, s_n). By our hypothesis, for each $n \in \mathbb{N}^*$, there exists a stationary solution p_n such that

$$u_n(t) \to p_n \text{ in } H, \text{ as } t \to \infty. \tag{3.2.12}$$

On the other hand, using for instance (3.2.11), we can see that (p_n) is a Cauchy sequence in H, and so (p_n) converges to some $p \in H$. We shall show that

$$u(t) \to p \text{ in } H, \text{ as } t \to \infty.$$

Indeed, by the inequality

$$\|u(t) - p\|_H \leq \|u(t) - u_n(t)\|_H + \|u_n(t) - p_n\|_H + \|p_n - p\|_H,$$

it follows that (see inequality (3.2.11))

$$\|u(t) - p\|_H \leq \|u_n(t) - p_n\|_H + \|p_n - p\|_H +$$

$$+ C_{11}\left(\left(\int_n^\infty \|s(\tau)\|_{\mathbb{R}^2}^2 \, d\tau\right)^{\frac{1}{2}} + \int_n^\infty \|s(\tau)\|_{\mathbb{R}^2} \, d\tau +$$

$$+ \int_n^\infty \|f(\tau)\|_H \, d\tau\right) < \epsilon + \|u_n(t) - p_n\|_H$$

for n large enough. Therefore, taking into account (3.2.12), we get

$$\limsup_{t\to\infty} \|u(t) - p\|_H \leq \epsilon \text{ for all } \epsilon > 0.$$

Hence $u(t) \to p$ in H, as $t \to \infty$. $\blacksquare$

3.3 Higher regularity of solutions

This is a special topic, less discussed in literature, but very important for some applications, including singularly perturbed nonlinear boundary value problems (see [BaMo3]), where higher regularity is needed for developing an asymptotic analysis.

Here, we restrict our investigation to the following simpler problem:

$$u_t(r,t) - u_{rr}(r,t) + au(r,t) = f(r,t), \ (r,t) \in (0,1) \times (0,T), \quad (3.3.1)$$

$$\begin{pmatrix} u_r(0,t) \\ -u_r(1,t) \end{pmatrix} \in \begin{pmatrix} \beta_1 u(0,t) \\ \beta_2 u(1,t) \end{pmatrix}, \ t \in (0,T), \quad (3.3.2)$$

$$u(r,0) = u_0(r), \ r \in (0,1), \quad (3.3.3)$$

where a is some nonnegative constant and β_1, β_2 are some nonlinear mappings (possibly multivalued). So, the present problem is still nonlinear and hence some difficulties are expected to appear.

Notice that the case of nonhomogeneous boundary conditions can be reduced to the homogeneous case (3.3.2) by a simple change of the unknown function:

$$\tilde{u}(r,t) = u(r,t) + r(1-r)\big(a(t)r + b(r)\big).$$

We have already proved some regularity properties of the solution. Let us reformulate Theorem 3.1.2 for the present particular case.

PROPOSITION 3.3.1
Assume that

$$f \in W^{1,1}(0,T; L^2(0,1)); \quad (3.3.4)$$

$$\beta_1, \beta_2 \subset \mathbb{R} \times \mathbb{R} \text{ are maximal monotone operators;} \qquad (3.3.5)$$

$$u_0 \in H^2(0,1); \qquad (3.3.6)$$

$$\begin{cases} u_0(0) \in D(\beta_1), \ u_0(1) \in D(\beta_2), \\ u_0'(0) \in \beta_1 u_0(0), \ -u_0'(1) \in \beta_2 u_0(1). \end{cases} \qquad (3.3.7)$$

Then, problem (3.3.1)-(3.3.3) has a unique strong solution

$$u \in W^{1,\infty}\big(0,T;L^2(0,1)\big) \cap L^\infty\big(0,T;H^2(0,1)\big) \cap$$
$$\cap W^{1,2}\big(0,T;H^1(0,1)\big). \qquad (3.3.8)$$

REMARK 3.3.1   The assumptions (3.3.7) are called *first order compatibility conditions.* ▌

It seems that, for multivalued β_1 and β_2, no essential improvement of the regularity result given by Proposition 3.3.1 is possible. This is an open problem. We are going to show, however, that any improvement of regularity is possible for single-valued and smooth β_1 and β_2. The main difficulty comes from the nonlinearity of β_1 and β_2. The linear case is well known from [LaSoU, Ch. VII, § 10], where the compatibility conditions, corresponding to our particular problem (3.3.1)-(3.3.3) with linear β_1 and β_2, consist in the fact that the time derivatives of $\partial^k u / \partial r^k$, $k = 0, 1, \ldots$, which can be calculated from (3.3.1) and (3.3.3), must satisfy boundary conditions (3.3.2) and equations obtained from them by differentiation with respect to t. Certainly, the case of nonlinear β_1 and β_2 is more delicate. We continue with the following result in this direction:

THEOREM 3.3.1
Assume that

$$f \in W^{2,1}\big(0,T;L^2(0,1)\big), \ f(\cdot,0) \in H^2(0,1); \qquad (3.3.9)$$

the operators β_1, β_2 are both everywhere defined on $\mathbb{R}$, single-valued, and

$$\beta_1, \beta_2 \in W^{2,\infty}_{loc}(\mathbb{R}), \ \beta_1' \geq 0, \ \beta_2' \geq 0; \qquad (3.3.10)$$

$$u_0 \in H^4(0,1); \qquad (3.3.11)$$

and u_0 satisfies (3.3.7), where "$\in$" is replaced by "$=$", as well as

$$z_0'(0) = \beta_1'\big(u_0(0)\big)z_0(0), \ -z_0'(1) = \beta_2'\big(u_0(1)\big)z_0(1), \qquad (3.3.12)$$

where $z_0 : [0,1] \to \mathbb{R}$ is defined by

$$z_0(r) = f(r,0) + u_0''(r) - a u_0(r) \ \text{ for all } r \in [0,1]. \qquad (3.3.13)$$

Then, the solution u of problem (3.3.1)-(3.3.3) belongs to

$$W^{2,2}\big(0,T;H^1(0,1)\big) \cap W^{2,\infty}\big(0,T;L^2(0,1)\big).$$

REMARK 3.3.2 The conditions (3.3.12) are called *second order compatibility conditions.* ∎

PROOF Clearly, all the conditions of Proposition 3.3.1 are met and so problem (3.3.1)-(3.3.3) has a unique solution u satisfying (3.3.8). Denote $V = H^1(0,1)$ and by V^* its dual space. First we show that $u_t \in W^{1,2}(0,T;V^*)$. To this purpose, it suffices to prove (cf. Theorem 1.1.2) that there exists a positive constant C such that

$$\int_0^{T-\delta} \|u_t(\cdot,t+\delta) - u_t(\cdot,t)\|_{V^*}^2 \, dt \le C\delta^2 \text{ for all } \delta \in (0,T). \qquad (3.3.14)$$

Indeed, we have for a.a. $t \in (0, T-\delta)$ and all $\phi \in V$,

$$\langle \phi, u_t(\cdot,t+\delta) - u_t(\cdot,t)\rangle_V = \langle \phi, u_{rr}(\cdot,t+\delta) - u_{rr}(\cdot,t)\rangle - $$
$$-a\langle \phi, u(t+\delta) - u(t)\rangle + \langle \phi, f(t+\delta) - f(t)\rangle, \qquad (3.3.15)$$

where $\langle \cdot, \cdot \rangle$ denotes the pairing between V and V^* (actually, in (3.1.15) this is just the inner product of $H = L^2(0,1)$). From (3.3.15) it follows that

$$\langle \phi, u_t(\cdot,t+\delta) - u_t(\cdot,t)\rangle + \langle \phi', u_r(\cdot,t+\delta) - u_r(\cdot,t)\rangle + $$
$$+\Big(\beta_1\big(u(0,t+\delta)\big) - \beta_1\big(u(0,t)\big)\Big)\phi(0) + $$
$$+\Big(\beta_2\big(u(1,t+\delta)\big) - \beta_2\big(u(1,t)\big)\Big)\phi(1) + $$
$$+a\langle \phi, u(t+\delta) - u(t)\rangle = \langle \phi, f(t+\delta) - f(t)\rangle. \qquad (3.3.16)$$

Taking into account (3.3.8) and (3.3.10), which in particular implies that β_1 and β_2 are Lipschitzian on bounded sets, one can easily derive from (3.3.16) the estimate

$$\|u_t(\cdot,t+\delta) - u_t(\cdot,t)\|_{V^*}^2 \le$$
$$\le C_1\Big(\|u(t+\delta) - u(t)\|_V^2 + \|f(t+\delta) - f(t)\|_H^2\Big), \qquad (3.3.17)$$

where C_1 is some positive constant. Now, (3.3.8), (3.3.9), (3.3.17), and Theorem 1.1.2 lead us to the desired estimate (3.3.14). Therefore, $z := u_t \in W^{1,2}(0,T;V^*)$ and hence we can differentiate with respect to t the obvious equation

$$\langle \phi, u_t(\cdot,t)\rangle + \langle \phi', u_r(\cdot,t)\rangle + \beta_1\big(u(0,t)\big)\phi(0) + $$
$$+\beta_2\big(u(1,t)\big)\phi(1) + a\langle \phi, u(t)\rangle = \langle \phi, f(t)\rangle$$

to obtain that

$$\langle \phi, z_t(\cdot, t)\rangle + \langle \phi', z_r(\cdot, t)\rangle + g_1(t)z(0, t)\phi(0) +$$
$$+ g_2(t)z(1, t)\phi(1) + a\langle \phi, z(t)\rangle = \langle \phi, f_t\rangle \tag{3.3.18}$$

for all $\phi \in V$ and a.a. $t \in (0, T)$, where

$$g_1(t) := \beta_1'\big(u(0, t)\big) \text{ and } g_2(t) := \beta_2'\big(u(1, t)\big).$$

In addition,

$$z(\cdot, 0) = z_0, \tag{3.3.19}$$

with z_0 defined by (3.3.13). It is easily seen that z is the unique solution of problem (3.3.18)-(3.3.19) in the class of u_t. Indeed, if we take in (3.3.18)-(3.3.19) $f_t \equiv 0$, $z_0 \equiv 0$, and $\phi = z(\cdot, t)$, then (cf. Theorem 1.1.3) we have

$$\frac{d}{dt}\|z(t)\|_H^2 \leq 0 \text{ for a.a. } t \in (0, T),$$

which clearly implies that $z \equiv 0$.

Notice that $z = u_t$ satisfies formally the problem

$$z_t(r, t) - z_{rr}(r, t) + az(r, t) = f_t(r, t), \ (r, t) \in (0, 1) \times (0, T), \tag{3.3.20}$$

$$\begin{pmatrix} z_r(0, t) \\ -z_r(1, t) \end{pmatrix} = \begin{pmatrix} g_1(t)z(0, t) \\ g_2(t)z(1, t) \end{pmatrix}, \ t \in (0, T), \tag{3.3.21}$$

$$z(r, 0) = z_0(r), \ r \in (0, 1). \tag{3.3.22}$$

By (3.3.8) and (3.3.10) it follows that $g_1, g_2 \in H^1(0, T)$ and $g_1 \geq 0$, $g_2 \geq 0$. Actually, problem (3.3.20)-(3.3.22) has a unique strong solution. To show this, let us notice that this problem can be expressed as the following Cauchy problem in $H = L^2(0, 1)$:

$$z'(t) + B(t)z(t) = f'(t), \ t \in (0, T), \ z(0) = z_0, \tag{3.3.23}$$

where $B(t): D\big(B(t)\big) \subset H \to H$ is defined by $B(t)v = -v'' + av$ and

$$D\big(B(t)\big) = \{v \in H^2(0, 1) \mid v'(0) = g_1(t)v(0), -v'(1) = g_2(t)v(1)\}.$$

As seen before (see Section 3.1), $B(t)$ is a linear maximal monotone operator, for each $t \in [0, T]$. Moreover, $B(t)$ is the subdifferential of the function $\phi(t, \cdot): H \to (-\infty, \infty]$, given by

$$\phi(t, v) = \begin{cases} \frac{1}{2}\int_0^1 \big(v'(x)^2 + av(x)^2\big)\, dx + \\ \qquad + \frac{1}{2}\big(g_1(t)v(0)^2 + g_2(t)v(1)^2\big) \text{ if } v \in H^1(0, 1), \\ \infty \text{ otherwise.} \end{cases}$$

For every $t \in [0, T]$, the effective domain $D\big(\phi(t, \cdot)\big) = H^1(0, 1)$, i.e., it does not depend on t. It turns out that the condition (1.5.15) of Theorem 1.5.8 is satisfied. Indeed, for every $v \in H^1(0, 1)$ and $0 \le s \le t \le T$, we have

$$\phi(t, v) - \phi(s, v) = \frac{1}{2}\big(g_1(t) - g_1(s)\big)v(0)^2 + \frac{1}{2}\big(g_2(t) - g_2(s)\big)v(1)^2 =$$

$$= \frac{1}{2}v(0)^2 \int_s^t g_1'(\tau)\, d\tau + \frac{1}{2}v(1)^2 \int_s^t g_2'(\tau)\, d\tau \le$$

$$\le \big(\|v\|_H^2 + \|v'\|_H^2\big) \int_s^t \big(|g_1'(\tau)| + |g_2'(\tau)|\big)\, d\tau,$$

since $H^1(0, 1)$ is embedded continuously into $C[0, 1]$. Therefore,

$$\phi(t, v) \le \phi(s, v) + \big(\gamma(t) - \gamma(s)\big)\Big(\phi(s, v) + \frac{1}{2}\|v\|_H^2\Big),$$

where

$$\gamma(t) := 2 \int_0^t \big(|g_1'(\tau)| + |g_2'(\tau)|\big)\, d\tau.$$

So, according to Theorem 1.5.8, problem (3.3.23) has a unique solution $z \in W^{1,2}(0, T; H) \cap L^\infty\big(0, T; H^1(0, 1)\big)$. Obviously, this z satisfies (3.3.18)-(3.3.19) and hence it coincides with u_t. So, we have already proved that

$$u \in W^{2,2}(0, T; H) \cap W^{1,\infty}\big(0, T; H^1(0, 1)\big).$$

We are now going to show that

$$z_t = u_{tt} \in L^\infty(0, T; H) \cap L^2\big(0, T, H^1(0, 1)\big).$$

To this purpose, we argue in a somewhat standard manner. So, we start from the obvious estimates

$$\frac{1}{2}\frac{d}{dt}\big\|z(t + h) - z(t)\big\|_H^2 + \big\|z_r(\cdot, t + h) - z_r(\cdot, t)\big\|_H^2 +$$

$$+ \big(g_1(t + h)z(0, t + h) - g_1(t)z(0, t)\big)\big(z(0, t + h) - z(0, t)\big) +$$

$$+ \big(g_2(t + h)z(1, t + h) - g_2(t)z(1, t)\big)\big(z(1, t + h) - z(1, t)\big) \le$$

$$\le \big\|f'(t + h) - f(t)\big\|_H \big\|z(t + h) - z(t)\big\|_H$$

for a.a. $0 \le t < t + h \le T$, and

$$\frac{1}{2}\frac{d}{dh}\big\|z(h) - z_0\big\|_H^2 + \big\|z_r(\cdot, h) - z_0'\big\|_H^2 +$$

$$+ \big(g_1(h)z(0, h) - g_1(0)z_0(0)\big)\big(z(0, h) - z_0(0)\big) +$$

$$+ \big(g_2(h)z(1, h) - g_2(0)z_0(1)\big)\big(z(1, h) - z_0(1)\big) \le$$

$$\le \big\|f_h(\cdot, h) - B(0)z_0\big\|_H \big\|z(h) - z_0\big\|_H \quad \text{for a.a. } h \in (0, T).$$

As $u \in W^{1,\infty}(0,T;H^1(0,1))$ and $\beta_1, \beta_2 \in W^{2,\infty}_{loc}(\mathbb{R})$, we can see that g_1, g_2 are both Lipschitz continuous. So, taking into account that $g_1 \geq 0$ and $g_2 \geq 0$ in $[0,1]$, we derive from the above inequalities that

$$\frac{1}{2}\frac{d}{dt}\left\|z(t+h) - z(t)\right\|^2_H + \left\|z_r(\cdot, t+h) - z_r(\cdot, t)\right\|^2_H \leq$$
$$\leq \left\|f_t(\cdot, t+h) - f_t(\cdot, t)\right\|_H \left\|z(t+h) - z(t)\right\|_H +$$
$$+ L_1 h\left|z(0, t+h) - z(0, t)\right| + L_2 h\left|z(1, t+h) - z(1, t)\right| \quad (3.3.24)$$

for a.a. $0 \leq t < t + h \leq T$, and

$$\frac{1}{2}\frac{d}{dh}\left\|z(h) - z_0\right\|^2_H + \left\|z_r(\cdot, h) - z_0'\right\|^2_H \leq$$
$$\leq \left\|f_h(\cdot, h) - B(0)z_0\right\|_H \left\|z(h) - z_0\right\|_H +$$
$$+ L_1 h\left|z(0, h) - z_0(0)\right| + L_2 h\left|z(1, h) - z_0(1)\right| \quad (3.3.25)$$

for a.a. $h \in (0,T)$. Since $H^1(0,1)$ is continuously embedded into $C[0,1]$, we get from (3.3.24)-(3.3.25) the following two estimates:

$$\frac{1}{2}\frac{d}{dt}\left\|z(t+h) - z(t)\right\|^2_H + \left\|z_r(\cdot, t+h) - z_r(\cdot, t)\right\|^2_H \leq$$
$$\leq \left\|f_t(\cdot, t+h) - f_t(\cdot, t)\right\|_H \left\|z(t+h) - z(t)\right\|_H + \frac{1}{2}C_2 h^2 +$$
$$+ \frac{1}{2}\left\|z_r(\cdot, t+h) - z_r(\cdot, t)\right\|^2_H + \frac{1}{2}\left\|z(t+h) - z(t)\right\|^2_H \quad (3.3.26)$$

for a.a. $0 \leq t < t + h \leq T$, and

$$\frac{1}{2}\frac{d}{dh}\left\|z(h) - z_0\right\|^2_H + \left\|z_r(\cdot, h) - z_0'\right\|^2_H \leq$$
$$\leq \left\|f_t(\cdot, h) - B(0)z_0\right\|_H \left\|z(h) - z_0\right\|_H + \frac{1}{2}\left\|z_r(\cdot, h) - z_0'\right\|^2_H +$$
$$+ \frac{1}{2}\left\|z(h) - z_0\right\|^2_H + \frac{1}{2}C_3 h^2 \text{ for a.a. } h \in (0,T), \quad (3.3.27)$$

where C_2 and C_3 are some positive constants. Therefore,

$$\frac{d}{dt}\left(e^{-t}\left\|z(t+h) - z(t)\right\|^2_H\right) + e^{-t}\left\|z_r(\cdot, t+h) - z_r(\cdot, t)\right\|^2_H \leq$$
$$\leq 2e^{-t}\left\|f_t(\cdot, t+h) - f_t(\cdot, t)\right\|_H \left\|z(t+h) - z(t)\right\|_H + C_2 h^2 \quad (3.3.28)$$

for a.a. $0 \leq t < t + h \leq T$, and

$$\frac{d}{dh}\left(e^{-h}\left\|z(h) - z_0\right\|^2_H\right) \leq C_3 h^2 +$$
$$+ 2e^{-h}\left\|f_t(\cdot, h) - B(0)z_0\right\|_H \left\|z(h) - z_0\right\|_H \quad (3.3.29)$$

for a.a. $h \in (0, T)$. We drop for the moment the second term of the left hand side of (3.3.28) and integrate the resulting inequality over $[0, t]$. Hence

$$\|z(t+h) - z(t)\|_H \leq C_4 \Big(\|z(h) - z_0\|_H + h +$$
$$+ \int_0^t \|f'(s+h) - f'(s)\|_H \, ds \Big) \tag{3.3.30}$$

for all $0 \leq t < t + h \leq T$. Similarly, starting from (3.3.29) we arrive at

$$\|z(h) - z_0\|_H \leq C_5 \Big(h^{3/2} + \int_0^h \|f'(s) - B(0)z_0\|_H \, ds \Big) \tag{3.3.31}$$

for all $h \in (0, T)$. Obviously, (3.3.30) and (3.3.31) yield $z_t \in L^\infty(0, T; H)$. Using this conclusion together with (3.3.26), we may easily obtain, by means of Theorem 1.1.2, that $z_r \in W^{1,2}(0, T; H)$. This concludes the proof. $\blacksquare$

REMARK 3.3.3   If in Theorem 3.3.1 f is assumed to be more regular with respect to r, then u is also more regular in r, since

$$u_{tt} = u_{rrt} - au_t + f_t = u_{rrrr} - au_{rr} + f_{rr} - au_t + f_t.$$

On the other hand, we can prove that

$$u \in W^{3,2}\big(0, T; H^1(0, 1)\big) \cap W^{3,\infty}\big(0, T; L^2(0, 1)\big),$$

under appropriate assumptions on the smoothness of β_1, β_2, u_0, f, and compatibility.

Indeed, one may use a reasoning similar to that of the proof of Theorem 3.3.1. This time, a problem of type (3.3.20)-(3.3.22) with *nonhomogeneous* boundary conditions must be solved. Fortunately, the nonhomogeneous terms in the boundary conditions are H^1-functions and so Theorem 1.5.8 can again be applied, with a slight modification of ϕ. The corresponding t-dependent operator is nonlinear, since its domain is an affine subset of $L^2(0, 1)$. We leave to the reader to formulate and prove this higher regularity result.

Of course, higher regularity with respect to r can also be obtained at the expense of additional assumptions. Actually, the above procedure can be repeated as many times as we want and so any regularity of the solution, with respect to t and r, can be reached under enough smoothness of the data and compatibility conditions. More precisely, the $H^k(Q_T)$-regularity can be reached for every $k \in \mathbb{N}^*$ and, hence, the $C^k(\overline{Q_T})$-regularity can be established as well for any $k \in \mathbb{N}^*$, due the Sobolev's embedding theorem (see, e.g., [Adams]). $\blacksquare$

REMARK 3.3.4 We have formulated Theorem 3.3.1 having in mind the idea of getting enough information that allows us to go to the next level of regularity. Notice, however, that our problem is parabolic and so in order to get, for instance, that

$$u \in W^{2,2}(0,T;L^2(0,1)) \cap W^{1,\infty}(0,T;H^1(0,1)),$$

we can relax our assumptions: It suffices that β_1, β_2 satisfy (3.3.10), $u_0 \in H^3(0,1)$ satisfies (3.3.7), and

$$f \in W^{1,2}(0,T;L^2(0,1)), \quad f(\cdot,0) \in H^1(0,1).$$

The reader may check this assertion by inspecting the proof of Theorem 3.3.1. Notice also, that some nonlinear cases of (3.3.1) can be studied by similar arguments. Even higher order parabolic problems could be investigated by the same technique, but we do not intend to go into further details.

References

[Adams] R.A. Adams, *Sobolev Spaces*, Academic Press, New York, 1975.

[BaMo3] L. Barbu & Gh. Moroşanu, *Analiza asimptotică a unor probleme la limită cu perturbaţii singulare*, Editura Academiei, Bucharest, 2000.

[GolLin] J.A. Goldstein & C.-Y. Lin, An L^p semigroup approach to degenerate parabolic boundary value problems, *Ann. Mat. Pura Appl.* (IV), Vol. CLIX (1991), 211–227.

[Kato] T. Kato, Nonlinear semigroups and evolution equations, *J. Math. Soc. Japan.*, 19 (4) (1967), 508–520.

[LaSoU] O.A. Ladyzhenskaya, V.A. Solonnikov & N.N. Ural'ceva, *Linear and Quasilinear Equations of Parabolic Type*, Translations of Math. Monographs 23, AMS, Providence, Rhode Island, 1968.

[Lin1] C.-Y. Lin, Fully nonlinear parabolic boundary value problems in one space dimensions, *J. Diff. Eq.*, 87 (1990), 62–69.

[LinFan] C.-Y. Lin & L.-C. Fan, A class of degenerate totally nonlinear parabolic equations, *J. Math. Anal. Appl.*, 203 (1996), 812–827.

[Moro1] Gh. Moroşanu, *Nonlinear Evolution Equations and Applications*, Editura Academiei and Reidel, Bucharest and Dordrecht, 1988.

[Moro5] Gh. Moroşanu, Parabolic problems associated to integrated circuits with time-dependent sources in *Differential Equations and Control Theory*, ed. by V. Barbu, Pitman Res. Notes in Math 250, Longman, London, 1991, 208–216.

[MMN1] Gh. Moroşanu, C.A. Marinov & P. Neittaanmäki, Well-posed nonlinear problems in integrated circuits modeling, *Circuits System Signal Process*, 10 (1991), 53–69.

[MMN2] Gh. Moroşanu, C.A. Marinov & P. Neittaanmäki, Well-posed nonlinear problems in the theory of electrical networks with distributed and lumped parameters in *Proceedings of the Seventeenth IASTED Int. Symposium Simulation and Modelling*, Acta Press, Zürich, 1989.

Chapter 4

Parabolic boundary value problems with algebraic-differential boundary conditions

We consider in this chapter the same doubly nonlinear parabolic equation as in Chapter 3, namely:

$$u_t(r,t) - \frac{\partial}{\partial r}G\big(r, u_r(r,t)\big) + K\big(r, u(r,t)\big) = f(r,t) \qquad (4.0.1)$$

for $r \in (0,1)$ and $t > 0$. We associate with it the boundary conditions

$$-G\big(0, u_r(0,t)\big) + \beta_1 u(0,t) \ni s_1(t), \ t > 0, \qquad (4.0.2)$$

$$u_t(1,t) + G\big(1, u_r(1,t)\big) + \beta_2 u(1,t) \ni s_2(t), \ t > 0, \qquad (4.0.3)$$

as well as the initial condition

$$u(r,0) = u_0(r), \ r \in (0,1). \qquad (4.0.4)$$

The same assumptions governing the nonlinear G and K will be required, i.e., (I.1) and (I.2'). We further assume that

(I.3') The operators $\beta_1, \beta_2 \subset \mathbb{R} \times \mathbb{R}$ are maximal monotone.

Notice that the boundary condition (4.0.2) is an algebraic one, while (4.0.3) is of a different character, because it contains the time derivative $u_t(1,t)$ and hence it is natural to call it a *differential boundary condition*. So, (4.0.2)-(4.0.3) will be called *algebraic-differential boundary conditions*. The present problem (4.0.1)-(4.0.4) is essentially different from the problem (3.0.1)-(3.0.3) investigated in Chapter 3. In order to study this new boundary value problem, we need another framework. More precisely, problem (4.0.1)-(4.0.4) will be expressed as a Cauchy problem in the product space $H_1 := L^2(0,1) \times \mathbb{R}$, which is a Hilbert space with the inner product defined by

$$\big((v_1,a),(v_2,b)\big)_{H_1} = \int_0^1 v_1(r)v_2(r)\,dr + ab.$$

A product space setup has previously been used in different contexts by R.B. Vinter [Vinter] and J. Zabczyk [Zabcz1], [Zabcz2] (see also Chapter 6 below). In the present situation, we can again see that the choice of an appropriate framework is an essential step toward the solvability and qualitative analysis of a problem. Due to the special character of problem (4.0.1)-(4.0.4), we study it in a separate chapter. However, the technique we shall use here has many similarities to that of the preceding chapter.

As noticed before, (4.0.1) represents a nonlinear variant of the one-dimensional heat equation. Also, it serves as a model for dispersion or diffusion phenomena. Notice also that a differential boundary condition like (4.0.3) may appear in the case of diffusion in chemical substances. More precisely, this condition describes a reaction at the boundary, and the term including $u_r(1, t)$ is responsible for the diffusive transport of materials to the boundary (see [Cannon, Ch. 7] for the linear case).

Let us also mention that problems of type (4.0.1)-(4.0.4) may appear as a degenerate case of some hyperbolic problems with algebraic-differential boundary conditions. More precisely, if in the telegraph system, with algebraic-differential boundary conditions, the inductance is negligible and set equal to zero, then a parabolic model of type (4.0.1)-(4.0.4) is obtained. Of course, this simplified model is an ideal one, but it still describes sufficiently well the corresponding propagation phenomenon, in the case of small inductances, as proved by some recent results of asymptotic analysis [Barbu], [BaMo1], [BaMo2], [BaMo3], [MoPer]. So, the study of the model (4.0.1)-(4.0.4) is strongly motivated.

The boundary condition (4.0.2) is a nonlinear algebraic condition of Robin-Steklov type. If $s_1(t) \equiv 0$, it can be written in a simpler form. Indeed, if (I.1) is satisfied, then clearly $G(0, \cdot)$ is invertible and $G(0, \cdot)^{-1}$ is nondecreasing, Lipschitzian, and continuously differentiable on $\mathbb{R}$. Moreover, it can easily be seen that the operator

$$\beta_3 \subset \mathbb{R} \times \mathbb{R}, \ \beta_3 x := G(0, \cdot)^{-1} \beta_1 x,$$

is maximal monotone. So, (4.0.2) can be written as a usual nonlinear Robin-Steklov boundary condition

$$-u_r(0, t) + \beta_3 u(0, t) \ni 0, \ t > 0.$$

However, for convenience, we shall keep the initial form, taking also into account that the above transformation is not possible in the general non-homogeneous case.

The case where both the boundary conditions (4.0.2)-(4.0.3) contain derivatives with respect to t can also be considered. In this situation, the natural framework will be the space $H_2 := L^2(0, 1) \times \mathbb{R}^2$, as we shall see later. Also, a system of n parabolic equations with some mixed boundary conditions (algebraic-differential boundary conditions) could be investigated. Such problems are more complex, but they can be solved by a similar technique.

We mention that the model (4.0.1)-(4.0.4), with $G(r, \cdot)$ a linear function, has been investigated in [Barbu], [BaMo1], [BaMo2], [MoPer].

In the next section we shall consider the case in which (4.0.2) is a *homogeneous* algebraic nonlinear Robin-Steklov boundary condition, i.e., $s_1(t) \equiv 0$. In Section 4.2 we allow this relation to be *nonhomogeneous*, i.e., $s(t) \not\equiv \text{const}$.

4.1 Homogeneous algebraic boundary condition

Here we suppose that $s_1(t)$ is a constant. Actually, by slightly modifying β_1, we can assume that $s_1(t) \equiv 0$. Define the operator $A_1 \colon D(A_1) \subset H_1 \to H_1$, $H_1 = L^2(0,1) \times \mathbb{R}$, by

$$D(A_1) = \{(v, a) \in H^2(0,1) \times \mathbb{R} \mid a = v(1) \in D(\beta_2),$$
$$v(0) \in D(\beta_1), -G\big(0, v'(0)\big) + \beta_1 v(0) \ni 0\}, \quad (4.1.1)$$

$$A_1(v, a) = \Big(-\frac{d}{dr} G(\cdot, v') + K(\cdot, v), G\big(1, v'(1)\big) + \beta_2 a \Big). \quad (4.1.2)$$

The operator A_1 will help us to write problem (4.0.1)-(4.0.4) as a Cauchy problem in H_1.

PROPOSITION 4.1.1
Assume (I.1), (I.2'), and (I.3'). Then the operator A_1 defined above is maximal cyclically monotone. More precisely, A_1 is the subdifferential of $\psi_1 \colon H_1 \to (-\infty, \infty]$ defined by

$$\psi_1(v, a) = \begin{cases} \int\limits_0^1 \Big(g\big(r, v'(r)\big) + k\big(r, v(r)\big) \Big)\, dr + j_1\big(v(0)\big) + j_2(a) \\ \qquad \textit{if } v \in H^1(0,1), \ v(0) \in D(j_1), \\ \qquad \qquad a = v(1) \in D(j_2) \ \textit{and } g(\cdot, v') \in L^1(0,1), \\ \infty \ \textit{otherwise,} \end{cases} \quad (4.1.3)$$

where $j_1, j_2 \colon \mathbb{R} \to (-\infty, \infty]$ are proper, convex, and lower semicontinuous functions such that $\beta_1 = \partial j_1$ and $\beta_2 = \partial j_2$, whilst $g(r, \cdot)$ and $k(r, \cdot)$ are given by

$$g(r, \xi) = \int_0^\xi G(r, \tau)\, d\tau \ \textit{ and } \ k(r, \xi) = \int_0^\xi K(r, \tau)\, d\tau.$$

In addition, the closure of $D(A_1)$ in H_1 is the set $L^2(0,1) \times \overline{D(\beta_2)}$.

PROOF   Obviously, the function $G(0, \cdot)\colon \mathbb{R} \to \mathbb{R}$ is surjective. Therefore, for every $b \in D(\beta_1)$ there exists a $c \in \mathbb{R}$ such that

$$-G(0, c) + \beta_1(b) \ni 0.$$

So, for every $d \in D(\beta_2)$ there exists a function $v_d\colon [0, 1] \to \mathbb{R}$,

$$v_d(r) = b + cr + (d - b - c)r^2,$$

such that $(v_d, d) \in D(A_1)$, i.e., $D(A_1) \neq \emptyset$. Now, we define the set

$$M = \left\{ (v_d + \phi, d) \mid \phi \in C_0^\infty(0, 1), d \in D(\beta_2) \right\}.$$

Obviously, we have

$$M \subset D(A_1) \subset L^2(0, 1) \times D(\beta_2). \tag{4.1.4}$$

Thus the closure of $D(A_1)$ in H_1 is $L^2(0, 1) \times \overline{D(\beta_2)}$, as asserted. Now, an easy computation shows that A_1 is a monotone operator, with respect to the scalar product of H_1. In order to prove the maximality of A_1, we shall use the equivalent Minty condition: $R(I + A_1) = H_1$; see Theorem 1.2.2. So, let $(p, y) \in H_1$ be arbitrary but fixed and consider the equation

$$(v, a) + A_1(v, a) \ni (p, y). \tag{4.1.5}$$

We shall prove that (4.1.5) has a solution $(v, a) \in D(A_1)$. In other words, we shall show that there exists a $v \in H^2(0, 1)$ such that

$$v(r) - \frac{d}{dr} G\big(r, v'(r)\big) + K\big(r, v(r)\big) = p(r) \ \text{ for a.a. } \ r \in (0, 1),$$
$$-G\big(0, v'(0)\big) + \beta_1 v(0) \ni 0,$$
$$G\big(1, v'(1)\big) + \beta_2 v(1) + v(1) \ni 0.$$

By choosing
$$\beta \subset \mathbb{R}^2 \times \mathbb{R}^2, \ \ \beta(\xi, \eta) = (\beta_1 \xi, \eta + \beta_2 \eta),$$

and K to be the mapping $(r, \xi) \mapsto \xi + K(r, \xi)$, Theorem 2.1.1 implies that the above problem has a solution $v \in H^2(0, 1)$. Hence A_1 is indeed a maximal monotone operator.

On the other hand, it is easily seen that the function ψ_1, given by (4.1.3), is proper, convex, and lower semicontinuous (see also the proof of Proposition 2.1.3). We shall show that $A_1 \subset \partial \psi_1$, which implies by the maximality of A_1 that $A_1 = \partial \psi_1$. Let functions v and w satisfy $(v, v(1)) \in D(A_1)$ and $(w, w(1)) \in D(\psi_1)$. Then

$$\psi_1\big(v, v(1)\big) - \psi_1\big(w, w(1)\big) = \int_0^1 \Big(g\big(r, v'(r)\big) - g\big(r, w'(r)\big) \Big)\, dr +$$

$$+ \int_0^1 \Big(k\big(r,v(r)\big) - k\big(r,w(r)\big) \Big)\, dr + j_1\big(v(0)\big) - j_1\big(w(0)\big) +$$

$$+ j_2\big(v(1)\big) - j_2\big(w(1)\big) \leq \int_0^1 G\big(r,v'(r)\big)\big(v'(r) - w'(r)\big)\, dr +$$

$$+ \int_0^1 K\big(r,v(r)\big)\big(v(r) - w(r)\big)\, dr + \theta_1\big(v(0) - w(0)\big) +$$

$$+ \theta_2\big(v(1) - w(1)\big) \ \text{ for all } \theta_1 \in \beta_1 v(0),\ \theta_2 \in \beta_2 v(1).$$

Now, by integrating by parts, we get

$$\psi_1\big(v,v(1)\big) - \psi\big(w,w(1)\big) \leq$$

$$\leq \left(\begin{pmatrix} -\frac{d}{dr}G(\cdot,v') + K(\cdot,v) \\ G\big(1,v'(1)\big) + \theta_2 \end{pmatrix}, \begin{pmatrix} v \\ v(1) \end{pmatrix} - \begin{pmatrix} w \\ w(1) \end{pmatrix} \right)_{H_1}$$

for all $\big(w,w(1)\big) \in D(\psi_1)$, $\theta_2 \in \beta_2 v(1)$. Therefore, $A_1 \subset \partial\psi_1$ as claimed.

∎

REMARK 4.1.1 Obviously, an alternative proof of Proposition 4.1.1 is to show directly that the subdifferential of ψ_1 coincides with A_1. However, the above proof works also in the vectorial case, in which the unknown function u takes its values in $\mathbb{R}^n$, each component of u satisfies an equation of type (4.0.1), and β_1 is defined on $\mathbb{R}^k$ and it is not necessarily a subdifferential. Of course, in this vectorial case A_1 is still maximal monotone, but not necessarily a subdifferential. On the other hand, the above proof reveals that the maximality of A_1 reduces to the maximality of an operator of type (3.1.1)-(3.1.2) defined in $L^2(0,1)$. ∎

In what follows, we are going to discuss the relationship between A_1 and problem (4.0.1)-(4.0.4). So, we consider in the space H_1 the following Cauchy problem

$$\frac{d}{dt}\begin{pmatrix} u(t) \\ \xi(t) \end{pmatrix} + A_1 \begin{pmatrix} u(t) \\ \xi(t) \end{pmatrix} \ni \begin{pmatrix} f(t) \\ s_2(t) \end{pmatrix},\ t > 0, \tag{4.1.6}$$

$$\begin{pmatrix} u(0) \\ \xi(0) \end{pmatrix} = \begin{pmatrix} u_0 \\ \xi_0 \end{pmatrix}. \tag{4.1.7}$$

It is easily seen that for every strong solution of problem (4.1.6)-(4.1.7), its first component u satisfies our original problem (4.0.1)-(4.0.4). Notice, however, that ξ_0 does not appear in the original problem and so, for some fixed u_0, f, s_2, we have the possibility to choose ξ_0 to be any number in $D(\beta_2)$ or even in $\overline{D(\beta_2)}$. Hence, it seems that problem (4.0.1)-(4.0.4) is not well posed. This question will be discussed later. Anyway, it is obvious that problem (4.0.1)-(4.0.4) is essentially different from problem (3.0.1)-(3.0.3).

Let us first discuss the existence of a strong or weak solution for problem (4.1.6)-(4.1.7); we call the first component u a strong, respectively weak, solution of problem (4.0.1)-(4.0.4).

THEOREM 4.1.1

Assume (I.1), (I.2'), and (I.3'). Let $T > 0$ be fixed, $(u_0, \xi_0) \in L^2(0,1) \times \overline{D(\beta_2)}$, $f \in L^1(0,T; L^2(0,1))$, and $s_2 \in L^1(0,T)$. Then the Cauchy problem (4.1.6)-(4.1.7) has a unique weak solution $(u, \xi) \in C([0,T]; H_1)$ with $u \in L^2(0,T; H^1(0,1))$.

If, in addition, $f \in L^2(Q_T)$, $Q_T = (0,1) \times (0,T)$, and $s_2 \in L^2(0,T)$, then (u, ξ) is a strong solution and it satisfies

$$\xi(t) = u(1,t) \ \text{ for a.a. } \ t \in (0,T),$$

and the mappings $(r,t) \mapsto \sqrt{t}\, u_t(r,t)$, $t \mapsto \sqrt{t}\, \xi'(t)$ belong to $L^2(Q_T)$ and $L^2(0,T)$, respectively. If, moreover, $(u_0, \xi_0) \in D(\psi_1)$, then $u_t \in L^2(Q_T)$, $u_r \in L^\infty(0,T; L^2(0,1))$, and $\xi \in H^1(0,T)$.

If, in addition, $u_0 \in H^2(0,1)$, $(u_0, u_0(1)) \in D(A_1)$, $s_2 \in W^{1,1}(0,T)$, and $f \in W^{1,1}(0,T; L^2(0,1))$, then

$$u \in W^{1,\infty}(0,T; L^2(0,1)) \cap L^\infty(0,T; H^2(0,1)) \cap W^{1,2}(0,T; H^1(0,1)),$$

and $u(1, \cdot) \in W^{1,\infty}(0,T)$.

PROOF   All the conclusions are either direct consequences of known theorems (see Theorems 1.5.1-1.5.3) or follow by arguments already used in the proofs of Theorems 3.1.1 and 3.1.2. Of course, some effort is needed but no essential difficulty appears and so the reader is encouraged to complete the proof.

Notice, however, that the fact that $u \in L^2(0,T; H^1(0,1))$ follows by a standard reasoning starting from the last part of the theorem. Indeed, if $(u_0, \xi_0) \in L^2(0,1) \times \overline{D(\beta_2)}$, $f \in L^1(0,T; L^2(0,1))$, and $s_2 \in L^1(0,T)$, we approximate them with respect to corresponding topologies by sequences $((u_{0n}, \xi_{0n}))$, (f_n), and (s_{2n}) of elements from $D(A_1)$, $W^{1,1}(0,T; L^2(0,T))$, and $W^{1,1}(0,T)$, respectively. Obviously, $\xi_{0n} = u_{0n}(1)$. The Cauchy problem (4.1.6)-(4.1.7), with $(f_n, s_{2n}, u_{0n}, \xi_{0n})$ instead of (f, s_2, u_0, ξ_0), has a unique strong solution (u_n, ξ_n). A standard calculation yields

$$\frac{1}{2}\frac{d}{dt} \left\| \begin{pmatrix} u_n(t) - u_m(t) \\ \xi_n(t) - \xi_m(t) \end{pmatrix} \right\|^2_{H_1} + k_0 \int_0^1 (u_{n,r}(r,t) - u_{m,r}(r,t))^2 \, dr \leq$$

$$\leq \left\| \begin{pmatrix} f_n(t) - f_m(t) \\ s_{2n}(t) - s_{2m}(t) \end{pmatrix} \right\|_{H_1} \left\| \begin{pmatrix} u_n(t) - u_m(t) \\ \xi_n(t) - \xi_m(t) \end{pmatrix} \right\|_{H_1}$$

for a.a. $t \in (0,T)$, and therefore

$$\left\| \begin{pmatrix} u_n(t) - u_m(t) \\ \xi_n(t) - \xi_m(t) \end{pmatrix} \right\|_{H_1} + \|u_{n,r} - u_{m,r}\|_{L^2(Q_t)} \leq$$

$$\leq C\left(\left\| \begin{pmatrix} u_{0n} - u_{0m} \\ \xi_{0m} - \xi_{0n} \end{pmatrix} \right\|_{H_1} + \right.$$

$$\left. + \int_0^t \left\| \begin{pmatrix} f_n(\theta) - f_m(\theta) \\ s_{2n}(\theta) - s_{2m}(\theta) \end{pmatrix} \right\|_{H_1} d\theta \right) \qquad (4.1.8)$$

for all $t \in [0,T]$, where C is some positive constant and $Q_t = (0,1) \times (0,t)$. Clearly, (4.1.8) proves the assertion above. $\blacksquare$

REMARK 4.1.2 Let (u,ξ) and $(\tilde{u},\tilde{\xi})$ be two weak solutions of (4.1.6)-(4.1.7), corresponding to some

$$(u_0, \xi_0, f, s_2), (\tilde{u}_0, \tilde{\xi}_0, \tilde{f}, \tilde{s}_2) \in L^2(0,1) \times \overline{D(\beta_2)} \times L^1\left(0,T; L^2(0,1)\right) \times L^1(0,T),$$

respectively. Then

$$\left\| \begin{pmatrix} u(t) - \tilde{u}(t) \\ \xi(t) - \tilde{\xi}(t) \end{pmatrix} \right\|_{H_1} + \|u_r - \tilde{u}_r\|_{L^2(Q_t)} \leq$$

$$\leq C\left(\left\| \begin{pmatrix} u_0 - \tilde{u}_0 \\ \xi_0 - \tilde{\xi}_0 \end{pmatrix} \right\|_{H_1} + \int_0^t \left\| \begin{pmatrix} f(\theta) - \tilde{f}(\theta) \\ s_2(\theta) - \tilde{s}_2(\theta) \end{pmatrix} \right\|_{H_1} d\theta \right) \qquad (4.1.9)$$

for all $t \in [0,T]$. $\blacksquare$

On the link between problems (4.0.1)-(4.0.4) and (4.1.6)-(4.1.7)

For simplicity, let us consider a particular case of problem (4.0.1)-(4.0.4) on the interval $[0,T]$:

$$u_t(r,t) - u_{rr}(r,t) = f(r,t), \ (r,t) \in (0,1) \times (0,T), \qquad (4.1.10)$$
$$u_r(0,t) = 0, \ t \in (0,T), \qquad (4.1.11)$$
$$u_t(1,t) + u_r(1,t) = 0, \ t \in (0,T), \qquad (4.1.12)$$
$$u(r,0) = u_0(r), \ r \in (0,1), \qquad (4.1.13)$$

where $u_0 \in L^2(0,1)$ and $f \in L^1\left(0,T; L^2(0,1)\right)$. According to the above theory we can associate with this problem the following Cauchy problem in $H_1 = L^2(0,1) \times \mathbb{R}$:

$$\left(u'(t), \xi'(t)\right) + A_1\left(u(t), \xi(t)\right) = \left(f(t), 0\right), \ t \in (0,T), \qquad (4.1.14)$$
$$\left(u(0), \xi(0)\right) = (u_0, \xi_0), \qquad (4.1.15)$$

where $\xi_0 \in \mathbb{R}$ and A_1 is given by

$$D(A_1) = \{(v, v(1)) \mid v \in H^2(0,1), v'(0) = 0\},$$
$$A_1(v, v(1)) = (-v'', v'(1)).$$

So, (4.1.14)-(4.1.15) can formally be expressed as

$$\frac{d}{dt}\begin{pmatrix} u(t) \\ \xi(t) \end{pmatrix} + \begin{pmatrix} -u_{rr}(\cdot, t) \\ u_r(1, t) \end{pmatrix} = \begin{pmatrix} f(t) \\ 0 \end{pmatrix}, \quad t \in (0, T), \qquad (4.1.16)$$
$$u_r(0, t) = 0, \quad t \in (0, T), \qquad (4.1.17)$$
$$(u(0), \xi(0)) = (u_0, \xi_0). \qquad (4.1.18)$$

The existence and uniqueness of a solution for problem (4.1.16)-(4.1.18) is assured by Theorem 4.1.1. In particular, for every $u_0 \in L^2(0,1)$, $\xi_0 \in \mathbb{R}$ and $f \in L^2(Q_T)$, this problem has a unique strong solution (u, ξ) with $\xi(t) = u(1, t)$ for a.a. $t \in (0, T)$. Of course, if f is more regular, for instance $f \in W^{1,1}(0, T; L^2(0, 1))$, then $(u(t), \xi(t)) \in D(A_1)$ for all $t \in (0, T]$, since A_1 is a subdifferential, and hence $\xi(t) = u(1, t)$ for all $t \in (0, T]$. Anyway, the first component u of the solution of (4.1.14)-(4.1.15) satisfies the original problem (4.1.10)-(4.1.13). However, ξ_0 does not appear in the original problem and so we may take any $\xi_0 \in \mathbb{R}$ in problem (4.1.14)-(4.1.15). It seems that problem (4.1.10)-(4.1.13) is not well posed. Indeed, if $u_0 \in L^2(0, 1)$ and $f \in L^2(Q_T)$ are fixed and ξ_{01}, ξ_{02} are two distinct real numbers, then the strong solutions (u_1, ξ_1) and (u_2, ξ_1) corresponding to (u_0, ξ_{01}, f) and (u_0, ξ_{02}, f), respectively, are distinct. Otherwise we would have

$$0 = u_1(1, t) - u_2(1, t) = \xi_1(t) - \xi_2(t) \text{ for a.a. } t \in (0, T),$$

and this implies by the continuity of ξ_1 and ξ_2 that $0 = \xi_1(0) - \xi_2(0) = \xi_{10} - \xi_{20}$, which is impossible. Therefore, problem (4.1.10)-(4.1.13) has an infinite number of solutions corresponding to (u_0, f). How do we explain this?

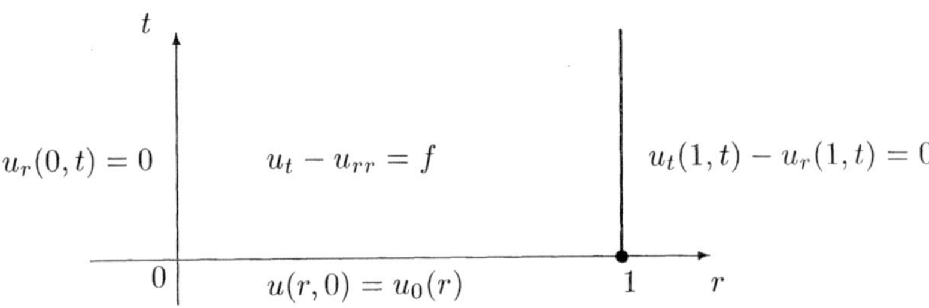

Actually, the boundary condition of (4.1.12) describes the evolution in time of $u(1,t)$. So, it is natural to prescribe an initial state for this evolution at the boundary of the domain. Moreover, we are led to consider a Cauchy problem in the space $H_1 = L^2(0,1) \times \mathbb{R}$ and the differential boundary condition is included in the corresponding evolution equation in H_1. This time, we have a complete initial condition, i.e., we prescribe as an initial state at time $t = 0$ the pair (u_0, ξ_0). Therefore, the model (4.1.14)-(4.1.15) (which is a particular case of problem (4.1.6)-(4.1.7)) is the right model, while the original model (4.1.10)-(4.1.13) is incomplete and it exhibits the nonuniqueness property. In classical textbooks one considers that u_0 is a smooth function and so $u_0(1)$ makes sense. Actually, this is considered as an initial value for $\xi(t) = u(1,t)$ without a precise statement. So, the compatibility condition at the corner $(r, t) = (1, 0)$ of the domain is satisfied, and if f is also smooth, then a classical solution of (4.1.10)-(4.1.13) does exist.

If the compatibility condition $\xi_0 = u_0(1)$ is satisfied then a good regularity is also given by Theorem 4.1.1. Indeed, if $u_0 \in H^2(0,1)$ and $\xi_0 = u_0(1)$, then $(u_0, \xi_0) \in D(A_1)$ and so for smooth f we have a strong solution, with some additional properties, which obviously coincides to the classical solution. However, as $A_1 = \partial\psi_1$, it is possible to have strong solutions even if the compatibility condition is not satisfied. This case may arise in concrete applications. Moreover, if $u_0 \in L^2(0,1)$, then $u_0(1)$ does not make sense, and so the compatibility is meaningless. Now, it is clear that the correct model is (4.1.14)-(4.1.15) and it admits a strong solution for every $u_0 \in L^2(0,1)$, $\xi_0 \in \mathbb{R}$, and $f \in L^2(Q_T)$. For a strong solution (u, ξ), we have $\xi(t) = u(1,t)$ for a.a. t and so the necessity of associating the Cauchy problem (4.1.14)-(4.1.15) is not so evident. But, the equality $\xi(t) = u(1,t)$ is not true for weak solutions (for example, in the case $f \in L^1(0, T; L^2(0,1))$). This means that the evolution at the boundary, given by the function $\xi(t)$, is not so dependent on the evolution in the interior of the domain, given by $u(\cdot, t)$. This is a strong motivation for the necessity of introducing the new unknown function $\xi(t)$ and the new model, i.e., the Cauchy problem (4.1.6)-(4.1.7), which describes more completely the physical phenomenon. So, the evolution of the phenomenon is given by a pair of functions $(u(t), \xi(t))$, which at time $t = 0$ starts from the initial state (u_0, ξ_0).

Taking into account the above comments, we shall next study only the Cauchy problem (4.1.6)-(4.1.7). In order to investigate the asymptotic behavior of the solutions, as $t \to \infty$, we continue with a result on the compactness of the resolvent of A_1:

PROPOSITION 4.1.2

Assume (I.1), (I.2'), and (I.3'). For every $\lambda > 0$, the operator $(I + \lambda A_1)^{-1}$

maps the bounded subsets of H_1 into bounded subsets of $H^2(0,1) \times \mathbb{R}$.

PROOF　　We essentially follow the proof of Proposition 3.1.1.　∎

THEOREM 4.1.2

Assume (I.1), (I.2'), and (I.3'). Let $F_1 := A_1^{-1}(0,0) \neq \emptyset$, $(u_0, \xi_0) \in L^2(0,1) \times \overline{D(\beta_2)}$, and $(f, s_2) \in L^1(\mathbb{R}_+; H_1)$. Then there exists a $p \in H^2(0,1)$ such that $\big(p, p(1)\big) \in F_1$ and the corresponding weak solution (u, ξ) of (4.1.6)-(4.1.7) satisfies

$$u(t) \to p \ in \ L^2(0,1), \tag{4.1.19}$$

$$\xi(t) \to p(1) \ in \ \mathbb{R}, \tag{4.1.20}$$

as $t \to \infty$. If, in addition, $(f, s_2) \in W^{1,1}(\mathbb{R}_+; H_1)$, then $\xi(t) = u(1,t)$ for all $t > 0$ and $u(t) \to p$ weakly in $H^2(0,1)$ and, hence, strongly in $C^1[0,1]$, as $t \to \infty$.

PROOF　　We can reason as in the proof of Theorem 3.1.3. Notice first that if $(f, s_2) \in W^{1,1}(\mathbb{R}_+; H_1)$, then the right derivative of $\big(u(t), \xi(t)\big) = \big(u(t), u(1,t)\big)$ is bounded for $t \geq \epsilon$, where ϵ is a fixed positive number. On the other hand, the condition $F_1 \neq \emptyset$ implies the boundedness in H_1 of the trajectory of the solution. Therefore, using a formula of type (3.1.26), we can see that the set $\{u(t) \mid t \geq \epsilon\}$ is bounded in $H^2(0,1)$ (cf. Proposition 4.1.2). The rest of the proof is similar to that of Theorem 3.1.3.　∎

REMARK 4.1.3　　The condition $F_1 \neq \emptyset$ is not superfluous. For example, if $G(r, \xi) = \xi$, $K(r, \xi) \equiv 0$, $\beta_1 \xi \equiv 0$, and $\beta_2 \xi \equiv -1$, then (I.1), (I.2') and (I.3') are all satisfied, but $F_1 = A_1^{-1}(0,0)$ is an empty set.　∎

4.2　Nonhomogeneous algebraic boundary condition

In this section we assume that $s_1(t)$ is not a constant function. So, we have a time dependent problem, which in turn leads to the following Cauchy problem in $H_1 = L^2(0,1) \times \mathbb{R}$:

$$\big(u'(t), \xi'(t)\big) + A_1(t)(u(t), \xi(t)) \ni \big(f(t), s_2(t)\big), \ t > 0, \tag{4.2.1}$$

$$\big(u(0), \xi(0)\big) = (u_0, \xi_0), \tag{4.2.2}$$

where $A_1(t): D\big(A(t)\big) \subset H_1 \to H_1$ is defined by

$$A_1(t)(v,a) = \Big(-\frac{d}{dr}G(\cdot,v') + K(\cdot,v), G\big(1,v'(1)\big) + \beta_2(a) \Big),$$

$$D\big(A_1(t)\big) = \big\{ (v,a) \mid v \in H^2(0,1),\ a = v(1) \in D(\beta_2),$$
$$v(0) \in D(\beta_1), G\big(0,v'(0)\big) + s_1(t) \in \beta_1 v(0) \big\}.$$

The Cauchy problem (4.2.1)-(4.2.2) is quite natural, more complete than the classical problem (4.0.1)-(4.0.4).

THEOREM 4.2.1

(Existence and uniqueness of a strong solution). Assume (I.1), (I.2'), and (I.3'). Let $T > 0$ be fixed, $(f,s_2) \in L^2(0,T;H_1)$, $H_1 = L^2(0,1) \times \mathbb{R}$, $(u_0,\xi_0) = \big(u_0,u_0(1)\big) \in D(\psi_1)$, and $s_1 \in W^{1,1}(0,T)$. Then the Cauchy problem (4.2.1)-(4.2.2) has a unique strong solution $(u,\xi) \in W^{1,2}(0,T;H_1)$ with $u_r \in L^\infty\big(0,T;L^2(0,1)\big)$. If, in addition, $\big(u_0,u_0(1)\big) \in D\big(A_1(0)\big)$, $s_1 \in W^{1,2}(0,T)$, and $(f,s_2) \in W^{1,1}(0,T;H_1)$, then $(u,\xi) \in W^{1,\infty}(0,T;H_1)$ and $u \in L^\infty\big(0,T;H^2(0,1)\big) \cap W^{1,2}\big(0,T;H^1(0,1)\big)$.

PROOF   To prove the first assertion of the theorem, we shall apply Theorem 1.5.8. Indeed, it is easily seen that $A_1(t) = \partial\psi_1(t,\cdot)$, $t \in [0,T]$, where $\psi_1(t,\cdot): H \to (-\infty,\infty]$ is defined by

$$\psi_1\big(t,(v,a)\big) = \begin{cases} \displaystyle\int_0^1 \Big(g\big(r,v'(r)\big) + k\big(r,v(r)\big) \Big)\,dr + j_1\big(v(0)\big) + \\ \qquad +j_2(a) - s_1(t)v(0) \text{ if } v \in H^1(0,1), \\ \qquad\quad v(0) \in D(j_1),\ a = v(1) \in D(j_2), \\ \qquad\quad g(\cdot,v') \in L^1(0,1), \\ \infty \text{ otherwise.} \end{cases} \tag{4.2.3}$$

According to Proposition 4.1.1, $A_1(t)$ is a maximal cyclically monotone operator, for all $t \in [0,T]$. We have just to notice that j_1 is here replaced by the function $\theta \mapsto j_1(\theta) - s_1(t)\theta$. The effective domain of $\psi_1(t,\cdot)$ is the same as that of ψ_1 defined in (4.1.3). Hence it does not depend on t. Now, for some $\big(v,v(1)\big) \in D(\psi_1)$ and $0 \le s \le t \le T$, we have

$$\psi_1\Big(t,\big(v,v(1)\big)\Big) - \psi_1\Big(s,\big(v,v(1)\big)\Big) =$$

$$= -\big(s_1(t) - s_1(s)\big)v(0) \le |v(0)| \int_s^t |s_1'(\tau)|\,d\tau. \tag{4.2.4}$$

On the other hand, it follows from (2.1.28) and the usual properties of proper, convex, lower semicontinuous functions that

$$\psi_1\Big(s,\big(v,v(1)\big)\Big) \ge \frac{k_0}{3}\|v'\|_H^2 - C_1\|v\|_H - C_2|v(0)| - C_3|v(1)| - C_4,$$

where C_1, C_2, C_3, and C_4 are positive constants and $\| \cdot \|_H$ denotes the norm of $H = L^2(0,1)$. Therefore,

$$|v(0)| \leq \psi_1\Big(s, \big(v, v(1)\big)\Big) + C_5 \big\| \big(v, v(1)\big) \big\|_{H_1} + C_6, \tag{4.2.5}$$

where C_5 and C_6 are positive constants. By (4.2.4)-(4.2.5) it follows that the key condition (1.5.15) of Theorem 1.5.8 is satisfied with

$$\gamma(t) = \int_0^t |s_1'(\tau)|\, d\tau.$$

So, according to Theorem 1.5.8, there is a unique strong solution $(u, \xi) \in W^{1,2}(0,T; H_1)$; hence, $\xi(t) = u(1,t)$ for a.a. $t \in (0,T)$. We also have that $u_r \in L^\infty\big(0,T; L^2(0,1)\big)$ by the estimate (1.5.17).

The last part of the theorem follows by standard arguments (see also the proof of Theorem 3.2.1). ∎

REMARK 4.2.1   The last part of Theorem 4.2.1 can also be derived as a consequence of Theorem 1.5.9 (see also Theorem 3.2.1). This alternative procedure is appropriate for the vectorial case, where $A_1(t)$ is not necessarily a subdifferential. ∎

THEOREM 4.2.2
(Existence and uniqueness of a weak solution). Assume (I.1), (I.2'), and (I.3'). Let $T > 0$ be fixed, $f \in L^1\big(0,T; L^2(0,1)\big)$, $s_1 \in L^2(0,T)$, $s_2 \in L^1(0,T)$, $u_0 \in L^2(0,1)$, and $\xi_0 \in \overline{D(\beta_2)}$. Then the Cauchy problem (4.2.1)-(4.2.2) has a unique weak solution $(u, \xi) \in C\big([0,T]; H_1\big)$ with $u_r \in L^2(Q_T)$, $Q_T = (0,1) \times (0,T)$.

PROOF   We adapt the proof of Theorem 3.2.2. So, take sequences $\big((u_{0n}, \xi_{0n})\big)$, (s_{1n}), (s_{2n}), and (f_n) of elements of $D\big(A_{1n}(0)\big)$, $W^{1,2}(0,T)$, $W^{1,1}(0,T)$, and $W^{1,1}\big(0,T; L^2(0,1)\big)$, respectively, such that

$$u_{0n} \to u_0 \text{ in } L^2(0,1), \;\; \xi_{0n} \to \xi_0 \text{ in } \mathbb{R}, \;\; s_{1n} \to s_1 \text{ in } L^2(0,T),$$
$$s_{2n} \to s_2 \text{ in } L^1(0,T) \text{ and } f_n \to f \text{ in } L^1\big(0,T; L^2(0,1)\big),$$

as $n \to \infty$. Above $A_{1n}(t)$ is obtained from $A_1(t)$ by replacing (f, s_1, s_2) by (f_n, s_{1n}, s_{2n}). We denote by (u_n, ξ_n) the strong solutions, given by Theorem 4.2.1, of the Cauchy problem (4.2.1)-(4.2.2), in which $(u_0, \xi_0, s_1, s_2, f)$ is replaced by $(u_{0n}, \xi_{0n}, s_{1n}, s_{2n}, f_n)$. By performing some computations that are comparable with those in the proof of Theorem 3.2.2, we arrive at

$$\big\| \big(u_n(\cdot,t), u_n(1,t)\big) - \big(u_m(\cdot,t), u_m(1,t)\big) \big\|_{H_1} +$$

$$+\left(\int_0^T \|u_n(\cdot,\tau)-u_m(\cdot,\tau)\|^2_{H^1(0,1)}\,d\tau\right)^{1/2} \leq$$

$$\leq C\Big(\|u_{0n}-u_{0m}\|_{L^2(0,1)}+|\xi_{0n}-\xi_{0m}|+\|s_{1n}-s_{1m}\|_{L^2(0,T)}+$$

$$+\|s_{2n}-s_{2m}\|_{L^1(0,T)}+\|f_n-f_m\|_{L^1\left(0,T;L^2(0,1)\right)}\Big)$$

for all $t \in [0,T]$, where C is a positive constant, independent of n, m, and T. This clearly concludes the proof. ∎

THEOREM 4.2.3

(Asymptotic behavior). Assume (I.1), (I.2'), and (I.3'). Let $F_1 = A_1^{-1}(0,0)$ be nonempty, $(u_0,\xi_0) \in L^2(0,1) \times \overline{D(\beta_2)}$, $(f,s_2) \in L^1(\mathbb{R}_+;H_1)$, $s_1 \in L^2(\mathbb{R}_+)$, and let $(u,\xi) \in C(\mathbb{R}_+;H_1)$ with $u \in L^2_{loc}(\mathbb{R}_+;H^1(0,1))$ be the corresponding weak solution of problem (4.2.1)-(4.2.2). Then $u(t)$ converges strongly in H_1 to an element of F_1, as $t \to \infty$.

PROOF The basic idea of the proof is the same as in the proof of Theorem 3.2.3, in the sense that the functions f, s_1, s_2 are approximated by their truncated functions. We skip the details. ∎

REMARK 4.2.2 One may also consider the case where both boundary conditions are of differential type, that is

$$u_t(0,t) - G\big(0,u_r(0,t)\big) + \beta_1 u(0,t) \ni s_1(t),\ t > 0, \qquad (4.2.6)$$

$$u_t(1,t) + G\big(1,u_r(1,t)\big) + \beta_2 u(1,t) \ni s_2(t),\ t > 0. \qquad (4.2.7)$$

In this case, we choose as a basic setup the space $H_2 := L^2(0,1) \times \mathbb{R}^2$, and associate with the problem (4.0.1), (4.2.6)-(4.2.7), (4.0.4) the following Cauchy problem in H_2

$$\big(u'(t),\xi_1'(t),\xi_2'(t)\big) + A_2\big(u(t),\xi_1(t),\xi_2(t)\big) \ni$$

$$\ni \big(f(t),s_1(t),s_2(t)\big),\ t > 0, \qquad (4.2.8)$$

$$\big(u(0),\xi_1(0),\xi_2(0)\big) = (u_0,\xi_{10},\xi_{20}). \qquad (4.2.9)$$

The operator $A_2 : D(A_2) \subset H_2 \to H_2$ is defined by

$$D(A_2) = \big\{(v,a,b)\mid v \in H^2(0,1),\ a = v(0) \in D(\beta_1),\ b = v(1) \in D(\beta_2)\big\},$$

$$A_2(v,a,b) = \begin{pmatrix} -\frac{d}{dr}G(\cdot,v') + H(\cdot,v) \\ -G\big(0,v'(0)\big) + \beta_1 a \\ G\big(1,v'(1)\big) + \beta_2 b \end{pmatrix}.$$

Notice that in this case we have a quasiautonomous equation and we can take advantage of this situation. Our comments concerning the well-posedness of problem (4.1.14)-(4.1.15) are still applicable here and the correct model to be taken into consideration is the Cauchy problem (4.2.8)-(4.2.9), in which we prescribe as an initial state the triple $(u_0, \xi_{10}, \xi_{20})$ (even if ξ_{10} and ξ_{20} do not appear in the classical original problem). We do not insist more on this subject, because its investigation would rely on arguments we have already used before. $\blacksquare$

4.3   Higher regularity of solutions

We consider next a simpler case in which the parabolic equation is a linear one. More precisely, we are going to investigate the problem

$$u_t(r,t) - u_{rr}(r,t) + au(r,t) = f(r,t), \ (r,t) \in Q_T, \tag{4.3.1}$$

$$u_r(0,t) \in \beta_1 u(0,t), \ t \in (0,T), \tag{4.3.2}$$

$$u_t(1,t) + u_r(1,t) + \beta_2 u(1,t) \ni s(t), \ t \in (0,T), \tag{4.3.3}$$

$$u(r,0) = u_0(r), \ r \in (0,1). \tag{4.3.4}$$

Here a is a nonnegative constant and $Q_T = (0,1) \times (0,T)$. The algebraic boundary condition can be homogenized and so we have assumed $s_1(t) \equiv 0$ and redenoted $s_2(t)$ by $s(t)$. Now, let us reformulate the last part of Theorem 4.1.1 for the present situation:

PROPOSITION 4.3.1
Assume that

$$f \in W^{1,1}(0,T;L^2(0,1)), \ s \in W^{1,1}(0,T), \tag{4.3.5}$$

$$\beta_1, \beta_2 \subset \mathbb{R} \times \mathbb{R} \ \textit{are maximal monotone operators}, \tag{4.3.6}$$

$$u_0 \in H^2(0,1), u_0(0) \in D(\beta_1), \ u_0(1) \in D(\beta_2), \tag{4.3.7}$$

$$u_0'(0) \in \beta_1\big(u_0(0)\big). \tag{4.3.8}$$

Then, problem (4.3.1)-(4.3.4) has a unique strong solution u in the sense that $\big(u(t), u(1,t)\big)$ is a strong solution of the associated Cauchy problem with $(u_0, u_0(1))$ as the initial value. Moreover,

$$u \in W^{1,\infty}(0,T;L^2(0,1)) \cap L^\infty(0,T;H^2(0,1)) \cap$$
$$\cap W^{1,2}(0,T;H^1(0,1)), \tag{4.3.9}$$

$$u(1,\cdot) \in W^{1,\infty}(0,T). \tag{4.3.10}$$

We continue with the following higher regularity result:

THEOREM 4.3.1

Assume that

$$f \in W^{1,2}\big(0,T;L^2(0,1)\big), \ s \in W^{1,2}(0,T); \tag{4.3.11}$$

the mappings β_1, β_2 are everywhere defined on $\mathbb{R}$, single valued,

$$\beta_1,\beta_2 \in W^{2,\infty}_{loc}(\mathbb{R}) \ \text{with} \ \beta_1' \geq 0 \ \text{and} \ \beta_2' \geq 0; \tag{4.3.12}$$

$$\big(u_0,u_0(1)\big) \in D(A_1); \tag{4.3.13}$$

$$\big(f(0),s(0)\big) - A_1\big(u_0,u_0(1)\big) \in V_1 :=$$
$$= \big\{ \big(\phi,\phi(1)\big) \mid \phi \in H^1(0,1) \big\}. \tag{4.3.14}$$

Then, the solution u of (4.3.1)-(4.3.4) belongs to $W^{2,2}\big(0,T;L^2(0,1)\big) \cap W^{1,\infty}\big(0,T;H^1(0,1)\big)$, and $u(1,\cdot)$ belongs to $W^{1,\infty}(0,T)$.

PROOF Notice that here we have (see (4.1.1)),

$$D(A_1) = \big\{ (v,b) \mid v \in H^2(0,1), b = v(1), \ v'(0) \in \beta_1 v(0) \big\},$$

and so the condition (4.3.13) is equivalent to (4.3.7) and (4.3.8). According to Proposition 4.3.1 our problem (4.3.1)-(4.3.4) has a unique solution u satisfying (4.3.9)-(4.3.10). Clearly, V_1 is a real Hilbert space, whose inner product is the sum of the usual inner product of $H^1(0,1)$ and the ordinary multiplication in $\mathbb{R}$. Denote by V_1^* the dual of V_1 and by $\langle \cdot,\cdot \rangle$ the pairing between them. Obviously,

$$\Big\langle \big(\phi,\phi(1)\big), \big(u'(t),u_t(1,t)\big) \Big\rangle + \big(\phi',u_r(\cdot,t)\big)_{L^2(0,1)} +$$
$$+ a\big(\phi,u(\cdot,t)\big)_{L^2(0,1)} + \beta_1\big(u(0,t)\big)\phi(0) + \beta_2\big(u(1,t)\big)\phi(1) =$$
$$= \big(\phi,f(t)\big)_{L^2(0,1)} + s(t)\phi(1) \tag{4.3.15}$$

for all $\phi \in H^1(0,1)$ and a.a. $t \in (0,T)$. Using (4.3.15), we easily obtain (see also the proof of Theorem 3.3.1) that

$$\int_0^{T-\delta} \big\| \big(u'(t+\delta),u_t(1,t+\delta)\big) - \big(u'(t),u_t(1,t)\big) \big\|_{V_1^*}^2 \, dt \leq C\delta^2$$

for all $\delta \in (0,T]$, where C is a positive constant. This implies (cf. Theorem 1.1.2) that the function $t \mapsto \big(u'(t),u_t(1,t)\big)$ belongs to $W^{1,2}(0,T;V_1^*)$. So, we can differentiate (4.3.15) to conclude that $z = u_t$ satisfies the equation

$$\Big\langle \big(\phi,\phi(1)\big), \big(z'(t),z_t(1,t)\big) \Big\rangle + \big(\phi',z_r(\cdot,t)\big)_{L^2(0,1)} + a\big(\phi,z(t)\big)_{L^2(0,1)} +$$
$$+ \beta_1'\big(u(0,t)\big)z(0,t)\phi(0) + \beta_2'\big(u(1,t)\big)z(1,t)\phi(1) =$$
$$= \big(\phi,f_t(\cdot,t)\big)_{L^2(0,1)} + s'(t)\phi(1) \tag{4.3.16}$$

for all $\phi \in H^1(0,1)$ and a.a. $t \in (0,T)$, as well as the initial condition

$$\big(z(0), z(1,0)\big) = \big(f(0), s(0)\big) - A_1(u_0, u_0(1)). \qquad (4.3.17)$$

It is easily seen that the problem (4.3.16)-(4.3.17) has a unique solution. Formally, $\big(z(t), z(1,t)\big)$ satisfies the following Cauchy problem:

$$\frac{d}{dt}\big(z(t), z(1,t)\big) + B_1(t)\big(z(t), z(1,t)\big) =$$
$$= (f'(t), s'(t)), \ \ t \in (0,T), \qquad (4.3.18)$$
$$\big(z(0), z(1,0)\big) = \big(f(0), s(0)\big) - A_1(u_0, u_0(1)), \qquad (4.3.19)$$

where $B_1(t) \colon D\big(B_1(t)\big) \subset H_1 \to H_1$ is defined by

$$D\big(B_1(t)\big) = \big\{(v,b) \mid v \in H^2(0,1), \ b = v(1), \ v'(0) = g_1(t)v(0)\big\},$$
$$B_1(t)(v,b) = \big(-v'' + av, v'(1) + g_2(t)b\big),$$

with $g_1(t) := \beta_1'\big(u(0,t)\big)$ and $g_2(t) := \beta_2'\big(u(1,t)\big)$. According to Proposition 4.1.1, $B_1(t)$ is a maximal cyclically monotone operator for all $t \in [0,T]$, since it is the subdifferential of the function $\tilde{\psi}_1(t, \cdot) \colon H_1 \to (-\infty, \infty]$,

$$\tilde{\psi}_1\big(t, (v,b)\big) = \begin{cases} \dfrac{1}{2}\int_0^1 \big(v'(r)^2 + av(r)^2\big)\,dr + \frac{1}{2}g_1(t)v(0)^2 + \frac{1}{2}g_2(t)b^2 \\ \qquad\qquad \text{if } v \in H^1(0,1) \text{ and } b = v(1), \\ \infty \text{ otherwise.} \end{cases}$$

Clearly, the effective domain of $\tilde{\psi}_1(t, \cdot)$ is the space V_1. Actually, problem (4.3.18)-(4.3.19) has a unique strong solution $\big(z, z(1, \cdot)\big)$ satisfying

$$\big(z, z(1, \cdot)\big) \in W^{1,2}(0,T;H_1), \ \ z_r \in L^\infty\big(0,T;L^2(0,1)\big), \qquad (4.3.20)$$

since Theorem 1.5.8 is again applicable. Indeed, the initial value is in V_1 (see (4.3.14)), the right hand side of (4.3.17) belongs to $L^2(0,T;H_1)$, and the key condition (1.5.15) of the quoted result is fulfilled, as the reader can easily verify. Obviously, z also satisfies problem (4.3.16)-(4.3.17) and hence $z = u_t$. This fact and (4.3.20) complete the proof. $\blacksquare$

REMARK 4.3.1   If f is also regular with respect to r, then higher regularity of u with respect to r can also be derived by using (4.3.1) and the conclusion of the above theorem.

On the other hand, we can continue the procedure to deduce that $u_{ttt} \in L^2(Q_T)$, under additional assumptions concerning the regularity of the data and their compatibility with (4.3.2). Actually, we can obtain as much regularity as we want, under appropriate requirements on the data. $\blacksquare$

REMARK 4.3.2   If the boundary condition (4.3.2) is linear, then an alternative approach can be used to derive higher regularity of the solution of (4.3.1)-(4.3.4). In this case the Cauchy problem associated to (4.3.1)-(4.3.4) is a semilinear problem and its solution can be represented by a variation of constants formula, which is the basic tool for this alternative method (see [BaMo3, pp. 72-78]).   ∎

References

[Barbu]   L. Barbu, On a singularly perturbed problem for the telegraph equations, *Comm. Appl. Anal.*, 3 (1999), no. 4, 529–546.

[BaMo1]   L. Barbu & Gh. Moroşanu, Asymptotic analysis of the telegraph equations with non-local boundary conditions, *PanAmer. Math. J.*, 8 (1998), 13–22.

[BaMo2]   L. Barbu & Gh. Moroşanu, A first order asymptotic expansion of the solution of a singularly perturbed problem for the telegraph equations, *Appl. Anal.*, 72 (1999), no. 1-2, 111–125.

[BaMo3]   L. Barbu & Gh. Moroşanu, *Analiza asimptotică a unor probleme la limită cu perturbaţii singulare*, Editura Academiei, Bucharest, 2000.

[Cannon]   J.R. Cannon, *The One-Dimensional Heat Equation*, Encyclopedia of Math. and its Appl., vol. 23, Addison-Wesley, Menlo Park, California, 1984.

[MoPer]   Gh. Moroşanu & A. Perjan, The singular limit of telegraph equations, *Comm. Appl. Nonlin. Anal.*, 5 (1998), no. 1, 91–106.

[Vinter]   R.B. Vinter, *Semigroups on product spaces with applications to initial value problems with non-local boundary conditions*, Report 1978, Imperial College, Dept. of Computing and Control.

[Zabcz1]   J. Zabczyk, *On semigroups corresponding to non-local boundary conditions with applications to system theory*, Report 49, Control Theory Centre, University of Warwick, Coventry, 1976.

[Zabcz2]   J. Zabczyk, On decomposition of generators, *SIAM J. Control Optim.*, 16 (1978), 523–534.

Chapter 5

Hyperbolic boundary value problems with algebraic boundary conditions

This chapter is dedicated to a class of partial differential systems of the form

$$u_t(r,t) + v_r(r,t) + K_1\big(r,u(r,t)\big) = f_1(r,t), \qquad (5.0.1)$$

$$v_t(r,t) + u_r(r,t) + K_2\big(r,v(r,t)\big) = f_2(r,t), \ 0 < r < 1, \ t > 0, \quad (5.0.2)$$

to which we associate the following algebraic boundary condition

$$\big(-u(0,t),u(1,t)\big) \in L\big(v(0,t),v(1,t)\big), \ t > 0, \qquad (5.0.3)$$

and initial conditions

$$u(r,0) = u_0(r), \ v(r,0) = v_0(r), \ r \in (0,1). \qquad (5.0.4)$$

Here $L \subset \mathbb{R}^2 \times \mathbb{R}^2$ is a nonlinear operator. Let us now introduce our main hypotheses:

(H.1) Functions $K_1, K_2 \colon [0,1] \times \mathbb{R} \mapsto \mathbb{R}$ satisfy:

$$K_1(\cdot,\xi) \text{ and } K_2(\cdot,\xi) \text{ belong to } L^2(0,1) \text{ for all } \xi \in \mathbb{R}, \quad (5.0.5)$$

$$\xi \mapsto K_1(r,\xi) \text{ and } \xi \mapsto K_2(r,\xi) \text{ are continuous and}$$

$$\text{nondecreasing for a.a. } r \in (0,1). \qquad (5.0.6)$$

(H.2) The operator $L \subset \mathbb{R}^2 \times \mathbb{R}^2$ is maximal monotone.

Notice that the boundary condition (5.0.3) is very general so that many classical boundary conditions can be derived as particular cases of it. For example, if L is the subdifferential of the function $j \colon \mathbb{R}^2 \to (-\infty, \infty]$ defined by

$$j(x,y) = \begin{cases} 0 & \text{if } x = a \text{ and } y = b, \\ \infty & \text{otherwise,} \end{cases}$$

where $a, b \in \mathbb{R}$ are fixed, then (5.0.3) becomes

$$v(0, t) = a, \quad v(1, t) = b, \quad t > 0,$$

(*bilocal boundary conditions*). Another case we may consider is that in which L is the subdifferential of the function $j_1 \colon \mathbb{R}^2 \to (-\infty, \infty]$ defined by

$$j_1(x, y) = \begin{cases} 0 & \text{if } x = y, \\ \infty & \text{otherwise.} \end{cases}$$

An easy computation reveals that in this case (5.0.3) reads as

$$u(0, t) = u(1, t), \quad v(0, t) = v(1, t), \quad t > 0$$

(*space periodic boundary conditions*). It should be mentioned that many practical problems arising from physics and engineering can be regarded as particular cases of problem (5.0.1)-(5.0.4). For instance, the unsteady fluid flow through pipes (possibly with nonlinear pipe friction) [LiWH], [StrWy] and electrical transmission line phenomena (see, e.g., [CooKr] and the references therein) can be described by mathematical models of type (5.0.1)-(5.0.4). Also, the vectorial case of (5.0.1)-(5.0.4) (i.e., the case of n partial differential systems of the form (5.0.1)-(5.0.2) connected by some algebraic boundary conditions) is an appropriate model for integrated circuits (see, e.g., [GhaKe] and [MarNe]) and for more complex problems of hydraulics, as for instance, the fluid flow through a tree-structured system of transmission pipelines, [Barbu3], [Iftimie], [Moro1, p. 316].

We consider here the simplest case of such problems in order to make our exposition clear and to better illustrate our methods. Some extensions will however be discussed later.

5.1   Existence, uniqueness, and long-time behavior of solutions

Our basic framework here will be the product space $H_3 := L^2(0, 1)^2$, which is a real Hilbert space with the scalar product defined by

$$\big((p_1, p_2), (q_1, q_2)\big)_{H_3} = \int_0^1 p_1(r) q_1(r) \, dr + \int_0^1 p_2(r) q_2(r) \, dr,$$

and the corresponding norm $\| \cdot \|_{H_3}$,

$$\big\| (p_1, p_2) \big\|_{H_3}^2 = \| p_1 \|_{L^2(0,1)}^2 + \| p_2 \|_{L^2(0,1)}^2.$$

We associate with our boundary value problem the operator $A_3: D(A_3) \subset H_3 \to H_3$ defined by

$$D(A_3) = \{(p, q) \in H^1(0, 1)^2 \mid \big(-p(0), p(1)\big) \in L\big(q(0), q(1)\big)\}, \quad (5.1.1)$$
$$A_3(p, q) = \big(q' + K_1(\cdot, p), p' + K_2(\cdot, q)\big). \quad (5.1.2)$$

PROPOSITION 5.1.1

Assume (H.1) and (H.2). Then A_3 is a maximal monotone operator and $D(A_3)$ is dense in H_3.

PROOF   Notice first that $D(A_3)$ is not empty. Indeed, it contains at least (p^*, q^*),

$$p^*(r) = (a + b)r - a, \; q^*(r) = (d - c)r + c \text{ for all } r \in [0, 1], \quad (5.1.3)$$

where a, b, c, and d are some real numbers such that $\big((c, d), (a, b)\big) \in L$.

Proving the monotonicity of A_3 is just a simple exercise, involving the monotonicity of $K_1(r, \cdot)$, $K_2(r, \cdot)$, and L. In order to prove the maximality of A_3, we suppose in the first stage that $K_1 \equiv 0$ and $K_2 \equiv 0$. In this particular case A_3 will be denoted by A_3'. We know that the maximality of A_3' is equivalent to the surjectivity of $I + A_3'$, where I is the identity operator on H_3 (see Theorem 1.2.2). So, let $(\tilde{p}, \tilde{q}) \in H_3$ be arbitrary and consider the equation

$$(p, q) + A_3'(p, q) = (\tilde{p}, \tilde{q}). \quad (5.1.4)$$

But (5.1.4) is equivalent to the problem of finding $(p, q) \in H^1(0, 1)^2$, which satisfies

$$p + q' = \tilde{p}, \; q + p' = \tilde{q}, \quad (5.1.5)$$
$$\big(-p(0), p(1)\big) \in L\big(q(0), q(1)\big). \quad (5.1.6)$$

The general solution of the system (5.1.5) is given by

$$p(r) = c_1 e^r + c_2 e^{-r} + \hat{p}(r), \; q(r) = -c_1 e^r + c_2 e^{-r} + \hat{q}(r), \quad (5.1.7)$$

where $(\hat{p}, \hat{q}) \in H^1(0, 1)^2$ is a fixed particular solution of (5.1.5), and c_1, c_2 are real constants. So, the question is whether there are some $c_1, c_2 \in \mathbb{R}$ such that (p, q) given by (5.1.7) satisfy (5.1.6). But this reduces to an algebraic equation in $\mathbb{R}^2$, which has a solution. We leave the details to the reader. Now, let us return to the maximality of A_3 in the general case. Let $(\tilde{p}, \tilde{q}) \in H_3$ and $\lambda > 0$ be arbitrary. We consider the equation

$$(p, q) + A_3'(p, q) + \big(K_{1\lambda}(\cdot, p), K_{2\lambda}(\cdot, q)\big) = (\tilde{p}, \tilde{q}), \quad (5.1.8)$$

where $K_{1\lambda}(r,\cdot)$ and $K_{2\lambda}(r,\cdot)$ are the Yosida approximations of $K_1(r,\cdot)$ and $K_2(r,\cdot)$, respectively. Obviously, for every $w \in L^2(0,1)$ the function $t \to K_{1\lambda}(r,w(r))$ is measurable in $(0,1)$ and since

$$\left| K_{1\lambda}(r,w(r)) \right| \le \left| K_{1\lambda}(r,w(r)) - K_{1\lambda}(r,0) \right| + \left| K_{1\lambda}(r,0) \right| \le$$

$$\le \frac{1}{\lambda}|w(r)| + |K_1(r,0)|,$$

it follows that $r \mapsto K_{1\lambda}(r,w(r))$ belongs to $L^2(0,1)$. Moreover, according to Theorem 1.2.7, the canonical extension (composition) $w \mapsto K_{1\lambda}(r,w(r))$ is maximal monotone in $L^2(0,1)$, since it is everywhere defined, monotone, and Lipschitz continuous. Similarly, $w \mapsto K_{2\lambda}(\cdot,w)$ is maximal monotone and everywhere defined on $L^2(0,1)$. So, according to Theorems 1.2.7 and 1.2.2, there exists a couple $(p_\lambda, q_\lambda) \in D(A_3)$ satisfying (5.1.8), i.e.,

$$(p_\lambda, q_\lambda) + A_3'(p_\lambda, q_\lambda) + \left(K_{1\lambda}(\cdot,p_\lambda), K_{2\lambda}(\cdot,q_\lambda) \right) = (\tilde{p}, \tilde{q}). \tag{5.1.9}$$

We intend to pass to the limit in (5.1.9), as $\lambda \to 0^+$, in order to derive the maximality of A_3. Notice first that

$$\{(p_\lambda, q_\lambda) \mid \lambda > 0\} \text{ is bounded in } H_3. \tag{5.1.10}$$

Indeed, if we denote

$$(p^*, q^*) + A_3'(p^*, q^*) + \left(K_{1\lambda}(\cdot,p^*), K_{2\lambda}(\cdot,q^*) \right) =: (p_\lambda^*, q_\lambda^*), \tag{5.1.11}$$

subtract (5.1.11) from (5.1.9), and multiply the resulting equation by $(p_\lambda - p^*, q_\lambda - q^*)$ with respect to the inner product of H_3, we get

$$\left\| (p_\lambda, q_\lambda) - (p^*, q^*) \right\|_{H_3} \le \left\| (\tilde{p}, \tilde{q}) - (p_\lambda^*, q_\lambda^*) \right\|_{H_3} \text{ for all } \lambda > 0. \tag{5.1.12}$$

Here (p^*, q^*) is the couple defined by (5.1.3). On the other hand,

$$\left| K_{1\lambda}(r,p^*(r)) \right| \le \left| K_1(r,p^*(r)) \right|, \ \left| K_{2\lambda}(r,q^*(r)) \right| \le \left| K_2(r,q^*(r)) \right|$$

for all $r \in [0,1]$. Therefore, according to (H.1), the set

$$\left\{ \left(K_{1\lambda}(\cdot,p^*), K_{2\lambda}(\cdot,q^*) \right) \mid \lambda > 0 \right\}$$

is bounded in H_3, i.e., $\{p_\lambda^*, q_\lambda^*) \mid \lambda > 0\}$ is also bounded in H_3. This and (5.1.12) imply (5.1.10).

Now, we multiply (5.1.9) by $(p_\lambda - p^* + \operatorname{sgn} q_\lambda', q_\lambda - q^* + \operatorname{sgn} p_\lambda')$ to obtain that (see also (5.1.10))

$$\|p_\lambda'\|_{L^1(0,1)} + \|q_\lambda'\|_{L^1(0,1)} \le -((q_\lambda', p_\lambda'), (p_\lambda - p^*, q_\lambda - q^*))_{H_3} -$$

$$- \left(\begin{pmatrix} K_{1\lambda}(\cdot,p_\lambda) \\ K_{2\lambda}(\cdot,q_\lambda) \end{pmatrix}, \begin{pmatrix} p_\lambda - p^* + \operatorname{sgn} q_\lambda' \\ q_\lambda - q^* + \operatorname{sgn} p_\lambda' \end{pmatrix} \right)_{H_3} + C,$$

where C is a positive constant. Therefore, by making use of the monotonicity of A_3', $K_{1\lambda}(r,\cdot)$, and $K_{2\lambda}(r,\cdot)$, we get that

$$\|p_\lambda'\|_{L^1(0,1)} + \|q_\lambda'\|_{L^1(0,1)} \leq -\left(A_3'(p^*,q^*),(p_\lambda - p^*, q_\lambda - q^*)\right)_{H_3} -$$
$$-\left(\begin{pmatrix} K_{1\lambda}(\cdot,p^* - \operatorname{sgn} q_\lambda') \\ K_{2\lambda}(\cdot,q^* - \operatorname{sgn} p_\lambda') \end{pmatrix}, \begin{pmatrix} p_\lambda - p^* + \operatorname{sgn} q_\lambda' \\ q_\lambda - q^* + \operatorname{sgn} p_\lambda' \end{pmatrix}\right)_{H_3} + C.$$

By (H.1) and (5.1.10), the last inequality implies that the sets $\{p_\lambda \mid \lambda > 0\}$ and $\{q_\lambda \mid \lambda > 0\}$ are bounded in $W^{1,1}(0,1)$. Hence, in particular,

$$\{p_\lambda \mid \lambda > 0\} \text{ and } \{q_\lambda \mid \lambda > 0\} \text{ are bounded in } C[0,1]. \tag{5.1.13}$$

Since, for all $r \in [0,1]$,

$$\left|K_{1\lambda}(r,p_\lambda(r))\right| \leq \left|K_1(r,p_\lambda(r))\right|, \quad \left|K_{2\lambda}(r,q_\lambda(r))\right| \leq \left|K_2(r,q_\lambda(r))\right|,$$

it follows by (H.1) and (5.1.13) that the set

$$\left\{\left(K_{1\lambda}(\cdot,p_\lambda), K_{2\lambda}(\cdot,q_\lambda)\right) \mid \lambda > 0\right\} \text{ is bounded in } H_3. \tag{5.1.14}$$

Now, (5.1.9), (5.1.10), and (5.1.14) imply that

$$\{(p_\lambda,q_\lambda) \mid \lambda > 0\} \text{ is bounded in } H^1(0,1)^2. \tag{5.1.15}$$

So, at least on a subsequence, there exists

$$\lim_{\lambda \to 0+} (p_\lambda, q_\lambda) = (p,q) \text{ in } C[0,1]^2. \tag{5.1.16}$$

Since no danger of confusion exists, in the following we shall simply write $\lambda \to 0+$ to indicate that some subsequence of (λ) tends to zero.

Let us prove now that

$$\left(K_{1\lambda}(\cdot,p_\lambda), K_{2\lambda}(\cdot,q_\lambda)\right) \to \left(K_1(\cdot,p), K_2(\cdot,q)\right) \text{ in } H_3, \tag{5.1.17}$$

as $\lambda \to 0+$. To this purpose, notice first that for a.a. $r \in (0,1)$ and every $\lambda > 0$,

$$\left|\left(I + \lambda K_1(r,\cdot)\right)^{-1} p_\lambda(r) - p(r)\right| \leq$$
$$\leq \left|\left(I + \lambda K_1(r,\cdot)\right)^{-1} p_\lambda(r) - \left(I + \lambda K_1(r,\cdot)\right)^{-1} p(r)\right| +$$
$$+\left|\left(I + \lambda K_1(r,\cdot)\right)^{-1} p(r) - p(r)\right| \leq \left|p_\lambda(r) - p(r)\right| +$$
$$+\left|\left(I + \lambda K_1(r,\cdot)\right)^{-1} p(r) - p(r)\right|.$$

So, from (5.1.16) we have for a.a. $r \in (0,1)$,

$$\left(I + \lambda K_1(r,\cdot)\right)^{-1} p_\lambda(r) \to p(r) \text{ in } L^2(0,1), \text{ as } \lambda \to 0+.$$

Therefore, for a.a. $r \in (0, 1)$, as $\lambda \to 0+$,

$$K_{1\lambda}\big(r, p_\lambda(r)\big) = K_1\Big(r, \big(I + \lambda K_1(r, \cdot)\big)^{-1} p_\lambda(r)\Big) \to K_1\big(r, u(r)\big). \quad (5.1.18)$$

Notice also that for a.a. $r \in (0, 1)$ and $\lambda > 0$,

$$\big|K_{1\lambda}\big(r, p_\lambda(r)\big)\big| \leq \big|K_1\big(r, p_\lambda(r)\big)\big| \leq \qquad\qquad (5.1.19)$$
$$\leq \max\big\{\, |K_1(r, c_0)|, |K_1(r, -c_0)| \,\big\} =: \eta(r),$$

where $c_0 = \sup\big\{ |p_\lambda(r)| \mid \lambda > 0,\, 0 \leq r \leq 1 \big\} < \infty$ (see (5.1.16)) and $r \mapsto \eta(r)$ belongs to $L^2(0, 1)$. From (5.1.18)-(5.1.19) and by the Lebesgue Dominated Convergence Theorem, we get

$$K_{1\lambda}(\cdot, p_\lambda) \to K_1(\cdot, p) \text{ in } L^2(0, 1), \text{ as } \lambda \to 0+.$$

Similarly,
$$K_{2\lambda}(\cdot, q_\lambda) \to K_2(\cdot, q) \text{ in } L^2(0, 1), \text{ as } \lambda \to 0+,$$

i.e., (5.1.17) holds. Now, on account of (5.1.15)-(5.1.17), we can take the limit in (5.1.9) to obtain that $(p, q) \in D(A_3)$ and

$$(p, q) + A_3(p, q) = (\tilde{p}, \tilde{q}).$$

We have also used the fact that A_3' is maximal monotone. So, A_3 is maximal monotone, too. In order to prove the density of $D(A_3)$ in H_3, it suffices to notice that the set

$$\big\{ (p^* + \phi_1, q^* + \phi_2) \mid \phi_1, \phi_2 \in C_0^\infty(0, 1) \big\}$$

(which is dense in H_3) is included in $D(A_3)$. Here (p^*, q^*) is again the couple defined by (5.1.3). The proof is now complete. ∎

REMARK 5.1.1   The above proof contains some technicalities, but we could not avoid them, since no classical perturbation results are applicable here directly. ∎

REMARK 5.1.2   It is easy to show that A_3 is not cyclically monotone, i.e., A_3 is not a subdifferential. This is to make sure that we are handling a hyperbolic problem. ∎

As usual, we associate with problem (5.0.1)-(5.0.4) a Cauchy problem,

$$\big(u'(t), v'(t)\big) + A_3\big(u(t), v(t)\big) = \big(f_1(t), f_2(t)\big), \ t > 0, \qquad (5.1.20)$$
$$\big(u(0), v(0)\big) = (u_0, v_0). \qquad\qquad (5.1.21)$$

THEOREM 5.1.1

Assume (H.1) and (H.2). Let $T > 0$ be fixed. Then, for every $(u_0, v_0) \in L^2(0,1)$, $(u_0, v_0) \in H_3 = L^2(0,1)^2$, and $(f_1, f_2) \in L^1(0,T;H_3)$, there exists a unique weak solution $(u,v) \in C\big([0,T];H_3\big)$ of (5.1.20)-(5.1.21). If, in addition, $(u_0, v_0) \in D(A_3)$ and $(f_1, f_2) \in W^{1,1}(0,T;H_3)$, then (u,v) is a strong solution, $(u,v) \in W^{1,\infty}(0,T;H_3)$ and $u_r, v_r \in L^\infty\big(0,T;L^2(0,1)\big)$.

PROOF By Proposition 5.1.1, $D(A_3)$ is dense in H_3, and so known results (see Theorems 1.5.1 and 1.5.2) imply all the conclusions except the regularity property $u_r, v_r \in L^\infty\big(0,T;L^2(0,1)\big)$. To prove that, we start from the fact that

$$A_3\big(u(t), v(t)\big) = A_3'\big(u(t), v(t)\big) + \tag{5.1.22}$$

$$+ \Big(K_1\big(\cdot, u(t)\big), K_2\big(\cdot, v(t)\big)\Big) = \big(f_1(t) - u_t(\cdot, t), f_2(t) - v_t(\cdot, t)\big)$$

and so $t \mapsto A_3\big(u(t), v(t)\big)$ belongs to $L^\infty(0,T;H_3)$. We multiply (5.1.22) by

$$\big(u(t) - p^* + \operatorname{sgn} v_r(\cdot, t), v(t) - q^* + \operatorname{sgn} u_r(\cdot, t)\big)$$

and continue with a reasoning similar to that already used in the proof of Proposition 5.1.1. So, we can easily conclude the proof. Recall that p^* q^* are the functions defined by (5.1.3). We leave the details to the reader, as an easy but useful exercise. ∎

REMARK 5.1.3 A strong (weak) solution of (5.1.20)-(5.1.21) will be called a *strong* (respectively, *weak*) *solution* of problem (5.0.1)-(5.0.4). ∎

Comments and extensions

1. The above existence theory also works if the time derivatives u_t, v_t in (5.0.1)-(5.0.2) are multiplied by some $L^\infty(0,1)$-functions α_1 and α_2, which are bounded from below by positive constants. In this case, we have just to replace H_3 by the following product of weighted L^2-spaces:

$$\tilde{H}_3 := L^2\big(0,1;\alpha_1(r)\,dr\big) \times L^2\big(0,1;\alpha_2(r)\,dr\big).$$

So, after dividing (5.0.1) and (5.0.2) by α_1 and α_2, respectively, we see that the corresponding operator $\tilde{A}_3$, defined by

$$D(\tilde{A}_3) = D(A_3), \quad \tilde{A}_3(p,q) = \begin{pmatrix} 1/\alpha_1 & 0 \\ 0 & 1/\alpha_2 \end{pmatrix} A_3(p,q),$$

is maximal monotone in $\tilde{H}_3$ with respect to the weighted inner product

$$\big((p_1,p_2),(q_1,q_2)\big)_{\tilde{H}_3} = \int_0^1 p_1(r)q_1(r)\alpha_1(r)\,dr + \int_0^1 p_2(r)q_2(r)\alpha_2(r)\,dr,$$

provided (H.1) and (H.2) are satisfied. Indeed, the monotonicity of $\tilde{A}_3$ as well as its maximality in $\tilde{H}_3$ reduce to the corresponding properties of A_3 in H_3.

2. Moreover, the existence theory still works if $K_1(r,\cdot)$ and $K_2(r,\cdot)$ are perturbed by some Lipschitz continuous perturbations, even t-dependent.

3. We have not considered the case of a nonhomogeneous boundary condition, say

$$\big(-u(0,t),u(1,t)\big) \in L\big(v(0,t),v(1,t)\big) + \big(s_1(t),s_2(t)\big), \ t > 0,$$

since such a condition can be homogenized by a simple transformation, such as

$$\tilde{u}(r,t) = u(r,t) + (1-r)s_1(t) - rs_2(t), \ \tilde{v}(r,t) = v(r,t).$$

By this trick, (5.0.1) will become t-dependent, but such a situation can be handled by known t-dependent existence results, for instance by Kato's result (see Theorem 1.5.7). Indeed, we may assume in a first stage, in addition to (H.1) and (H.2), that $K_1(r,\cdot)$ is Lipschitz continuous for a.a. $r \in (0,1)$ and f_1, f_2, s_1, s_2 are sufficiently smooth so that Kato's key condition (1.5.12) is satisfied. Then, replacing $K_1(r,\cdot)$ by its Yosida approximation $K_{1\lambda}(r,\cdot)$, $\lambda > 0$, and arguing as in the proof of Theorem 5.1.1 (see also Proposition 5.1.1), we can develop a procedure of passage to limit as $\lambda \to 0+$ so that the Lipschitz condition on $K_1(r,\cdot)$ can be dropped. The details are left to the reader.

4. Partial differential systems with higher derivatives of u and v with respect to r can also be treated by a similar, but more complex theory. More precisely, instead of v_r and u_r we can consider some n-th order differential operators Dv and D^*u, where D^* is the Lagrange adjoint of D, and some boundary conditions are associated accordingly (see [MoPe1]).

5. Multivalued nonlinearities K_1 and K_2 can also be studied, but some difficulties do appear and a more elaborated theory is needed in this case [Moro1, p. 245]. However, the treatment is interesting and useful for applications involving distributed feedback control [Moro1, p. 323], as considered by Duvaut and Lions [DuvLi] for different models.

6. With some nonessential generalizations, the above existence theory covers the vectorial case, in which the system (5.0.1)-(5.0.2) is replaced by n partial differential systems of type (5.0.1)-(5.0.2), with the unknowns u_i, v_i, $i = 1, 2, \ldots, n$, coupled by algebraic nonlinear boundary conditions, [Barbu3], [Iftimie], and [Moro1, § III.4]. As said before, this case fits some applications in hydraulics and electronic engineering.

7. Here we discuss only the case of initial conditions. The time-periodic problem associated with (5.0.1)-(5.0.3) is investigated in [MoPe2].

Long-time behavior of solutions

We start with the following auxiliary result:

PROPOSITION 5.1.2

Assume (H.1) and (H.2). Then, for every $\lambda > 0$, the operator $(I + \lambda A_3)^{-1}$ maps bounded subsets of H_3 into bounded subsets of $H^1(0,1)^2$, where I is the identity operator of H_3, and A_3 is given by (5.1.1)-(5.1.2).

PROOF   Let $\lambda > 0$ be fixed and let $M \subset H_3$ be a bounded set. Denote

$$(u_{pq}, v_{pq}) := (I + \lambda A_3)^{-1}(p, q) \text{ for all } (p, q) \in M. \tag{5.1.23}$$

Obviously, the set $\{(u_{pq}, v_{pq}) \mid (p, q) \in M\}$ is bounded in H_3. On the other hand, (5.1.23) can be written as

$$A_3(u_{pq}, v_{pq}) = \frac{1}{\lambda}\big((p, q) - (u_{pq}, v_{pq})\big) \text{ for all } (p, q) \in M. \tag{5.1.24}$$

But (5.1.24) is like (5.1.22) with t replaced by (p, q). So, we can repeat the same reasoning and deduce that the set

$$(I + \lambda A_3)^{-1} M := \big\{(u_{pq}, v_{pq}) \mid (p, q) \in M\big\}$$

is bounded in $H^1(0,1)^2$.   ∎

THEOREM 5.1.2

Assume (H.1) and (H.2). Let $(u_0, v_0) \in H_3$, (f_1, f_2) belong to $L^1(\mathbb{R}_+; H_3)$, $F_3 := A_3^{-1}(0,0)$ be nonempty, $K_1(r, \cdot)$ and $K_2(r, \cdot)$ be strictly increasing for a.a. $r \in (0, 1)$. Let $t \mapsto \big(u(t), v(t)\big)$ denote the weak solution of problem (5.0.1)-(5.0.4). Then F_3 is a singleton, say $F_3 = \{(\tilde{p}, \tilde{q})\}$, and $\big(u(t), v(t)\big)$ converges in H_3 to $(\tilde{p}, \tilde{q})$, as $t \to \infty$. If, in addition, $(u_0, v_0) \in D(A_3)$ and $(f_1, f_2) \in W^{1,1}(\mathbb{R}_+; H_3)$, then $\big(u(t), v(t)\big) \to (\tilde{p}, \tilde{q})$ weakly in $H^1(0,1)^2$ and, hence, strongly in $C[0,1]^2$, as $t \to \infty$.

PROOF   Let (p_1, q_1) and (p_2, q_2) belong to F_3. Then $(p_i, q_i) \in D(A_3)$ and $A_3(p_i, q_i) = (0,0)$, $i = 1, 2$. It is easily seen that

$$\Big(K_1\big(r, p_1(r)\big) - K_1\big(r, p_2(r)\big) \Big) \big(p_1(r) - p_2(r)\big) = 0,$$

$$\Big(K_2\big(r, q_1(r)\big) - K_2\big(r, q_2(r)\big) \Big) \big(q_1(r) - q_2(r)\big) = 0$$

for a.a. $t \in (0, T)$. Since $K_1(r, \cdot)$ and $K_2(r, \cdot)$ are strictly increasing, $p_1 = p_2$ and $q_1 = q_2$. Thus F_3 is a singleton: $F_3 = \{(\tilde{p}, \tilde{q})\}$. Now, let us assume that $(u_0, v_0) \in D(A_3)$ and $(f_1, f_2) \in W^{1,1}(\mathbb{R}_+; H_3)$. Then $(u', v') \in L^\infty(\mathbb{R}_+; H_3)$ (cf. (1.5.7) of Theorem 1.5.1). On the other hand, as F_3 is nonempty, the trajectory $\{(u(t), v(t)) \mid t \geq 0\}$ is bounded in H_3. So, making use of the formula

$$\big(u(t), v(t)\big) = (I + A_3)^{-1} \Big(\big(f_1(t), f_2(t)\big) - \big(u'(t), v'(t)\big) + \big(u(t), v(t)\big) \Big),$$

we deduce that the trajectory is bounded in $H^1(0,1)^2$ (cf. Proposition 5.1.2). Therefore, it is clear that the last assertion of the theorem can be obtained as a consequence of the first one. In order to prove the first assertion, it suffices to show that for every $(u_0, v_0) \in D(A_3)$ and $f_1(t) \equiv 0$, $f_2(t) \equiv 0$ (cf. Theorem 1.5.4). Fix $(u_0, v_0) \in D(A_3)$. The trajectory $\{S(t)(u_0, v_0) \mid t \geq 0\}$ is bounded in $H^1(0,1)^2$, where $S(t)$, $t \geq 0$, denotes the contraction semigroup generated by $-A_3$. It follows that the ω-limit set

$$\omega(u_0, v_0) = \{(p, q) \in H_3 \mid \text{ there exist } t_n > 0, \ n \in \mathbb{N}, \text{ such that}$$
$$\lim_{n \to \infty} t_n = \infty \text{ and } \lim_{n \to \infty} S(t_n)(u_0, v_0) = (p, q) \text{ in } H_3\}$$

is nonempty. By Theorem 1.5.6, $\omega(u_0, v_0) \subset D(A_3)$. We shall show that actually $\omega(u_0, v_0)$ is a singleton. Let $(p, q) \in \omega(u_0, v_0)$ be arbitrary. We know that the trajectory $\{S(t)(p, q) \mid t \geq 0\}$ lies on a sphere centered at $(\tilde{p}, \tilde{q})$ (see Theorem 1.5.6), i.e.,

$$\big\| S(t)(p, q) - (\tilde{p}, \tilde{q}) \big\|_{H_3} = \big\| (p, q) - (\tilde{p}, \tilde{q}) \big\|_{H_3}, \ t \geq 0. \tag{5.1.25}$$

On the other hand, $t \mapsto S(t)(p, q)$ is a strong solution of the equation

$$\frac{d}{dt} S(t)(p, q) + A_3 S(t)(p, q) = 0, \ t > 0, \tag{5.1.26}$$

since $(p, q) \in D(A_3)$. Multiplying (5.1.26) by $S(t)(p, q) - (\tilde{p}, \tilde{q})$ and making use of (5.1.25), we get

$$\big(A_3 S(t)(p, q), S(t)(p, q) - (\tilde{p}, \tilde{q}) \big)_{H_3} = 0 \text{ for a.a. } t > 0.$$

In other words, for a.a. $t > 0$,

$$\left(A_3 S(t)(p,q) - A_3(\tilde{p}, \tilde{q}), S(t)(p,q) - (\tilde{p}, \tilde{q})\right)_{H_3} = 0. \qquad (5.1.27)$$

Using the strict monotonicity of $K_1(r, \cdot)$ and $K_2(r, \cdot)$, we obtain by (5.1.27) that

$$S(t)(p,q) = (\tilde{p}, \tilde{q}) \text{ for all } t \geq 0.$$

Hence $(p,q) = (\tilde{p}, \tilde{q})$. Therefore, $\omega(u_0, v_0) = \{(\tilde{p}, \tilde{q})\}$, as claimed. ∎

REMARK 5.1.4 The conclusions of Theorem 5.1.2 are still valid under some alternative assumptions. For instance, the strict monotonicity of $K_1(r, \cdot)$ and $K_2(r, \cdot)$ can be replaced by the following: $K_1(r, \cdot)$ *is strictly increasing for a.a.* $r \in (0,1)$ *and L is injective in one of its variables* (i.e., $L(y_1, y_2) \cap L(z_1, z_2) \neq \emptyset$ implies that either $y_1 = z_1$ or $y_2 = z_2$). A similar condition is: $K_2(r, \cdot)$ *is strictly increasing for a.a.* $r \in (0,1)$ *and L^{-1} is injective in one of its variables.* Such conditions are useful for applications, see [Moro1, p. 314]. The reader can easily reproduce the proof of the modified theorem for the case of these new conditions. ∎

5.2   Higher regularity of solutions

Different concepts of solution are important for the solvability of various models arising in applications. For example, the water-hammer problem in the case of a sudden stoppage of flow at the valve (see, e.g., [LiWH, p. 272], and [Moro1, p. 314]) can be solved only in the class of weak solutions. Indeed, this is a problem of type (5.0.1)-(5.0.4) with the boundary conditions

$$u(0, t) = u_1, \ v(1, t) = 0, \ t > 0,$$

and the initial data are not compatible with these boundary conditions (the initial velocity v_0 is a positive constant). This means that the couple (u_0, v_0) does not belong to the domain of the corresponding operator A_3 (see (5.1.1)-(5.1.2)). Notice that the above boundary conditions can be regarded as a particular case of (5.0.3), where L is the subdifferential of the function $j_2 \colon \mathbb{R}^2 \to (-\infty, \infty]$,

$$j_2(x, y) = \begin{cases} -u_1 x & \text{if } y = 0, \\ \infty & \text{otherwise.} \end{cases}$$

For other applications, in which the data are smooth and compatible with the boundary conditions, the concept of strong solution is more appropriate.

Moreover, the singular perturbation analysis of such problems [BaMo3] requires higher regularity of solutions. Of course, enough smoothness of the data and higher order compatibility are to be required. In the following, we are going to illustrate a method to derive higher regularity of solutions for a simpler case of problem (5.0.1)-(5.0.4), namely

$$u_t(r,t) + v_r(r,t) + Ru(r,t) = f_1(r,t), \tag{5.2.1}$$

$$v_t(r,t) + u_r(r,t) + Gv(r,t) = f_2(r,t),\ 0 < r < 1,\ 0 < t < T, \tag{5.2.2}$$

$$-u(0,t) \in \beta_1 v(0,t),\ u(1,t) \in \beta_2 v(1,t),\ r \in (0,1) \tag{5.2.3}$$

$$u(r,0) = u_0(r),\ v(r,0) = v_0(r),\ r \in (0,1), \tag{5.2.4}$$

where R and G are nonnegative constants. Let us first restate the second part of Theorem 5.1.1:

PROPOSITION 5.2.1

Assume that $\beta_1, \beta_2 \subset \mathbb{R} \times \mathbb{R}$ are maximal monotone operators, $f_1, f_2 \in W^{1,1}\big(0, T; L^2(0,1)\big)$, and $u_0, v_0 \in H^1(0,1)$ such that $v_0(0) \in D(\beta_1)$, $v_0(1) \in D(\beta_2)$, and

$$-u_0(0) \in \beta_1 v_0(0),\ u_0(1) \in \beta_2 v_0(1). \tag{5.2.5}$$

Then, problem (5.2.1)-(5.2.4) has a unique strong solution (u,v) such that

$$u, v \in W^{1,\infty}\big(0, T; L^2(0,1)\big) \cap L^\infty\big(0, T; H^1(0,1)\big).$$

REMARK 5.2.1   The conditions (5.2.5) are called *zeroth order compatibility conditions.* ∎

In order to get higher regularity results, we first consider the particular case when $R = G = 0$. In this case, the general solution of system (5.2.1)-(5.2.2) can be expressed by means of some formulae of D'Alembert type. This fact helps us to prove the following C^1-regularity result:

PROPOSITION 5.2.2

Assume that $R = G = 0$, and let $T > 0$ be fixed. Assume also that β_1, β_2 are defined on $\mathbb{R}$, single-valued, and

$$\beta_1, \beta_2 \in C^1(\mathbb{R}) \text{ with } \beta_1' \geq 0 \text{ and } \beta_2' \geq 0; \tag{5.2.6}$$

$$f_1, f_2 \in C^1\big([0,T]; C[0,1]\big); \tag{5.2.7}$$

$u_0, v_0 \in C^1[0,1]$, and satisfy (5.2.5) as well as

$$f_1(0,0) - v_0'(0) + \beta_1'\big(v_0(0)\big)\big(f_2(0,0) - u_0'(0)\big) = 0, \tag{5.2.8}$$

$$f_1(1,0) - v_0'(1) - \beta_2'\big(v_0(1)\big)\big(f_2(1,0) - u_0'(1)\big) = 0. \tag{5.2.9}$$

Then, the solution (u,v) of (5.2.1)-(5.2.4) belongs to $C^1\left(\overline{Q_T}\right)^2$, where $Q_T = (0,1) \times (0,T)$.

REMARK 5.2.2   The conditions (5.2.8)-(5.2.9) are called *first order compatibility conditions*. Those together with (5.2.5) are necessary conditions for the C^1-regularity of (u,v). Indeed, (5.2.5) are implied by the continuity of (u,v), while (5.2.8)-(5.2.9) follow from (5.2.3) by differentiating and taking then $t = 0$.   ∎

PROOF of Proposition 5.2.2.   The general solution of system (5.2.1)-(5.2.2) with $R = G = 0$ is given by the following formulae of D'Alembert type:

$$u(r,t) = \frac{1}{2}\int_0^t \left((f_1 + f_2)(r - s, t - s) + (f_1 - f_2)(r + s, t - s)\right) ds +$$

$$+ \frac{1}{2}\big(\phi(r - t) + \psi(r + t)\big), \tag{5.2.10}$$

$$v(r,t) = \frac{1}{2}\int_0^t \left((f_1 + f_2)(r - s, t - s) - (f_1 - f_2)(r + s, t - s)\right) ds +$$

$$+ \frac{1}{2}\big(\phi(r - t) - \psi(r + t)\big)+, \tag{5.2.11}$$

where ϕ and ψ are some arbitrary C^1-functions. We consider that in (5.2.10)-(5.2.11) f_1 and f_2 are extended to $\mathbb{R} \times [0,T]$ by

$$f_i(r,t) := f_i(2 - r, t) \text{ for all } i = 1, 2, \ r \in (1,2],$$
$$f_i(r,t) := f_i(-r, t) \text{ for all } i = 1, 2, \ r \in [-1,0),$$

and so on. Obviously, these extensions belong to $C^1\big([0,T]; C(\mathbb{R})\big)$. When necessary, every function defined on $\overline{Q_T}$ will be extended in a similar manner. The functions $\phi \colon [-T,1] \to \mathbb{R}$ and $\psi \colon [0,1+T] \to \mathbb{R}$ appearing in (5.2.10)-(5.2.11) can be determined from (5.2.3)-(5.2.4). Indeed, using (5.2.4), we get

$$\phi(r) = u_0(r) + v_0(r) \text{ and } \psi(r) = u_0 - v_0(r) \text{ for all } r \in [0,1]. \tag{5.2.12}$$

For simplicity, we shall assume that $T \leq 1$. Now, let us require that the functions u, v given by (5.2.10)-(5.2.11) satisfy (5.2.3):

$$\frac{1}{2}\big(\phi(-t) + \psi(t)\big) + \int_0^t f_1(s, t - s)\, ds +$$

$$+ \beta_1 \left(\frac{1}{2}\big(\phi(-t) - \psi(t)\big) + \int_0^t f_2(s, t - s)\, ds\right) = 0, \tag{5.2.13}$$

$$\frac{1}{2}\big(\phi(1-t)+\psi(1+t)\big)+\int_0^t f_1(1-s,t-s)\,ds =$$

$$= \beta_2\Big(\frac{1}{2}\big(\phi(1-t)-\psi(1+t)\big)+\int_0^t f_2(1-s,t-s)\,ds\Big), \quad (5.2.14)$$

for all $t \in [0,T]$. It is easily seen that ϕ, ψ can uniquely and completely be determined from (5.2.13)-(5.2.14). Indeed, setting

$$z_1(t) := \frac{1}{2}\big(\phi(-t)-\psi(t)\big)+\int_0^t f_2(s,t-s)\,ds,$$

$$z_2(t) := \frac{1}{2}\big(\phi(1-t)-\psi(1+t)\big)+\int_0^t f_2(1-s,t-s)\,ds,$$

the equations (5.2.13)-(5.2.14) can be written as

$$(I+\beta_1)z_1(t) = h_1(t) \ \text{and} \ (I+\beta_2)z_2(t) = h_2(t) \qquad (5.2.15)$$

for all $t \in [0,T]$, where

$$h_1(t) = -\psi(t) - \int_0^t (f_1-f_2)(s,t-s)\,ds,$$

$$h_2(t) = \phi(1-t) + \int_0^t (f_1+f_2)(1-s,t-s)\,ds.$$

As β_1 and β_2 are maximal monotone, it follows that z_1 and z_2 are uniquely determined from (5.2.15) and so also ϕ and ψ are uniquely determined. Moreover, the compatibility conditions (5.2.5) and (5.2.8)-(5.2.9) imply that $\phi \in C^1[-T,1]$ and $\psi \in C^1[0,1+T]$. Therefore, by virtue of (5.2.10)-(5.2.11), $(u,v) \in C^1\big([0,T];C[0,1]\big)^2$. Finally, as (u,v) satisfies system (5.0.1)-(5.0.2) with $R = G = 0$, we can easily see that $(u,v) \in C^1\big(\overline{Q_T}\big)^2.$ ∎

REMARK 5.2.3   Clearly, under the assumptions of Proposition 5.2.2, problem (5.2.1)-(5.2.4) is equivalent with the system (5.2.10)-(5.2.14). But, the problem (5.2.10)-(5.2.14) may have a solution under weaker assumptions and, in this situation, it is quite natural to call it a *generalized solution* of problem (5.2.1)-(5.2.4).   ∎

By revising the proof of the previous proposition, we can prove the following result:

PROPOSITION 5.2.3

Assume that β_1 and β_2 are single-valued and maximal monotone, $f_1, f_2 \in C(\overline{Q_T})$, and $u_0, v_0 \in C[0,1]$ and satisfy (5.2.5). Then, problem (5.2.1)-(5.2.4) has a unique generalized solution $(u,v) \in C(\overline{Q_T})^2$.

REMARK 5.2.4 Actually, the multivalued case for β_1, β_2 is still allowed in the last result, but we have in mind the higher regularity of (u, v) that requires even differentiability of β_1 and β_2. ∎

REMARK 5.2.5 Obviously, the generalized solution given by Proposition 5.2.3 satisfies (5.2.3) for every $t \in [0, T]$. This solution is stronger than the weak solution of problem (5.2.1)-(5.2.4). ∎

On the other hand, it is easily seen that Proposition 5.2.3 is still true if β_1, β_2 are t-dependent, but under some continuity assumptions. In particular, this does happen in the linear case

$$\beta_1(t, \xi) = a_1(t)\xi, \ \beta_2(t, \xi) = a_2(t)\xi \text{ for all } (t, \xi) \in [0, T] \times \mathbb{R},$$

where $a_1, a_2 \in C[0, T]$ are some nonnegative functions. Of course, instead of (5.2.3), we shall have the boundary conditions

$$u(0, t) + a_1(t)v(0, t) = 0, \tag{5.2.16}$$
$$u(1, t) - a_2(t)v(1, t) = 0, \tag{5.2.17}$$

and (5.2.5) must be replaced by

$$u_0(0) + a_1(0)v_0(0) = 0, \ u_0(1) - a_2(0)v_0(1) = 0. \tag{5.2.18}$$

PROPOSITION 5.2.4

Assume that β_1 and β_2 are single-valued and maximal monotone, $f_1, f_2 \in L^\infty\big(0, T; L^p(0, 1)\big)$, and $u_0, v_0 \in L^p(0, 1)$, where $p \in [1, \infty)$. Then, problem (5.2.1)-(5.2.4) with $R = G = 0$ has a unique generalized solution $(u, v) \in L^\infty\big(0, T; L^p(0, 1)\big)^2$. If, in addition, $f_1, f_2 \in C\big([0, T]; L^p(0, 1)\big)^2$, then $(u, v) \in C\big([0, T]; L^p(0, 1)\big)^2$.

PROOF We assume again, for simplicity and without any loss of generality, that $T \leq 1$. Let $f_1, f_2 \in C\big([0, T]; L^p(0, 1)\big)$. From (5.2.12) and (5.2.13)-(5.2.14) (or (5.2.15)) we have that $\phi \in L^p(-T, 1)$ and $\psi \in L^p(0, 1 + T)$. Indeed, $(I + \beta_1)^{-1}$ and $(I + \beta_2)^{-1}$ are Lipschitzian and so, taking into account that the functions $(t, s) \mapsto f_i(s, t-s), (t, s) \mapsto f_i(1-s, t-s)$, $i = 1, 2$, are both Lebesgue measurable (see, e.g., [HewSt], p. 395), we can apply the Fubini-Tonelli Theorem to deduce that $\phi \in L^p(-T, 1)$ and $\psi \in L^p(0, 1 + T)$ (see, e.g., [Nicole, p. 258] or [HewSt]). Similar arguments show that the integral terms in (5.2.10)-(5.2.11) belong to $C\big([0, T]; L^p(0, 1)\big)$

as well. The other situation of our proposition can be handled similarly. ∎

REMARK 5.2.6   The generalized solution (u,v) given by Proposition 5.2.4 does not depend on how f_1 and f_2 are extended to $C\big([0,T]; L^p_{loc}(\mathbb{R})\big)$ or $L^\infty\big(0,T; L^p_{loc}(\mathbb{R})\big)$. ∎

REMARK 5.2.7   Proposition 5.2.4 is still valid for the linear case

$$\beta_1(t,\xi) = a_1(t)\xi, \ \ \beta_2(t,\xi) = a_2(t)\xi,$$

where $a_1, a_2 \in C[0,T]$ are some nonnegative functions. ∎

Now, we are going to investigate the general case $R > 0$ and/or $G > 0$. Suppose that the conditions of Proposition 5.2.3 are satisfied. Then we know that problem (5.2.1)-(5.2.4) has a unique weak solution $(u,v) \in C\big([0,T]; L^2(0,1)\big)^2$ (see Theorem 5.1.1). Actually, this solution is more regular. To show that let us consider the operator B that assigns to each couple $\alpha = (\alpha_1, \alpha_2) \in C\big(\overline{Q_T}\big)^2$ the unique solution of system (5.2.10)-(5.2.14) with (f_1, f_2) replaced by $(f_1 - R\alpha_1, f_2 - G\alpha_2)$. By Proposition 5.2.3, this solution is in $C\big(\overline{Q_T}\big)^2$. Consider that $C\big(\overline{Q_T}\big)$ is endowed with the norm $\|\cdot\|_*$,

$$\|w\|_* := \sup_{(r,t)\in\overline{Q_T}} e^{-\gamma t} |w(r,t)|,$$

where γ is a positive constant. Obviously, $Z = C(\overline{Q_T})^2$ is a real Banach space with respect to the norm

$$\|(w_1, w_2)\|_Z := \max\big\{ \|w_1\|_*, \|w_2\|_* \big\}.$$

By a straightforward computation it turns out that $B: Z \to Z$ is a strict contraction:

$$\|B\alpha - B\tilde{\alpha}\|_Z \le \rho\|\alpha - \tilde{\alpha}\|_Z,$$

where $0 < \rho < 1$, provided γ is a sufficiently large constant. Indeed, we have, for instance,

$$\left| \int_0^t (f_1 - R\alpha_1 + f_2 - G\alpha_2)(r - s, t - s)\, ds - \right.$$

$$\left. - \int_0^t (f_1 - R\tilde{\alpha}_1 + f_2 - G\tilde{\alpha}_2)(r - s, t - s)\, ds \right| \le$$

$$\le 2\max\{R,G\}\|\alpha - \tilde{\alpha}\|_Z \int_0^t e^{\gamma(t-s)}\, ds \le$$

$$\le \frac{2}{\gamma}\max\{R,G\}\, e^{\gamma t}\, \|\alpha - \tilde{\alpha}\|_Z \ \text{ for all } t \in [0,T].$$

Therefore,

$$\left\| \int_0^t (f_1 - R\alpha_1 + f_2 - G\alpha_2)(r - s, t - s)\, ds - \right.$$

$$\left. - \int_0^t (f_1 - R\tilde{\alpha} + f_2 - G\tilde{\alpha}_2)(r - s, t - s)\, ds \right\|_Z \leq$$

$$\leq \frac{2}{\gamma} \max\{R, G\} \|\alpha - \tilde{\alpha}\|_Z.$$

Using (5.2.12)-(5.2.14), with $(f_1 - R\alpha_1, f_2 - G\alpha_2)$ instead of (f_1, f_2), we obtain similar estimates for $\|\phi(r - t) - \tilde{\phi}(r - t)\|_Z$ and $\|\psi(r + t) - \tilde{\psi}(r + t)\|_Z$. To do that we can use (5.2.15) and the fact that $(I + \beta_1)^{-1}$ and $(I + \beta_2)^{-1}$ are Lipschitz continuous. So, for a sufficiently large γ, B is indeed a strict contraction and therefore, by the Banach Fixed Point Theorem, B has a unique fixed point.

Summarizing what we have done so far, we can state the following result:

PROPOSITION 5.2.5

Assume the conditions of Proposition 5.2.3. Then, problem (5.2.1)-(5.2.4) has a unique generalized solution $(u, v) \in C\left(\overline{Q_T}\right)^2$. More precisely, (u, v) satisfies (5.2.10)-(5.2.14), where (f_1, f_2) is replaced by $(f_1 - Ru, f_2 - Gv)$.

REMARK 5.2.8 By similar arguments, Proposition 5.2.4 can be extended to the case of general coefficients $R, G \geq 0$. ∎

Now, we are prepared to state and prove the following C^1-regularity result:

THEOREM 5.2.1

Assume that $R \geq 0$, $G \geq 0$ and that f_1, f_2, β_1, β_2, u_0, and v_0 satisfy all assumptions of Proposition 5.2.2, except (5.2.8)-(5.2.9), which are replaced by

$$f_1(0, 0) - Ru_0(0) - v_0'(0) +$$
$$+ \beta_1'\big(v_0(0)\big)\big(f_2(0, 0) - Gv_0(0) - u_0(0)\big) = 0, \tag{5.2.19}$$
$$f_2(1, 0) - Ru_0(1) - v_0'(1) -$$
$$- \beta_2'\big(v_0(1)\big)\big(f_2(1, 0) - Gv_0(1) - u_0'(1)\big) = 0. \tag{5.2.20}$$

Then, the solution (u, v) of problem (5.2.1)-(5.2.4) belongs to $C^1\left(\overline{Q_T}\right)^2$.

PROOF The solution (u, v) exists and is unique, by Proposition 5.2.1. Moreover, $(u, v) \in C\left(\overline{Q_T}\right)^2$ (see Proposition 5.2.5). Proposition 5.2.5 can be extended to the case of t-dependent linear boundary conditions, in which

$$\beta_1(t,\xi) = a_1(t)\xi \text{ and } \beta_2(t,\xi) = a_2(t)\xi,$$

where $a_1, a_2 \in C[0,T]$ with $a_1 \geq 0$, $a_2 \geq 0$. Thus there exists a unique generalized solution $(\tilde{u}, \tilde{v}) \in C\left(\overline{Q_T}\right)^2$ of the problem

$$\tilde{u}_t(r,t) + \tilde{v}_r(r,t) = -R\tilde{u}(r,t) + f_{1t}(r,t), \tag{5.2.21}$$

$$\tilde{v}_t(r,t) + \tilde{u}_r(r,t) = -G\tilde{u}(r,t) + f_{2t}(r,t), \ 0 < r < 1, \ 0 < t < T, \tag{5.2.22}$$

$$\tilde{u}(0,t) + \beta_1'(v(0,t))\tilde{v}(0,t) = 0 \text{ for all } t \in [0,T], \tag{5.2.23}$$

$$\tilde{u}(1,t) - \beta_2'(v(1,t))\tilde{v}(1,t) = 0 \text{ for all } t \in [0,T], \tag{5.2.24}$$

$$\tilde{u}(r,0) = f_1(r,0) - Ru_0(r) - v_0'(r) \text{ for all } r \in [0,1], \tag{5.2.25}$$

$$\tilde{v}(r,0) = f_2(r,0) - Gv_0(r) - u_0'(r) \text{ for all } r \in [0,1]. \tag{5.2.26}$$

On the other hand, (u,v) satisfies (5.2.10)-(5.2.14) with $(f_1 - Ru, f_2 - Gv)$ instead of (f_1, f_2). By differentiating (5.2.10)-(5.2.14) with respect to t, we obtain that $u_t, v_t \in L^\infty\left(0, T; L^2(0,1)\right)$ (see Proposition 5.2.1) satisfy (5.2.21)-(5.2.26) in the generalized sense. By the uniqueness of the solution for (5.2.21)-(5.2.26) in the class $L^\infty\left(0, T; L^2(0,1)\right)^2$ (see Proposition 5.2.4 and Remark 5.2.8), we have that $(u_t, v_t) = (\tilde{u}, \tilde{v})$ and hence $(u,v) \in C^1\left([0,T]; C[0,T]\right)^2$. Actually, using (5.2.1)-(5.2.2), we can see that $(u,v) \in C^1\left(\overline{Q_T}\right)^2$. $\blacksquare$

REMARK 5.2.9   Analogous C^k-regularity results can be obtained for $k \geq 2$. $\blacksquare$

For instance, let us point out how to obtain the C^2-regularity. Of course, the data u_0, v_0, f_1, f_2, β_1, β_2 have to be more regular and satisfy, in addition to (5.2.5) and (5.2.19)-(5.2.20), some second order compatibility conditions, which can be derived in a natural way as necessary conditions for C^2-regularity. So, if $(u,v) \in C^2\left(\overline{Q_T}\right)^2$ is a solution of (5.2.1)-(5.2.4), then it satisfies, for all $t \in [0,T]$,

$$-u_{tt}(0,t) = \beta_1'\big(v(0,t)\big)v_{tt}(0,t) + \beta_1''\big(v(0,t)\big)\big(v_t(0,t)\big)^2, \tag{5.2.27}$$

$$u_{tt}(1,t) = \beta_2'\big(v(1,t)\big)v_{tt}(1,t) + \beta_1''\big(v(1,t)\big)\big(v_t(1,t)\big)^2. \tag{5.2.28}$$

On the other hand,

$$u_{tt} = f_{1t} - Ru_t - v_{rt} = f_{1t} - R\big(f_1 - Ru - v_r\big) - \big(f_{2r} - Gv_r - u_{rr}\big),$$

$$v_{tt} = f_{2t} - Gv_t - u_{tr} = f_{2t} - G\big(f_2 - Gv - u_r\big) - \big(f_{1r} - Ru_r - v_{rr}\big).$$

Therefore,

$$u_{tt}(i,0) = f_{1t}(i,0) - R\big(f_1(i,0) - Ru_0(i) - v_0'(i)\big) -$$

$$-\big(f_{2r}(i,0) - Gv_0'(i) - u_0''(i)\big) =: \omega_i, \;\; i = 0, 1, \qquad (5.2.29)$$

$$v_{tt}(i,0) = f_{2t}(i,0) - G\big(f_2(i,0) - Gv_0(i) - u_0'(i)\big) -$$
$$-\big(f_{1r}(i,0) - Ru_0'(i) - v_0''(i)\big) =: \theta_i, \;\; i = 0, 1. \qquad (5.2.30)$$

Taking $t = 0$ in (5.2.27)-(5.2.28) and using (5.2.29)-(5.2.30), one gets the following *second order compatibility conditions*:

$$-\omega_0 = \beta_1'\big(v_0(0)\big)\theta_0 + \beta_1''\big(v_0(0)\big)\big(f_2(0,0) - Gv_0(0) - u_0'(0)\big)^2, \quad (5.2.31)$$

$$\omega_1 = \beta_2'\big(v_0(1)\big)\theta_1 + \beta_1''\big(v_0(1)\big)\big(f_2(1,0) - Gv_0(1) - u_0'(1)\big)^2. \quad (5.2.32)$$

We are now in a position to state the following C^2-regularity result:

THEOREM 5.2.2
Assume that $R \geq 0$, $G \geq 0$, $\beta_1, \beta_2 \in C^2(\mathbb{R})$, $\beta_1' \geq 0$, $\beta_2' \geq 0$, $f_1, f_2 \in C^1\big(\overline{Q_T}\big)$, $f_{1tt}, f_{2tt} \in C\big(\overline{Q_T}\big)$, $u_0, v_0 \in C^2[0,1]$ and the compatibility conditions (5.2.5), (5.2.19)-(5.2.20), (5.2.31)-(5.2.32) are all fulfilled. Then, the solution (u,v) of problem (5.2.1)-(5.2.4) belongs to $C^2\big(\overline{Q_T}\big)^2$.

PROOF　By Theorem 5.2.1, problem (5.2.1)-(5.2.4) has a unique solution $(u,v) \in C^1\big(\overline{Q_T}\big)^2$. Recall that $(\tilde{u},\tilde{v}) = (u_t, v_t)$ is the unique generalized solution of class $C\big(\overline{Q_T}\big)^2$ to problem (5.2.21)-(5.2.26). Actually, by Theorem 5.2.1, extended to the case of linear t-dependent boundary conditions, $(\tilde{u},\tilde{v}) = (u_t, v_t) \in C^1\big(\overline{Q_T}\big)$. This implies that $(u,v) \in C^2\big(\overline{Q_T}\big)$, since (u,v) satisfies system (5.2.1)-(5.2.2) and $f_1, f_2 \in C^1\big(\overline{Q_T}\big)$. ∎

REMARK 5.2.10　We have used in the above treatment the C^k-spaces as our framework. It seems that they are more appropriate for this hyperbolic problem than the energetic H^k-spaces. On the other hand, it is clear that every level of regularity for the solution of problem (5.2.1)-(5.2.4) can be reached under enough regularity and compatibility of the data with the boundary conditions. ∎

References

[Barbu3]　V. Barbu, Nonlinear boundary value problems for a class of hyperbolic systems, *Rev. Roumaine Math. Pures Appl.*, 22 (1977), no. 2, 155–168.

[BaMo3]　L. Barbu & Gh. Moroşanu, *Analiza asimptotică a unor probleme la limită cu perturbaţii singulare*, Editura Academiei, Bucharest, 2000.

[CooKr]　K.L. Cooke & D.W. Krumme, Differential-difference equations and nonlinear initial boundary value problems for linear hyperbolic partial differential equations, *J. Math. Anal. Appl.*, 24 (1968), 372–387.

[DuvLi]　G. Duvaut & J.-L. Lions, *Inequalities in Mechanics and Physics*, Springer-Verlag, Berlin, 1976.

[GhaKe]　M.S. Ghausi & J.J. Kelly, *Introduction to Distributed Parameter Networks with Applications to Integrated Circuits*, R.E. Krieger Publ., Huntington, New York, 1977.

[HewSt]　E. Hewitt & K. Stromberg, *Real and Abstract Analysis*, Springer-Verlag, Berlin, 1965.

[Iftimie]　V. Iftimie, Sur un probleme non linéaire pour un système hyberbolique, *Rev. Roumaine Math. Pures. Appl.*, 21 (1976), no. 3, 303–328.

[LiWH]　W.-H. Li, *Differential Equations of Hydraulics, Transients, Dispersion, and Groundwater Flow*, Prentice-Hall, New Jersey, 1972.

[MarNe]　C.A. Marinov & P. Neittaanmäki, *Mathematical Models in Electrical Circuits: Theory and Applications*, Kluwer, Dordrecht, 1991.

[Moro1]　Gh. Moroşanu, *Nonlinear Evolution Equations and Applications*, Editura Academiei and Reidel, Bucharest and Dordrecht, 1988.

[MoPe1]　Gh. Moroşanu & D. Petrovanu, Nonlinear boundary value problems for a class of hyperbolic partial differential systems, *Atti Sem. Mat. Fis. Univ. Modena*, 34 (1987), no.2, 295–315.

[MoPe2]　Gh. Moroşanu & D. Petrovanu, Time periodic solutions for a class of partial differential hyperbolic systems, *An. Şt. Univ. "Al. I. Cuza", Secţ. Mat. (N.S.)*, 36 (1990), no. 2, 93–98.

[Nicole]　M. Nicolescu, *Analiză matematică, vol. III*, Editura Tehnică, Bucharest, 1960.

[StrWy]　L.V. Streeter & E.B. Wyle, *Hydraulic Transients*, McGraw-Hill, New York, 1967.

Chapter 6

Hyperbolic boundary value problems with algebraic-differential boundary conditions

In this chapter we consider the same hyperbolic partial differential system as in the preceding chapter, but relabeled for our convenience as

$$u_t(r,t) + v_r(r,t) + K_1\big(r, u(r,t)\big) = f_1(r,t), \tag{6.0.1}$$

$$v_t(r,t) + u_r(r,t) + K_2\big(r, v(r,t)\big) = f_2(r,t),\ 0 < r < 1,\ t > 0. \tag{6.0.2}$$

This time we are interested in boundary conditions of algebraic-differential type, i.e.,

$$-u(0,t) \in \beta_1 v(0,t), \tag{6.0.3}$$

$$v_t(1,t) - u(1,t) + \beta_2 v(1,t) \ni e(t),\ t > 0, \tag{6.0.4}$$

where β_1, β_2 are some given mappings (possibly multivalued) and $e\colon \mathbb{R}_+ \to \mathbb{R}$. Also, we associate with (6.0.1)-(6.0.2) the initial conditions

$$u(r,0) = u_0(r),\ v(r,0) = v_0(r),\ 0 < r < 1. \tag{6.0.5}$$

The case in which we have a nonhomogeneous algebraic boundary condition instead of (6.0.3) will be omitted. Actually, its treatment requires easy arguments, as we shall see later.

Such problems are suggested by some applications in electrical engineering (see, e.g., [CooKr], [Moro1, p. 322]). More precisely, a classical model from transmission line theory is given by the telegraph system

$$Lu_t + v_r + Ru = \tilde{e},\ Cv_t + u_r + Gv = 0,\ 0 < r < 1,\ t > 0,$$

with the boundary conditions

$$-v(0,t) = R_0 u(0,t),\ u(1,t) = C_1 v_t(1,t) + \beta\big(v(1,t)\big),\ t > 0,$$

and initial conditions at $t = 0$. Notice that if the inductance L is negligible (small parameter), then the corresponding reduced model (obtained for $L = 0$) consists of an algebraic equation,

$$u = \frac{1}{R}(-v_r + \tilde{e}),$$

as well as a parabolic problem of type (4.0.1)-(4.0.4) (see Chapter 4). The coefficients multiplying u_t, v_t, and $v_t(1,t)$ that appear in applications are taken to be equal to 1 in (6.0.1)-(6.0.4). Actually, such a situation can be solved by choosing an appropriate weighted space.

The assumption (H.1) of the preceding chapter is kept here and we further assume that:

(H.2') The operators $\beta_1, \beta_2 \subset \mathbb{R} \times \mathbb{R}$ are both *maximal monotone*.

6.1   Existence, uniqueness, and long-time behavior of solutions

Consider as a basic framework the space $H_4 = L^2(0,1)^2 \times \mathbb{R}$. This is a real Hilbert space with the scalar product

$$\left((p_1, q_1, c_1), (p_2, q_2, c_2)\right)_{H_4} := \int_0^1 \left(p_1(r)p_2(r) + q_1(r)q_2(r)\right) dr + c_1 c_2$$

and the associated Hilbertian norm. Define the operator $A_4 \colon D(A_4) \subset H_4 \mapsto H_4$ by

$$D(A_4) = \left\{(p,q,c) \in H^1(0,1)^2 \times \mathbb{R} \mid c = q(1) \in D(\beta_2), \quad (6.1.1)\right.$$
$$\left. q(0) \in D(\beta_1), -p(0) \in \beta_1 q(0)\right\},$$
$$A(p,q,c) = \left(q' + K_1(\cdot,p), p' + K_2(\cdot,q), -p(1) + \beta_2 c\right). \quad (6.1.2)$$

Of course, A_4 is naturally connected with our problem (6.0.1)-(6.0.5), as we shall see later.

PROPOSITION 6.1.1
If (H.1) and (H.2') hold, then A_4 is maximal monotone and

$$\overline{D(A_4)} = L^2(0,1) \times \overline{D(\beta_2)}, \quad (6.1.3)$$

with respect to the topology of H_4.

PROOF   The monotonicity of A_4 follows by an elementary computation. In order to prove the maximality of A_4, fix an arbitrary $(p_1, q_1, c_1) \in H_4$ and consider the equation

$$(p, q, c) + A_4(p, q, c) \ni (p_1, q_1, c_1). \tag{6.1.4}$$

But, (6.1.4) can be equivalently written as

$$(p, q) + \hat{A}_3(p, q) \ni (p_1, q_1), \tag{6.1.5}$$

where $\hat{A}_3$ is the operator A_3 defined by (5.1.1)-(5.1.2) with L given by

$$D(L) = D(\beta_1) \times D(\beta_2), \quad L(y_1, y_2) := \big(\beta_1 y_1, y_2 + \beta_2 y_2 - c_1\big).$$

Obviously, this L is maximal monotone in $\mathbb{R}^2$ and so, according to Proposition 5.1.1, there exists a pair $(p, q) \in D(\hat{A}_3)$ satisfying Eq. (6.1.5). This means that $\big(p, q, q(1)\big)$ belongs to $D(A_4)$ and satisfies (6.1.4). Hence A_4 is indeed maximal monotone. The proof of (6.1.3) is immediate. $\blacksquare$

Now, we consider in H_4 the Cauchy problem

$$\big(u'(t), v'(t), \xi'(t)\big) + A_4\big(u(t), v(t), \xi(t)\big) \ni \tag{6.1.6}$$
$$\ni \big(f_1(t), f_2(t), e(t)\big) \ \text{for a.a.} \ t > 0,$$
$$\big(u(0), v(0), \xi(0)\big) = (u_0, v_0, \xi_0). \tag{6.1.7}$$

THEOREM 6.1.1
Assume (H.1) and (H.2'). Let $T > 0$ be fixed. For every $(u_0, v_0, \xi_0) \in L^2(0,1)^2 \times \overline{D(\beta_2)}$, and $(f_1, f_2, e) \in L^1(0, T; H_4)$ there exists a unique weak solution $(u, v, \xi) \in C\big([0,T]; H_4\big)$ of problem (6.1.6)-(6.1.7). If, in addition, $u_0, v_0 \in H^1(0,1)$, $\xi_0 = v_0(1)$ such that $\big(u_0, v_0, v_0(1)\big) \in D(A_4)$ and $(f_1, f_2, e) \in W^{1,1}(0, T; H_4)$, then (u, v, ξ) belongs to $W^{1,\infty}(0, T; H_4)$, it is a strong solution of problem (6.1.6)-(6.1.7), $\xi(t) = v(1,t)$, $0 \le t \le T$, and $u_r, v_r \in L^\infty\big(0, T; L^2(0,1)\big)$.

PROOF   All the conclusions follow from general existence results (Theorems 1.5.1 and 1.5.2; see also (6.1.3)) with the exception of the last regularity property, i.e., $u_r, v_r \in L^\infty\big(0, T; L^2(0,1)\big)$. To prove this, we can apply a familiar technique (see Chapter 5). More precisely, we multiply (6.0.1) and (6.0.2) by $u(t) - u_0 + \operatorname{sgn} v_r(\cdot, t)$ and $v(t) - v_0 + \operatorname{sgn} u_r(\cdot, t)$, respectively, to obtain:

$$v_r(u - u_0) + |v_r| + K_1(\cdot, u)(u - u_0 + \operatorname{sgn} v_r) =$$
$$= (f_1 - u_t)(u - u_0 - \operatorname{sgn} v_r),$$
$$u_r(v - v_0) + |u_r| + K_2(\cdot, v)(v - v_0 + \operatorname{sgn} u_r) =$$
$$= (f_2 - v_t)(v - v_0 + \operatorname{sgn} u_r).$$

Therefore, by adding the last two equations and using the monotonicity of $K_1(r,\cdot)$ and $K_2(r,\cdot)$ we arrive at:

$$\int_0^1 \frac{\partial}{\partial r}\big(u(r,t)-u_0(r)\big)\big(v(r,t)-v_0(r)\big)\,dr + \|v_r(\cdot,t)\|_{L^1(0,1)} +$$

$$+\|u_r(\cdot,t)\|_{L^1(0,1)} \le C_1 \quad \text{for a.a. } t \in (0,T), \tag{6.1.8}$$

where C_1 is a positive constant. We have also used (H.1) and the facts $u,v \in W^{1,\infty}(0,T;L^2(0,1))$ and $f_1,f_2 \in C([0,T];L^2(0,1))$. On the other hand, by (6.0.3)-(6.0.4) and (H.2'), we have

$$\int_0^1 \frac{\partial}{\partial r}\big(u(r,t)-u_0(r)\big)\big(v(r,t)-v_0(r)\big)\,dr \ge$$

$$\ge \big(v_t(1,t)+\beta_2 v(1,t)-e(t)-u_0(1)\big)\big(v(1,t)-v_0(1)\big) \ge$$

$$\ge \big(v_t(1,t)+\beta_2 v_0(1)-e(t)-u_0(1)\big)\big(v(1,t)-v_0(1)\big) \ge -C_2 \tag{6.1.9}$$

for a.a. $t \in (0,T)$, where C_2 is another positive constant. We have also used that $e \in C[0,T]$ and $v_t \in L^\infty(0,T)$. Now, clearly, (6.1.8) and (6.1.9) imply that $u_r, v_r \in L^\infty(0,T;L^1(0,1))$ and hence $u,v \in L^\infty(0,T;C[0,1])$. By virtue of (H.1), this implies that $K_1(\cdot,u), K_2(\cdot,v) \in L^\infty(0,T;L^2(0,1))$ and so, as u and v satisfy the system (6.0.1)-(6.0.2), we may conclude that $u_r, v_r \in L^\infty(0,T;L^2(0,1))$. $\blacksquare$

Comments and extensions

1. Let us first discuss the relation between the original problem (6.0.1)-(6.0.5) and the Cauchy problem (6.1.6)-(6.1.7). If the data u_0, v_0, f_1, f_2 are smooth functions and $\xi_0 = v_0(1)$ then problem (6.1.6)-(6.1.7) has a unique strong solution and so $\xi(t) = v(1,t)$, $t \ge 0$, and the first two components u, v satisfy the original problem (6.0.1)-(6.0.5). Notice that ξ_0 does not appear in (6.0.1)-(6.0.5), because this is a classical model, for which usually one looks for a smooth solution and the condition $\xi_0 = v_0(1)$ is implicitly assumed (even if it is not explicitly stated). But, in the case of our Cauchy problem (6.1.6)-(6.1.7) this condition may not be satisfied. Moreover, if (u_0, v_0, ξ_0) belongs to the closure of $D(A_4)$ in H_4, which is $L^2(0,1)^2 \times \overline{D(\beta_2)}$, then $v_0(1)$ may not make sense. In both cases the Cauchy problem (6.1.6)-(6.1.7) admits just a weak solution. The same situation does appear when f_1, f_2, e are not sufficiently smooth (see Theorem 6.1.1 above).

   In such cases, $\xi(t)$ does not coincide with $v(1,t)$ anymore, i.e., it is not the trace of v on the half-line $\{1\} \times \{t \mid t > 0\}$, which is a part of the boundary of the domain. However, $\xi(t)$ still represents an evolution on that part of the boundary and, of course, an initial condition is needed, i.e., $\xi(0) = \xi_0$ Therefore, it is clear that the Cauchy problem (6.1.6)-(6.1.7) is well posed and represents a more complete model, since it covers situations that are beyond the classical framework.

2. If instead of (6.0.3) we have a nonhomogenous algebraic boundary condition, say

$$-u(0,t) \in \beta_1 v(0,t) + s(t), \ t > 0,$$

then we can homogenize such a condition by a simple change, namely

$$\tilde{u}(r,t) = u(r,t) + s(t), \ \tilde{v}(r,t) = v(r,t).$$

Equation (6.0.1) will become t-dependent. However, this is not so bad, because we are able to proceed as explained before, in Chapter 5 (see Comments and extensions). As a matter of fact, those comments and extensions are still valid for problem (6.0.1)-(6.0.5) with slight modifications.

Long-time behavior of solutions

We need again a compactness result for the resolvent of the operator appearing in the Cauchy problem.

PROPOSITION 6.1.2
If (H.1) and (H.2') hold, then, for every $\lambda > 0$, the operator $(I + \lambda A_4)^{-1}$ maps bounded subsets of H_4 into bounded subsets of $H^1(0,1)^2 \times \mathbb{R}$, where A_4 is the operator defined by (6.1.1)-(6.1.2) and I is the identity operator of H_4.

PROOF   Fix a $\lambda > 0$ and consider a bounded subset of H_4, say $M = \{(p_j, q_j, c_j) \in H_4 \mid j \in J\}$, where J is some nonempty set. Let

$$(u_j, v_j, y_j) := (I + \lambda A_4)^{-1}(p_j, q_j, c_j) \text{ for all } j \in J. \tag{6.1.10}$$

Obviously, the set $\{(u_j, v_j, y_j) \mid j \in J\}$ is bounded in H_4. So, it remains to show that the set $\{(u_j, v_j) \mid j \in J\}$ is bounded in $H^1(0,1)^2$. Notice that (6.1.10) can equivalently be written as

$$v'_j + K_1(\cdot, u_j) = \frac{1}{\lambda}(p_j - u_j) \text{ for all } j \in J, \tag{6.1.11}$$

$$u'_j + K_2(\cdot, v_j) = \frac{1}{\lambda}(q_j - v_j) \text{ for all } j \in J, \tag{6.1.12}$$

$$-u_j(0) \in \beta_1 v_j(0) \text{ for all } j \in J, \tag{6.1.13}$$

$$u_j(1) \in \frac{1}{\lambda}v_j(1) + \beta_2 v_j(1) - \frac{1}{\lambda}c_j \text{ for all } j \in J. \tag{6.1.14}$$

Employing again a standard device (that has been used the last time in the proof of Theorem 6.1.1), we can infer from (6.1.11)-(6.1.14) that the sets $\{u_j' \mid j \in J\}$ and $\{v_j' \mid j \in J\}$ are bounded in $L^2(0,1)$. ∎

THEOREM 6.1.2

Assume that (H.1) and (H.2') hold, $(u_0, v_0, \xi_0) \in \overline{D(A_4)} = L^2(0,1)^2 \times \overline{D(\beta_2)}$, $(f_1, f_2, e) \in L^1(\mathbb{R}_+; H_4)$, and $F_4 := A_4^{-1}(0,0,0)$ is nonempty. Moreover, assume that:

$$K_1(r,\cdot) \text{ and } K_2(r,\cdot) \text{ are both strictly increasing} \qquad (6.1.15)$$

for a.a. $r \in (0,1)$. Let $(u,v,\xi) : [0,\infty) \to H_4$ be the weak solution of problem (6.1.6)-(6.1.7). Then F_4 is a singleton, say $F_4 = \{(\hat{p}, \hat{q}, \hat{q}(1))\}$, and $(u(t), v(t), \xi(t))$ converges strongly in H_4 to $(\hat{p}, \hat{q}, \hat{q}(1))$, as $t \to \infty$.

If, in addition, $(u_0, v_0, \xi_0) \in D(A_4)$ and $(f_1, f_2, e) \in W^{1,1}(\mathbb{R}_+; H_4)$, then $(u(t), v(t))$ converges weakly in $H^1(0,1)^2$ and, hence, strongly in $C[0,1]^2$ to $(\hat{p}, \hat{q})$, as $t \to \infty$.

The proof relies on Proposition 6.1.2 and is very similar to the proof of Theorem 5.1.2. So, we leave it to the reader, as an exercise.

REMARK 6.1.1   Theorem 6.1.2 still holds if (6.1.15) is replaced by alternative conditions, such as:

(i) The mapping $K_1(r,\cdot)$ is strictly increasing for a.a. $r \in (0,1)$ and at least one of β_1, β_2 is injective;

(ii) The mapping $K_2(r,\cdot)$ is strictly increasing for a.a. $r \in (0,1)$ and at least one of $\beta_1^{-1}, \beta_2^{-1}$ is injective.

We again encourage the reader to prove Theorem 6.1.2 under each of the new assumptions (i) and (ii). ∎

6.2 Higher regularity of solutions

In this section we consider a special case of problem (6.0.1)-(6.0.5), namely

$$u_t(r,t) + v_r(r,t) + Ru(r,t) = f_1(r,t), \qquad (6.2.1)$$

$$v_t(r,t) + u_r(r,t) + Gv(r,t) = f_2(r,t), \ 0 < r < 1, \ t > 0, \quad (6.2.2)$$

$$-v(0,t) = R_0 u(0,t), \ t > 0, \qquad (6.2.3)$$

$$v_t(1,t) - u(1,t) + \beta v(1,t) \ni e(t), \ t > 0, \tag{6.2.4}$$

$$u(r,0) = u_0(r), \ v(r,0) = v_0(r), \ 0 < r < 1, \tag{6.2.5}$$

where R, G, R_0 are some nonnegative constants, and $\beta: D(\beta) \subset \mathbb{R} \mapsto R$ is a nonlinear mapping. So, this problem is still nonlinear. We are going to illustrate a new method to derive higher regularity. For the time being, let us partly restate Theorem 6.1.1 for the case of problem (6.2.1)-(6.2.5):

PROPOSITION 6.2.1

Assume that $\beta \subset \mathbb{R} \times \mathbb{R}$ is maximal monotone, $(f_1, f_2, e) \in W^{1,1}(0,T; H_4)$ with $H_4 = L^2(0,1) \times \mathbb{R}$, $u_0, v_0 \in H^1(0,1)$ such that $v_0(1) \in D(\beta)$, and the zero-th order compatibility condition

$$-v_0(0) = R_0 u_0(0) \tag{6.2.6}$$

is satisfied. Then, problem (6.2.1)-(6.2.5) has a unique strong solution (u,v) with

$$u, v \in W^{1,\infty}\big(0,T; L^2(0,1)\big), \ v(1,\cdot) \in W^{1,\infty}(0,T),$$

$$u_r, v_r \in L^\infty\big(0,T; L^2(0,1)\big),$$

in the sense that $t \mapsto \big(u(t), v(t), v(1,t)\big)$ is a strong solution of Cauchy problem (6.1.6)-(6.1.7) with $\xi_0 = v_0(1)$ as adapted to the special case (6.2.1)-(6.2.5).

REMARK 6.2.1 Clearly, the assumptions on u_0 and v_0 in Proposition 6.2.1 say nothing else but that

$$z_0 := \big(u_0, v_0, v_0(1)\big) \in D(A_4) =$$
$$= \big\{(p,q,c) \in H^1(0,1)^2 \times \mathbb{R} \mid c = q(1) \in D(\beta), -q(0) = R_0 p(0)\big\}.$$

∎

In order to obtain higher regularity results we can use the classical basic idea of formally differentiating our problem with respect to t, establishing some regularity for the new problem, and then returning to the original problem to derive higher regularity for its solution. Unfortunately, this is not a trivial task in the nonlinear case. To do that, let us first denote

$$w_0 := \big(f_1(0), f_2(0), e(0)\big) - A_4\big(u_0, v_0, v_0(1)\big) =$$
$$= \big(f_1(0), f_2(0), e(0)\big) - \big(v_0' + R u_0, u_0' + G v_0, -u_0(1) + \beta v_0(1)\big).$$

Here β is assumed to be single-valued.

THEOREM 6.2.1

Let $T > 0$ be fixed and assume that the following conditions are satisfied:
$\beta\colon D(\beta) = \mathbb{R} \to \mathbb{R}$ *is single valued;*

$$\beta \in W^{2,\infty}_{loc}(\mathbb{R}) \ and \ \beta' \geq 0 \ in \ \mathbb{R}; \tag{6.2.7}$$

$$f_1, f_2 \in W^{2,\infty}(0,T;L^2(0,1)), \ e \in W^{2,\infty}(0,T); \tag{6.2.8}$$

$$z_0, w_0 \in D(A_4). \tag{6.2.9}$$

Then the strong solution (u,v) of problem (6.2.1)-(6.2.5) satisfies

$$u,v \in W^{2,\infty}(0,T;L^2(0,1)) \cap W^{1,\infty}(0,T;H^1(0,1)), \ v(1,\cdot) \in W^{2,\infty}(0,T).$$

PROOF   By Proposition 6.2.1 we know that problem (6.2.1)-(6.2.5) has a unique strong solution (u,v); that means that $z(t) = \big(u(t),v(t),v(1,t)\big)$ is the strong solution of the Cauchy problem (6.1.6)-(6.1.7), i.e.,

$$z'(t) + A_4 z(t) = F(t), \ t \in (0,T), \ z(0) = z_0, \tag{6.2.10}$$

where

$$F(t) = \big(f_1(t), f_2(t), e(t)\big).$$

Denote by A_4' the operator A_4 in the case $\beta = 0$. Clearly, A_4' is a linear maximal monotone operator, with

$$D(A_4') = D(A_4) = \{(p,q,c) \in H^1(0,1)^2 \times \mathbb{R} \mid c = q(1), -q(0) = R_0 p(0)\}.$$

Consider the operators $B(t)\colon D\big(B(t)\big) = H_4 \mapsto H_4$, defined by

$$B(t)(p,q,c) = b(t)(0,0,c) - F'(t),$$

where $b(t) := \beta'\big(v(1,t)\big)$. Obviously, the operators $E(t) := A_4' + B(t)$, $0 \leq t \leq T$, with $D\big(E(t)\big) = D(A_4)$, are all maximal monotone (see Theorem 1.2.7). Moreover, since $b \in W^{1,\infty}(0,T)$ and $F \in W^{2,\infty}(0,T;H_4)$, there exists a positive constant L_0 such that

$$\|E(t)x - E(s)x\|_{H_4} \leq L_0|t-s|\big(1 + \|x\|_{H_4}\big),$$

for all $x \in D(A_4)$ and all $s,t \in [0,T]$. Therefore, according to Theorem 1.5.7, the following Cauchy problem

$$w'(t) + E(t)w(t) = 0, \ t \in (0,T), \ w(0) = w_0, \tag{6.2.11}$$

has a unique strong solution $w \in W^{1,\infty}(0,T;H_4)$ such that $w(t) \in D(A_4)$, for all $t \in [0,T]$. So, $w(t)$ has the form $w(t) = \big(\tilde{u}(t), \tilde{v}(t), \tilde{v}(1,t)\big)$, $t \in [0,T]$.

If we denote by $\{S(t)\colon H_4 \to H_4 \mid t \geq 0\}$ the contraction semigroup generated by $-A_4'$, then the respective solutions of problems (6.2.10), (6.2.11) satisfy

$$z(t) = S(t)z_0 + \int_0^t S(t-s)F_1(s)\,ds \ \text{ for all } t \in [0,T], \quad (6.2.12)$$

$$w(t) = S(t)w_0 + \int_0^t S(t-s)F_2(s)\,ds \ \text{ for all } t \in [0,T], \quad (6.2.13)$$

where

$$F_1(t) := F(t) - \Big(0,0,\beta\big(v(1,t)\big)\Big),$$

$$F_2(t) := F'(t) - \big(0,0,b(t)\tilde{v}(1,t)\big).$$

Since $z_0 \in D(A_4') = D(A_4)$ and $F_1 \in W^{1,\infty}(0,T;H_4)$, we can differentiate (6.2.12) (see Section 1.4) to obtain

$$z'(t) = S(t)w_0 + \int_0^t S(t-s)F_1'(s)\,ds \ \text{ for all } t \in [0,T]. \quad (6.2.14)$$

Now, by (6.2.13) and (6.2.14) one has

$$\|w(t) - z'(t)\|_{H_4} \leq \int_0^t \big\|F_2(s) - F_1'(s)\big\|_{H_4}\,ds \leq$$

$$\leq C \int_0^t \|w(s) - z'(s)\|_{H_4}\,ds \ \text{ for all } t \in [0,T], \quad (6.2.15)$$

where

$$C := \|b\|_{C[0,T]} = \sup\Big\{\big|\beta'(v(1,t))\big| \ \Big|\ t \in [0,T]\Big\}.$$

Gronwall's inequality applied to (6.2.15) yields $w(t) = z'(t)$ for all $t \in [0,T]$. So $z \in W^{2,\infty}(0,T;H_4)$, i.e., $u,v \in W^{2,\infty}(0,T;L^2(0,1))$, $v(1,\cdot) \in W^{2,\infty}(0,T)$. Notice that $w = z'$ satisfies

$$w'(t) + A_4'w(t) = F_3(t), \ \ t \in (0,T), \quad (6.2.16)$$

$$w(0) = w_0, \quad (6.2.17)$$

where

$$F_3(t) := -B(t)w(t) = F'(t) - \Big(0,0,\beta'\big(v(1,t)\big)v_t(1,t)\Big).$$

Since $w_0 \in D(A_4') = D(A_4)$ and $F_3 \in W^{1,\infty}(0,T;H_4)$, we can derive by Proposition 6.2.1 that $\tilde{u}_r, \tilde{v}_r \in L^\infty(0,T;L^2(0,1))$. Therefore, $u,v \in W^{1,\infty}(0,T;H^1(0,1))$. ∎

REMARK 6.2.2   If, in addition to the assumptions of Theorem 6.2.1, we suppose that $f_{1r}, f_{2r} \in L^2(Q_T)$, $Q_T = (0,1) \times (0,T)$, then $u, v \in H^2(Q_T)$. To see that, it suffices to use the system (6.2.1)-(6.2.2). Moreover, we can reach as much regularity for (u,v) as we want, provided the data are sufficiently smooth and satisfy suitable compatibility conditions.   ∎

REMARK 6.2.3   By using a similar technique we can investigate the case in which the boundary conditions are both of differential type. The basic framework for this case will be the product space $L^2(0,1)^2 \times \mathbb{R}^2$. On the other hand, the theory of the last two chapters applies partly to some variants of problems (5.0.1)-(5.0.4) and (6.0.1)-(6.0.5) corresponding to the cases $r \in \mathbb{R}_+ = [0,\infty)$ and $r \in \mathbb{R}$ (instead of $r \in [0,1]$); see [Moro1, p. 309]. ∎

References

[CooKr]   K.L. Cooke & D.W. Krumme, Differential-difference equations and nonlinear initial boundary value problems for linear hyperbolic partial differential equations, *J. Math. Anal. Appl.*, 24 (1968), 372–387.

[Moro1]   Gh. Moroşanu, *Nonlinear Evolution Equations and Applications*, Editura Academiei and Reidel, Bucharest and Dordrecht, 1988.

Chapter 7

The Fourier method for abstract differential equations

In this chapter we shall consider abstract differential equations of first and second order. They are generalizations of parabolic and hyperbolic partial differential equations, respectively. We shall show that the well-known method of Fourier can be extended to some abstract cases. Most of our results on existence and regularity of the solution are not new, but our point is to prove them by the extended Fourier method. Of course, the representation of the solution in the form of Fourier series is a great advantage.

Let $T > 0$ be fixed and let H be a real separable Hilbert space with the inner product $(\cdot, \cdot)_H$, which induces the norm $\| \cdot \|_H$, $\|u\|_H^2 = (u, u)_H$. We shall consider the following abstract differential equations

$$y'(t) + p(t)By(t) = f(t), \ 0 < t < T, \tag{7.0.1}$$

$$y''(t) + q(t)y'(t) + r(t)By(t) = g(t), \ 0 < t < T, \tag{7.0.2}$$

where $B \colon D(B) \subset H \to H$ is a linear operator satisfying the hypotheses (B.1)-(B.4) of Section 1.3 and $p, q, r \colon [0, T] \to [0, \infty)$, $f, g \colon [0, T] \to H$ are given functions.

In order to apply a more general notion of solution for these equations, we replace the operator B by its *energetic extension* B_E, which is the duality mapping of the *energetic space* $H_E \subset H$ (see Section 1.3 and, for further details, [Zeidler]).

7.1 First order linear equations

In this section we give some results, based on the Fourier method, for the Cauchy problem consisting of the first order equation (7.0.1) and the initial condition $y(0) = y_0$. First we show that the extended Cauchy problem, which is associated with the energetic extension B_E of B, has a solution. Then we assume more on B, y_0, and f, and obtain further regularity on the solution. Also existence and regularity for solutions of the periodic problem will be investigated. Finally, an example will be given. For further details, see [MoGeG].

THEOREM 7.1.1
Assume (B.1)-(B.4) of Section 1.3 and let $p \in L^\infty(0,T)$ with $p(t) \geq p_0 > 0$ for a.a. $t \in (0,T)$. If $y_0 \in H$ and $f \in L^2(0,T;H_E^*)$, then there exists a unique y satisfying

$$y \in L^2(0,T;H_E) \cap C([0,T];H) \cap H^2(0,T;H_E^*), \tag{7.1.1}$$

$$y'(t) + p(t)B_E\, y(t) = f(t) \ \text{ for a.a. } t \in (0,T), \tag{7.1.2}$$

$$y(0) = y_0. \tag{7.1.3}$$

PROOF   Let $e_1, e_2, \ldots$ and $\lambda_1, \lambda_2 \ldots$ be given by 1.3.1. We write

$$y_0 = \sum_{n=1}^\infty y_{0n} e_n \ \text{ and } \ f(t) = \sum_{n=1}^\infty f_n(t) e_n \ \text{ for all } t \in [0,T]. \tag{7.1.4}$$

Since $y_0 \in H$ and $f \in L^2(0,T;H_E^*)$, we have, for a.a. $t \in (0,T)$,

$$\|y_0\|_H^2 = \sum_{n=1}^\infty \lambda_n^{-1} y_{0n}^2, \ \ \|f(t)\|_{H_E^*}^2 = \|B_E^{-1} f(t)\|_{H_E}^2 = \sum_{n=1}^\infty \lambda_n^{-2} f_n^2(t),$$

$$y_{0n} = \lambda_n(y_0, e_n)_H, \ \ f_n(t) = \lambda_n\big(B_E^{-1} f(t), e_n\big)_{H_E}.$$

We are seeking a solution in the form

$$y(t) = \sum_{n=1}^\infty b_n(t) e_n, \ t \in [0,T]. \tag{7.1.5}$$

Formally, the coefficients $b_n(t)$ satisfy the following Cauchy problems, for $n \in \mathbb{N}^*$,

$$b_n'(t) + \lambda_n p(t) b_n(t) = f_n(t) \ \text{ for a.a. } t \in (0,T), \tag{7.1.6}$$

$$b_n(0) = y_{0n}. \tag{7.1.7}$$

The problems (7.1.6)-(7.1.7) have unique solutions $b_1, b_2, \ldots \in H^1(0,T)$. We multiply (7.1.6) by $b_n(t)$ and integrate over $[0,t]$. Then by $p(t) \geq p_0$ and (7.1.7),

$$\frac{1}{2\lambda_n} b_n^2(t) + p_0 \int_0^t b_n^2(s)\, ds \leq \frac{1}{2\lambda_n} y_{0n}^2 + \frac{1}{\lambda_n} \int_0^t |f_n(s)| \cdot |b_n(s)|\, ds \leq$$

$$\leq \frac{1}{2\lambda_n} y_{0n}^2 + \frac{\epsilon}{2} \int_0^t b_n^2(s)\, ds + \frac{1}{2\epsilon} \int_0^t \lambda_n^{-2} f_n^2(s)\, ds \tag{7.1.8}$$

for all $t \in [0, t]$ and $\epsilon > 0$. Choosing $\epsilon = p_0$ we arrive at

$$p_0 \frac{1}{\lambda_n} b_n^2(t) + p_0^2 \int_0^t b_n^2(s)\, ds \leq p_0 \frac{1}{\lambda_n} y_{0n}^2 + \int_0^t \lambda_n^{-2} f_n(s)^2\, ds \qquad (7.1.9)$$

for all $\in (0, T]$. Since $\int_0^T b_n^2(s)\, ds$ are bounded by terms of convergent series, the series in (7.1.5) converges in $L^2(0, T; H_E)$ toward some $y \in L^2(0, T; H_E)$. Using again (7.1.9) we conclude by the Weierstrass M-test that the series $\sum_n \lambda^{-1} b_n(t)$ converges uniformly on $[0, T]$. Hence $y \in C([0, T]; H)$. Now, we observe from (7.1.6) that the series $\sum_n \lambda_n^{-2} b_n'(t)^2$ converges in $L^1(0, T)$. This implies that the series $\sum_n b_n'(t) e_n$ converges in $L^2(0, T; H_E^*)$ toward some $y^* \in L^2(0, T; H_E^*)$. Let $\xi \in C_0^\infty(0, T)$. Then in H_E^*, as $N \to \infty$,

$$\int_0^T \xi(t) y^*(t)\, dt \leftarrow \int_0^T \xi(t) \sum_{n=1}^N b_n'(t) e_n\, dt =$$

$$= -\int_0^T \xi'(t) \sum_{n=1}^N b_n(t) e_n\, dt \to -\int_0^T \xi'(t) y(t)\, dt.$$

Thus, $y \in H^1(0, T; H_E^*)$ and $y^* = y'$. Since $B_E: H_E \to H_E^*$ is the duality mapping, then

$$\limsup_{N \to \infty} \left\| \sum_{n=1}^N \lambda_n p\, b_n e_n - p\, B_E y \right\|_{L^2(0,T;H_E^*)} \leq$$

$$\leq \limsup_{N \to \infty} \|p\|_{L^\infty(0,T)} \left\| \sum_{n=1}^N b_n e_n - y \right\|_{L^2(0,T;H_E)} = 0$$

and

$$\left\| \sum_{n=1}^N f_n e_n - f \right\|_{L^2(0,T;H_E^*)} = \left\| \sum_{n=1}^N f_n B_E^{-1} e_n - B_E^{-1} f \right\|_{L^2(0,T;H_E)} =$$

$$= \left\| \sum_{n=1}^N (B_E^{-1} f(t), e_n)_{H_E} e_n - B_E^{-1} f \right\|_{L^2(0,T;H_E)} \to 0,$$

as $N \to \infty$. Thus y, given by (7.1.5)-(7.1.7), is the desired solution.

It remains to prove the uniqueness of the solution. Let y_1 and y_2 be two solutions of our Cauchy problem. We denote $\tilde{y} = y_1 - y_2$. Then by the linearity of B_E,

$$\tilde{y}'(t) + p(t) B_E \tilde{y}(t) = 0 \text{ for a.a. } t \in (0, T), \text{ and } \tilde{y}(0) = 0. \qquad (7.1.10)$$

By Theorem 1.1.3, the mapping $t \mapsto \left\|\tilde{y}(t)\right\|_H^2$ is differentiable and

$$\frac{1}{2}\frac{d}{dt}\left\|\tilde{y}(t)\right\|_H^2 = \langle \tilde{y}'(t), \tilde{y}(t)\rangle = -p(t)\left\|\hat{y}(t)\right\|_{H_E}^2 \le 0 \tag{7.1.11}$$

for a.a. $t \in (0,T)$. Here $\langle \cdot, \cdot \rangle$ is the pairing between H_E and its dual H_E^*. Since $\hat{y}(0) = 0$, then $\hat{y}(t) \equiv 0$.

Theorem 7.1.1 is proved.   ∎

REMARK 7.1.1   By (7.1.9) for all $t \in (0,T]$,

$$p_0\|y(t)\|_H^2 + p_0^2 \int_0^t \|y(s)\|_{H_E}^2\, ds + p_1^2 \int_0^t \|y'(s)\|_{H_E^*}^2\, ds \le \tag{7.1.12}$$

$$\le 2p_0\|y_0\|_H^2 + 2\int_0^t \|f(s)\|_{H_E^*}^2\, ds, \text{ where } p_1^2 = \frac{p_0^2}{2\left(p_0^2 + \|p\|_{L^\infty(0,T)}^2\right)},$$

which implies the continuous dependence of the solution on y_0 and f.   ∎

THEOREM 7.1.2
Let $f \in L^2(0,T;H)$ and assume the conditions of Theorem 7.1.1 on B and p. If $y_0 \in H$ and y is given by Theorem 7.1.1, then $y \in C\bigl((0,T];H_E\bigr)$. Moreover, if $y_0 \in H_E$, then $y \in C\bigl([0,T];H_E\bigr)$.

PROOF   Let $t \in [0,T]$. From (7.1.6)-(7.1.7) we obtain

$$b_n(t) = y_{0n}\mathrm{e}^{-\lambda_n \int_0^t p(s)\,ds} + \int_0^t \mathrm{e}^{-\lambda_n \int_0^s p(\tau)\,d\tau} f_n(s)\, ds, \tag{7.1.13}$$

from which, by Hölder's inequality,

$$b_n(t)^2 \le 2\left(y_{0n}\mathrm{e}^{-\lambda_n p_0 t}\right)^2 + 2\left(\int_0^t \mathrm{e}^{-\lambda_n p_0(t-s)}|f_n(s)|\, ds\right)^2 \le$$

$$\le 2\left(y_{0n}\right)^2 \mathrm{e}^{-2\lambda_n p_0 t} + \frac{1}{\lambda_n p_0}\int_0^t f_n(s)^2\, ds. \tag{7.1.14}$$

Let $\delta \in (0,T]$ be fixed. Since $\mathrm{e}^{-x} \le x^{-1}$ for any $x > 0$, then

$$b_n(t)^2 \le \frac{y_{0n}^2}{\lambda_n p_0 \delta} + \frac{1}{\lambda_n p_0}\int_0^t f_n(s)^2\, ds \text{ for all } t \in (\delta, T]. \tag{7.1.15}$$

Since $y_0 \in H$, then $y \in C\bigl([\delta,T];H_E\bigr)$, for all $\delta \in (0,T)$. Hence $y \in C\bigl((0,T];H_E\bigr)$. If $y_0 \in H_E$, we obtain by (7.1.14) that $y \in C\bigl([0,T];H_E\bigr)$.

Theorem 7.1.2 is proved.   ∎

The next theorem is a regularity result. We shall use the Hilbert spaces $V_k \subset H$ introduced in (1.3.14), Section 1.3.

THEOREM 7.1.3

Let $k \in \mathbb{N}$, $f \in L^2(0, T; V_{k-1})$, and assume all the conditions of Theorem 7.1.1 on B and p. If $y_0 \in V_k$, then the solution y of (7.1.2)-(7.1.3) satisfies

$$y \in L^2(0, T; V_{k+1}) \cap C([0, T]; V_k) \cap H^1(0, T; V_{k-1}), \qquad (7.1.16)$$

$$p_0 \|y(t)\|_k^2 + p_0^2 \|y\|_{L^2(0,t;V_{k+1})}^2 + p_1^2 \|y'\|_{L^2(0,t;V_{k-1})}^2 \leq \qquad (7.1.17)$$

$$\leq 2p_0 \|y_0\|_k^2 + 2\|f\|_{L^2(0,t;V_{k-1})}^2 \quad \text{for all } t \in (0, T].$$

Moreover, if $y_0 \in V_{k-1}$ and $k > 0$, then $y \in C((0, T]; V_k)$.

PROOF   Let $y_0 \in V_k$. Multiplying (7.1.9) by λ_n^{k-2} we see that $y \in L^2(0, T; V_{k-1})$. Now, we multiply (7.1.8) by λ_n^k. We see that the series $\sum_n \lambda_n^{k-1} b_n(t)^2$ converges uniformly on $[0, T]$, whence $y \in C([0, T]; V_k)$. By (7.1.6), $y' \in L^2(0, T; V_{k-1})$. We also obtain (7.1.17).

Let $y_0 \in V_{k-1}$ and $k > 0$. We multiply (7.1.15) by λ_n^{k-1}. Then clearly $y \in C([\delta, T]; V_k)$ for any $\delta \in (0, T)$. Hence $y \in C((0, T]; V_k)$. Theorem 7.1.3 is proved.   ∎

Next we are seeking *periodic solutions* for (7.1.2). The following result holds.

THEOREM 7.1.4

Let $f \in L^2(0, T; H)$ and assume all the conditions of Theorem 7.1.1 on B and p. Then there exists a unique y, satisfying (7.1.2), $y(0) = y(T)$, and

$$y \in L^2(0, T; H_E) \cap C([0, T]; H) \cap H^1(0, T; H_E^*).$$

Moreover, if $f \in L^2(0, T; V_{k-1})$ for some $k \in \mathbb{N}^*$, then

$$y \in L^2(0, T; V_{k+1}) \cap C([0, T]; V_k) \cap H^1(0, T; V_{k-1}).$$

PROOF   We associate with equations (7.1.6) the periodic condition $b_n(0) = b_n(T)$. By an elementary calculation, we see that they are satisfied by $b_n \in H^1(0, T)$, given by

$$b_n(t) = D_n e^{-\lambda_n \int_0^t p(s)\, ds} + \int_0^t e^{-\lambda_n \int_s^t p(\tau)\, d\tau} f_n(s)\, ds,$$

$$D_n = \left(1 - e^{-\lambda_n \int_0^T p(s)\, ds}\right)^{-1} \int_0^T e^{-\lambda_n \int_s^T p(\tau)\, d\tau} f_n(s)\, ds.$$

Since $0 < p_0 \le p(t) \le \|p\|_{L^\infty(0,T)}$, there exists a constant $C_1 > 0$ such that

$$D_n^2 \le \left(\frac{\int_0^T e^{-\lambda_n p_0 (T-s)} |f_n(s)|\, ds}{1 - e^{-\lambda_1 p_0 T}} \right)^2 \le \frac{C_1}{\lambda_n} \int_0^T f_n(s)^2\, ds,$$

where again Hölder's inequality has been applied. Hence there is a constant $C_2 > 0$ such that

$$b_n(t)^2 \le \frac{C_2}{\lambda_n} \int_0^T f_n(s)^2\, ds, \ \text{ for all } t \in T, \ n \in \mathbb{N}^*. \tag{7.1.18}$$

Consequently the series in (7.1.5) converges in H, uniformly on $[0,T]$. Thus its limit $y \in C([0,T]; H)$. By Theorem 7.1.1, it also belongs to $L^2(0,T; H_E)$ with $y' \in L^2(0,T; H_E^*)$ and it satisfies (7.1.2). By the periodicity of functions b_n, it follows that $y(0) = y(T)$.

Let $f \in L^2(0,T; V_{k-1})$. By (7.1.18), the series in (7.1.5) converges uniformly in V_k on $[0,T]$. Thus its limit $y \in C([0,T]; V_k)$. Theorem 7.1.3 yields $y \in L^2(0,T; V_{k+1})$.

Theorem 7.1.4 is proved. ∎

Example 7.1.1

The following simple example, involving the heat equation, will clarify the difference between the extended problem and the original one. Let Ω be a nonempty bounded open set in $\mathbb{R}^N$ with a smooth boundary, $c, \kappa > 0$ constants and $Q_T = (0,T) \times \Omega$, where $N \in \mathbb{N}^*$. We choose H to be $L^2(\Omega)$ with the ordinary inner product and $D(B) = C_0^\infty(\Omega)$, $Bu = -\Delta u$, where Δu is the Laplacian of u. Clearly, B satisfies our hypotheses (B.1)-(B.3). The energetic space is now $H_E = H_0^1(\Omega)$, and $B_E \colon H_E \to H_E^*$, the energetic extension of B, is given by

$$B_E(u)(v) = \int_\Omega \nabla u(x) \cdot \nabla v(x)\, dx \ \text{ for all } u, v \in H_0^1(\Omega), \tag{7.1.19}$$

while the Friedrichs extension A of B is given by $D(A) = H_0^1(\Omega) \cap H^2(\Omega)$ and $Au = B_E u$. Let $p(t) \equiv \kappa/c$. Now, the problem $y' + pBy = f$ reads

$$cy_t(t,x) - \kappa \Delta y(x,t) = cf(t,x), \ (t,x) \in Q_T, \tag{7.1.20}$$

with the Dirchlet boundary condition

$$y(t,x) = 0, \ t \in (0,T), \ x \in \partial\Omega, \tag{7.1.21}$$

and the initial conditions

$$y(0,x) = y_0(x), \ x \in \Omega. \tag{7.1.22}$$

The system (7.1.20)-(7.1.21) describes heat conduction in Ω; $y(t, x)$ represents the temperature at point x at time moment t. Let us assume, for example, that $y_0 \in L^2(\Omega)$ and $f \in L^2(Q_T)$. Then there exists a unique solution y of (7.1.20)-(7.1.22) with $(-\Delta)_E$ instead of $-\Delta$, which satisfies

$$y \in C\big([0, T]; L^2(\Omega)\big) \cap L^2\big(0, T; H_0^1(\Omega)\big) \cap C\big((0, T]; H_0^1(\Omega)\big)$$

with $y' = y_t \in L^2\big(0, T; H^{-1}(\Omega)\big)$. One should also notice that y is the *generalized solution of (7.1.20) in Sobolev's sense:*

$$\int_{Q_T} y(t, x)\big(-c\phi_t(t, x) - \kappa \Delta \phi(t, x)\big)\, dx\, dt = \int_{Q_T} cf(t, x)\phi(t, x)\, dx\, dt$$

for each $\phi \in C_0^\infty(Q_T)$. If $f \in C_0^\infty(Q_T)$, $y_0 \in C_0^\infty(\Omega)$, and $\partial\Omega$ is sufficiently smooth, then y is a classical solution of $y' + pBy = f$ with $p(t) \equiv \kappa/c$, $D(B) = C_0^\infty(\Omega)$, and $B = -\Delta$. Hence the extended equation (7.1.2) is a natural generalization of the original problem (7.0.1). $\square$

7.2 Semilinear first order equations

Let us next consider the equation

$$y'(t) + p(t)By(t) = f(y)(t), \ t \in (0, T), \tag{7.2.1}$$

where $p\colon [0, T] \to [0, \infty)$ and B satisfies the hypotheses (B.1)-(B.4) in Section 1.3 and $y \mapsto f(y)$ is locally Lipschitzian and dominated by y' and By. That is why $f(y)$ is said to be a *perturbation.* In order to be more precise, we define for each $t \in (0, T]$ and $k \in \mathbb{N}$ a Banach space and its norm:

$$Y_{k,t}^1 = L^2(0, t; V_{k+1}) \cap H^1(0, t; V_{k-1}), \tag{7.2.2}$$

$$\|y\|_{Y_{k,t}^1} = \left(\int_0^t \big(p_0^2 \|y(s)\|_{k+1}^2 + p_1^2 \|y'(s)\|_{k-1}^2\big)\, ds \right)^{1/2}. \tag{7.2.3}$$

Let $k \in \mathbb{N}$, $l = 0, 1, \ldots, k$, $K \geq 0$, $y_0 \in V_{k-1}$, $\eta \in L^1(0, T)$ and denote $Z_M = \{y \in Y_{k,T}^1 \mid y(0) = y_0,\ \|y\|_{Y_T^1} \leq M\}$, for each $M > 0$. The following condition will be used throughout this section.

$(\mathbf{H}_f)$ The function f maps $Y_{k,T}^1$ into $L^2(0, T; V_{k-1})$. Moreover, for each $M > 0$, there exists $\eta_M \in L^1(0, T)$ such that for a.a. $t \in (0, T)$, and for each $\alpha, \beta \in Z_M$ and $\gamma \in Y_{k,T}^1$,

$$\big\|f(\alpha)(t) - f(\beta)(t)\big\|_{k-1}^2 \leq \eta_M(t)\|\alpha - \beta\|_{Y_{k,t}^1}^2 +$$

$$+K\left(p_0^2\left\|\alpha(t)-\beta(t)\right\|_{l+1}^2+p_1^2\left\|\alpha'(t)-\beta'(t)\right\|_{l-1}^2\right),$$

$$\left\|f(\gamma)(t)\right\|_{k-1}^2\le\eta(t)\left(\left\|\gamma\right\|_{Y_{k,t}^1}^2+1\right)+$$

$$+K\left(p_0^2\left\|\gamma(t)\right\|_{l+1}^2+p_1^2\left\|\gamma'(t)\right\|_{l-1}^2\right).$$

THEOREM 7.2.1
Assume (H_f) and all the conditions of Theorem 7.1.1 on B and p. Let λ_1 be the lowest eigenvalue of A. If $y_0\in V_k$ and $2K\lambda_1^{l-k}<1$, then there exists a unique $y\in Y_{k,T}^1\cap C\big([0,T];V_k\big)$ such that

$$y'(t)+p(t)B_E\,y(t)=f(y)(t)\ \ \text{for a.a. }t\in(0,T),\qquad(7.2.4)$$

$$y(0)=y_0.\qquad(7.2.5)$$

PROOF   Let $N>0$, $t\in(0,T]$, and $\xi\in L^1(0,T)$ be positive. We denote $\sigma=2K\lambda_1^{l-k}<1$ and define an equivalent norm $\|\cdot\|_{\xi,t}$ to $\|\cdot\|_{Y_{k,t}^1}$ by

$$\left\|u\right\|_{\xi,t}^2=\sup_{0\le s\le t}\ e^{-E_\xi(s)}\left\|u\right\|_{Y_{k,s}^1}^2,\ \ \text{where }E_\xi(t)=\frac{4}{1-\sigma}\int_0^t\xi(s)\,ds.\qquad(7.2.6)$$

Let $\tilde Z_N=\big\{y\in Y_{k,T}^1\mid y(0)=y_0,\ \|y\|_{\eta,T}\le N\big\}$ and $\alpha,\beta\in\tilde Z_N$. By Theorem 7.1.3 there exist unique $y_\alpha,y_\beta\in\tilde Z_N\cap C\big([0,T];V_k\big)$ such that

$$y_\alpha(0)=y_0\ \text{and}\ y_\alpha'(t)+p(t)B_H y_\alpha(t)=f(\alpha)(t),\ \text{a.e. on }(0,T),\quad(7.2.7)$$

$$y_\beta(0)=y_0\ \text{and}\ y_\beta'(t)+p(t)B_H y_\beta(t)=f(\beta)(t),\ \text{a.e. on }(0,T).\quad(7.2.8)$$

Since $\|u\|_l^2\le\lambda_1^{l-k}\|u\|_k^2$ for all $k\in\mathbb{N}$, $l=0,1,\dots,k$ and $u\in V_k$, we have

$$\left\|y_\alpha\right\|_{Y_{l,t}^1}^2\le\lambda_1^{l-k}\left\|y_\alpha\right\|_{Y_{k,t}^1}^2\ \ \text{for all }t\in(0,T].\qquad(7.2.9)$$

By (H_f) and (7.1.17), for all $t\in(0,T]$,

$$\left\|y_\alpha\right\|_{Y_{k,t}^1}^2\le 2p_0\left\|y_0\right\|_k^2+2K\left\|\alpha\right\|_{Y_{l,t}^1}^2+2\int_0^t\eta(s)\left(\left\|\alpha\right\|_{Y_{k,s}^1}^2+1\right)ds\le$$

$$\le 2p_0\left\|y_0\right\|_k^2+2\|\eta\|_{L^1(0,T)}+2K\lambda_1^{l-k}\left\|\alpha\right\|_{Y_{k,t}^1}+$$

$$+\frac{1}{2}(1-\sigma)\int_0^t E_\eta'(s)e^{E_\eta(s)}\left\|\alpha\right\|_{\eta,s}^2\,ds\le 2p_0\left\|y_0\right\|_k^2+$$

$$+2\|\eta\|_{L^1(0,T)}+\sigma\left\|\alpha\right\|_{Y_{k,t}^1}+\frac{1}{2}(1-\sigma)\left(e^{E_\eta(t)}-1\right)\left\|\alpha\right\|_{\eta,T}^2.$$

Now we see that

$$\|y_\alpha\|_{\eta,T}^2\le 2p_0\|y_0\|_k^2+2\|\eta\|_{L^1(0,T)}+\frac{1}{2}(1+\sigma)\|\alpha\|_{\eta,T}^2.\qquad(7.2.10)$$

We choose N large enough. Then $y_\alpha \in \tilde{Z}_N$, so $\alpha \mapsto y_\alpha$ maps the (nonempty) complete metric space $\tilde{Z}_N$ into itself. Next we choose $M = \mathrm{e}^{\,E_\eta(T)/2} N$ and use in $\tilde{Z}_N$ an equivalent norm $\|\cdot\|_{\eta_M,T}$. By (H_f) and (7.1.17), we have

$$\left\|y_\alpha - y_\beta\right\|^2_{Y^1_{k,t}} \le \int_0^t 2\eta_M(s)\|\alpha - \beta\|^2_{Y^1_{k,s}}\, ds + 2K\lambda_1^{l-k}\|\alpha - \beta\|^2_{Y^1_{k,t}} \le$$

$$\le \frac{1}{2}(1-\sigma)\left(\mathrm{e}^{\,E_{\eta_M}(t)} - 1\right)\|\alpha - \beta\|^2_{\eta_M,t} + \sigma\|\alpha - \beta\|^2_{Y^1_{k,t}}$$

for all $t \in (0,T]$. Thus

$$\left\|y_\alpha - y_\beta\right\|^2_{\eta_M,T} \le \frac{1}{2}(1+\sigma)\|\alpha - \beta\|^2_{\eta_M,T}. \tag{7.2.11}$$

Now, $\alpha \mapsto y_\alpha$ is a strict contraction. By Banach's fixed point theorem, there exists a unique $\alpha \in \tilde{Z}_N$ such that $\alpha = y_\alpha$. Thus we have found a solution $y \in \tilde{Z}_N$ for (7.2.4)-(7.2.5). It is unique in $\tilde{Z}_N$ but maybe not in the whole $Y^1_{k,T}$. So, let $\tilde{y} \in Y^1_{k,T}$ be another solution of (7.2.4)-(7.2.5). We choose in (7.2.11) $M = \max\left\{\|y\|_{Y^1_{k,T}}, \|\tilde{y}\|_{Y^1_{k,T}}\right\}$. Then $y = \tilde{y}$, since

$$\|y - \tilde{y}\|^2_{\eta_M,T} \le \frac{1}{2}(1+\sigma)\|y - \tilde{y}\|^2_{\eta_M,T}. \tag{7.2.12}$$

Theorem 7.2.1 is proved.   ∎

Example 7.2.1

We modify Example 7.1.1. The continuity equation of energy reads

$$cy_t(x,t) + \nabla \cdot \mathbf{q}(x,t) = 0 \ \text{ for a.a. } (x,t) \in Q_T, \tag{7.2.13}$$

where $\mathbf{q}(t,x)$ is the density of heat flow. Instead of the ordinary Fourier's law for the heat flow density $\mathbf{q}(x,t) = -\kappa\nabla y(x,t)$ we assume that

$$\mathbf{q}(x,t) = -\kappa\nabla y(x,t) + \int_0^t L\big(\nabla y(x,s), s\big)\, ds, \tag{7.2.14}$$

where $L\colon \mathbb{R}^N \times [0,T] \to \mathbb{R}^N$ is uniformly Lipschitzian with respect to its first variable and measurable with respect to its second variable. Hence $\mathbf{q}(x,t)$ depends on the whole history of the temperature gradient from 0 to t. A more general theory on heat conduction can be found in [GurPip]. We define $f\colon L^2(0,T;H_E) \to L^2(0,T;H_E^*)$ by

$$\langle v, f(u)(t)\rangle = \frac{1}{c}\int_0^t \int_\Omega L\big(\nabla u(x,s), s\big) \cdot \nabla v(x)\, dx\, ds \tag{7.2.15}$$

for all $v \in H_E$. Let $y_0 \in L^2(\Omega)$. Then all conditions of Theorem 7.2.1 are satisfied with $k = l = K = 0$. So the problem (7.2.13)-(7.2.14), (7.1.21)-(7.1.22) has in Sobolev's sense a unique generalized solution

$$y \in C\big([0,T]; L^2(\Omega)\big) \cap L^2\big(0,T; H_0^1(\Omega)\big) \cap H^1\big(0,T; H^{-1}(\Omega)\big).$$

$\square$

7.3   Second order linear equations

In this section we shall consider briefly the equation (7.0.2) with initial values. We state directly the following theorem.

THEOREM 7.3.1
Assume (B.1)-(B.4) and that $q \in L^2(0,T)$, $r \in W^{1,1}(0,T)$ and there exists $r_0 > 0$ such that $r(t) \geq r_0$, for all $t \in [0,T]$. If $y_0 \in H_E$, $y_1 \in H$, and $g \in L^2(0,T;H)$, then there exists a unique y such that

$$y \in C\big([0,T]; H_E\big) \cap C^1\big([0,T]; H\big) \cap H^2(0,T; H_E^*), \qquad (7.3.1)$$
$$y''(t) + q(t)y'(t) + r(t)B_E\, y(t) = g(t) \ \text{ for a.a. } t \in (0,T), \quad (7.3.2)$$
$$y(0) = y_0 \ \text{ and } \ y'(0) = y_1. \qquad (7.3.3)$$

PROOF   We are again seeking the solution in the series form $y(t) = \sum_n b_n(t)e_n$ and so we begin with the Cauchy problems

$$b_n''(t) + q(t)b_n'(t) + \lambda_n r(t)b_n(t) = g_n(t) \ \text{ for a.a. } t \in (0,T), \quad (7.3.4)$$
$$b_n(0) = y_{0n}, \ b_n'(0) = y_{1n}, \qquad (7.3.5)$$

where $n \in \mathbb{N}^*$ and

$$y_{0n} = (y_0, e_n)_{H_E} = \lambda_n(y_0, e_n)_H, \ y_{1n} = \lambda_n(y_1, e_n)_H, \qquad (7.3.6)$$
$$g_n(t) = \lambda_n(g(t), e_n)_H. \qquad (7.3.7)$$

By classical theory, these Cauchy problems have solutions $b_n \in H^2(0,T)$. We multiply (7.3.4) by $b_n'(t)$ and integrate over $[0,t]$. Then by (7.3.5)

$$\frac{1}{2}b_n'(t)^2 + \lambda_n \int_0^t r(s)b_n(s)b_n'(s)\, ds \leq \frac{1}{2}y_{1n}^2 +$$
$$+ \int_0^t \big(g_n(s)b_n'(s) - q(s)b_n'(s)^2\big)\, ds \ \text{ for all } t \in [0,T].$$

We integrate the second term by parts and use $r \geq r_0$. Then

$$b_n'(t)^2 + r_0 \lambda_n b_n(t)^2 \leq r(0)\lambda_n y_{0n}^2 + y_{1n}^2 + \int_0^t |g_n(s)|\,|b_n'(s)|\,ds +$$

$$+ \int_0^t \left(\frac{1}{r_0}|r'(s)| + 1 + |q(s)| \right)\left(r_0 \lambda_n b_n(s)^2 + b_n'(s)^2 \right) ds$$

for all $t \in [0,T]$. Using the Gronwall type inequality (Lemma 1.5.2), and $q, r' \in L^1(0,T)$, we conclude that there exists a constant $C_3 > 0$ such that

$$b_n'(t)^2 + r_0 \lambda_n b_n(t)^2 \leq C_3 \left(y_{1n}^2 + \lambda_n y_{0n}^2 + \left(\int_0^t |g_n(s)|\,ds \right)^2 \right) \qquad (7.3.8)$$

for each $t \in [0,T]$ and $n \in \mathbb{N}^*$. Combining (7.3.8) and (7.3.4) we obtain that there is a constant $C_4 > 0$ such that

$$\frac{b_n''(t)^2}{\lambda_n^2} \leq C_4 \left(1 + q(t)^2 \right) \left(y_{0n}^2 + \frac{y_{1n}^2}{\lambda_n} + \frac{1}{\lambda_n} \int_0^t |g_n(s)|^2\,ds \right) + \frac{g_n(t)^2}{\lambda_n^2} \qquad (7.3.9)$$

for a.a. $t \in (0,T)$ and each $n \in \mathbb{N}^*$. As in the proof of Theorem 7.1.1, the estimates (7.3.8) and (7.3.9) imply the existence of y, satisfying (7.3.1)-(7.3.3).

Let y and $\hat{y}$ be two solutions of (7.3.1)-(7.3.3) and denote $z = y - \hat{y}$. Then

$$z''(t) + q(t)z'(t) + r(t)B_E z(t) = 0 \text{ for a.a. } t \in (0,T), \qquad (7.3.10)$$

with $z(0) = 0$ and $z'(0) = 0$. Multiplying (7.3.10) by $z'(t)$ and integrating over $[0,t]$ yields, on account of the symmetry of B,

$$\|z'(t)\|_H^2 + r(t)\big(Bz(t), z(t)\big)_H =$$

$$= \int_0^t r'(s)\big(Bz(s), z(s)\big)_H\,ds + \int_0^t q(s)\|z'(s)\|_H^2\,ds \leq$$

$$\leq \int_0^t \left(\frac{|r'(s)|}{r_0} + |q(s)| \right)\left(\|z'(s)\|_H^2 + \big(Bz(s), z(s)\big)_H \right) ds$$

for all $t \in [0,T]$. By Gronwall's inequality and $r \geq r_0$, it follows that $\big(Bz(t), z(t)\big)_H \equiv 0$, whence $z(t) \equiv 0$, since B is coercive. Thus the solution is unique.

Theorem 7.3.1 is proved. $\blacksquare$

REMARK 7.3.1   Assume the conditions of Theorem 7.3.1. Then there exists a constant $C_5 > 0$ such that the solution y of (7.3.1)-(7.3.3) satisfies

$$\|y(t)\|_{H_E}^2 + \|y'(t)\|_H^2 \leq C_5 \left(\|y_0\|_{H_E}^2 + \|y_1\|_H^2 + \int_0^t \|g(s)\|_H^2\,ds \right) \qquad (7.3.11)$$

for all $t \in [0, T]$. Indeed, the Fourier coefficients of $z_n - u_n$ satisfy (7.3.8) with zero initial values, whence (7.3.11) follows. ∎

We can relax the assumptions on the initial values and $g(t)$, but then we must extend the notion of solution for (7.3.2), as in the case of first order equations; see, e.g., [Brézis1, p. 64]. Let $y_0 \in H$, $y_1 \in H_E^*$, and $g \in L^2(0, T; H_E^*)$. We call $z \in C\big([0, T]; H\big) \cap C^1\big([0, T]; H_E^*\big)$ a *weak solution* of (7.3.2), if there exist sequences $(\tilde{g}_n)$ of functions of $L^2(0, T; H)$ and (z_n) of functions $z_n \in C\big([0, T]; H_E\big) \cap C^1\big([0, T]; H\big) \cap H^2(0, T; H_E^*)$ such that

$$z_n''(t) + q(t)z_n'(t) + r(t)B_E z_n(t) = \tilde{g}_n(t) \text{ for a.a. } t \in (0, T), \quad (7.3.12)$$

$$\|z_n - z\|_{C\big([0,T];H\big)} + \|z_n' - z'\|_{C\big([0,T];H_E^*\big)} \to 0, \quad (7.3.13)$$

$$\text{and } \|g - \tilde{g}_n\|_{L^1(0,T;H_E^*)} \to 0, \text{ as } n \to \infty. \quad (7.3.14)$$

THEOREM 7.3.2

Assume the conditions of Theorem 7.3.1. If $y_0 \in H$, $y_1 \in H_E^*$, and $g \in L^2(0, T; H_E^*)$, then there exists a unique $y \in C\big([0, T]; H\big) \cap C^1\big([0, T]; H_E^*\big)$ that is a weak solution of (7.3.2) and satisfies $y(0) = y_0$ and $y'(0) = y_1$.

PROOF   We define functions $z_n \in H^2(0, T; H_E)$ and $\tilde{g}_n \in L^2(0, T; H_E)$ by

$$z_n(t) = \sum_{j=1}^{n} b_j(t)e_j, \qquad \tilde{g}_n(t) = \sum_{j=1}^{n} g_j(t)e_j, \quad (7.3.15)$$

$$g_n(t) = \lambda_n \big(B_E^{-1}g(t), e_n\big)_{H_E}, \quad (7.3.16)$$

$$y_{0n} = \lambda_n \big(y_0, e_n\big)_H, \text{ and } y_{1n} = \lambda_n \big(B_E^{-1}y_1, e_n\big)_{H_E} \quad (7.3.17)$$

such that (7.3.4) and (7.3.5) are satisfied. Then we still have the estimate (7.3.8). Hence there exists $z \in C\big([0, T]; H\big) \cap C^1\big([0, T]; H_E^*\big)$, which satisfies (7.3.13). Moreover, (7.3.12) is satisfied. Since $y_0 \in H$, $y_1 \in H_E^*$, and $g \in L^2(0, T; H_E^*)$, then z_n converges toward a weak solution of (7.3.2), because $\tilde{g}_n$ converges toward g even in $L^2(0, T; H_E^*)$. Indeed, by the Parseval equality

$$\big\|\tilde{g}_n(t) - g(t)\big\|_{H_E^*}^2 = \bigg\| \sum_{j=1}^{n} g_j(t)B_E^{-1}e_j - B_E^{-1}g(t) \bigg\|_{H_E}^2 \leq$$

$$\leq \sum_{j=n}^{\infty} (B_E^{-1}g(t), e_j)_{H_E}^2 \begin{cases} \leq \big\|B_E^{-1}g(t)\big\|_{H_E}^2, \\ \to 0, \text{ as } n \to \infty, \end{cases}$$

which allows the use of the Lebesgue Dominated Convergence Theorem. Let u be another weak solution satisfying $u(0) = y_0$ and $u'(0) = y_1$. So there are

(u_n, f_n) satisfying (7.3.1)-(7.3.3) instead of (y, g). Thus all $(z_n - u_n, \tilde{g}_n - f_n)$ satisfy (7.3.1)-(7.3.3) with zero initial values. Thus by the estimate (7.3.8),

$$r_0 \|z_n(t) - u_n(t)\|_H^2 + \|z_n'(t) - u_n'(t)\|_{H_E^*}^2 \leq$$

$$\leq C_5 \int_0^t \|\tilde{g}_n(s) - f_n(s)\|_{H_E^*}^2 \, ds \text{ for all } t \in [0, T].$$

As $n \to \infty$, we obtain $z(t) \equiv u(t)$. Hence the weak solution is unique. Theorem 7.3.2 is proved. ∎

THEOREM 7.3.3

Assume the conditions of Theorem 7.3.1. If $k \in \mathbb{N}$, $y_0 \in V_{k+1}$, $y_1 \in V_k$, and $g \in L^2(0, T; V_k)$, then there exists a constant C_6, independent of y_0, y_1, and g such that

$$y \in C\big([0, T]; V_{k+1}\big) \cap C^1\big([0, T]; V_k\big) \cap H^2(0, T; V_{k-1}), \quad (7.3.18)$$

$$\|y(t)\|_{k+1}^2 + \|y'(t)\|_k^2 + \|y''\|_{L^2(0,t;V_{k-1})}^2 \leq$$

$$\leq C_6 \Big(\|y_0\|_{k+1}^2 + \|y_1\|_k^2 + \|g\|_{L^2(0,t;V_{k-1})}^2 \Big) \quad (7.3.19)$$

for all $t \in [0, T]$, where y is the solution of (7.3.2)-(7.3.3).

PROOF   We repeat the proof of Theorem 7.3.1 or 7.3.2 with the following modification:

$$y_{0n} = \lambda_n^{1-k}(y_0, e_n)_{k+1}, \; y_{1n} = (y_1, e_n)_k \text{ and } g_n(t) = (g(t), e_n)_k. \quad (7.3.20)$$

Then the results follow from (7.3.8)-(7.3.9). ∎

Next we state a theorem on the existence of periodic solutions for (7.3.2). For the sake of simplicity we assume $q(t)$ to vanish and $r(t)$ to be a constant. Note that the resonance phenomenon will be excluded by an extra condition on the eigenvalues of A.

THEOREM 7.3.4

Assume that B satisfies (B.1)-(B.4) in Section 1.3 and, in addition, $r_0 > 0$, $\delta < 1$, $g \in L^2(0, T; H)$, and $\cos \sqrt{r_0 \lambda_n} T \leq \delta$ for all $n \in \mathbb{N}^*$. Then there exist a unique $y: [0, T] \to H_E$ such that

$$y \in C\big([0, T]; H_E\big) \cap C^1\big([0, T]; H\big) \cap H^2(0, T; H_E^*), \quad (7.3.21)$$

$$y''(t) + r_0 B_E \, y(t) = g(t) \text{ for a.a. } t \in (0, T), \quad (7.3.22)$$

$$y(0) = y(T) \text{ and } y'(0) = y'(T). \quad (7.3.23)$$

PROOF   We are seeking the solution as $y(t) = \sum_n b_n(t)e_n$, where the functions $b_n \colon [0,T] \to \mathbb{R}$ satisfy

$$b''(t) + r_0\lambda_n b_n(t) = g_n(t) \text{ for a.a. } t \in (0,T), \tag{7.3.24}$$

$$b_n(0) = b_n(T), \ b_n'(0) = b_n'(T) \text{ and } g_n(t) = \lambda_n\big(g(t), e_n\big)_H. \tag{7.3.25}$$

An elementary calculation by the method of variation of constants gives

$$b_n(t) = D_n \cos\omega_n t + E_n \sin\omega_n t + \frac{1}{\omega_n}\int_0^t g_n(s)\sin\omega_n(t-s)\,ds, \tag{7.3.26}$$

where $\omega_n = \sqrt{r_0\lambda_n}$ and

$$\begin{pmatrix} D_n \\ E_n \end{pmatrix} = \frac{1}{2\omega_n(1-\cos\omega_n T)} \begin{pmatrix} 1-\cos\omega_n T & \sin\omega_n T \\ -\sin\omega_n T & 1-\cos\omega_n T \end{pmatrix} \times$$

$$\times \int_0^T \begin{pmatrix} \sin\omega_n(T-s) \\ \cos\omega_n(T-s) \end{pmatrix} g_n(s)\,ds.$$

Thus there exists a constant $C_7 > 0$ such that

$$b_n'(t)^2 + \lambda_n b_n(t)^2 \leq C_7 \int_0^T g_n(s)^2\,ds \text{ for all } t \in [0,T]. \tag{7.3.27}$$

Hence $\sum_n b_n e_n$ converges in $C\big([0,T]; H_E\big)$ and in $C^1\big([0,T]; H\big)$ toward a function y. By Theorem 7.3.1, $y \in H^2(0,T; H_E^*)$.

Let $z \in H^2(0,T; H_E^*) \cap C\big([0,T]; H_E\big)$ be a solution of (7.3.22)-(7.3.23). Then $\big(z(t), e_n\big)_H$ satisfies (7.3.24)-(7.3.25). Since their solutions are unique in $H^2(0,T)$, $\big(z(t), e_n\big)_H = b_n(t)$ for each $n \in \mathbb{N}^*$. Hence $z = y$.

Theorem 7.3.4 is proved.   ∎

REMARK 7.3.2   If $q(t)$ and $r(t)$ are constants, an explicit formula for the solutions of (7.3.4) can be calculated, cf. [MorSb].   ∎

Example 7.3.1

We consider an application in acoustics (see [HMJLB] for details). Let Ω be a nonempty open bounded set in $\mathbb{R}^N$, $N \in \mathbb{N}^*$, with a sufficiently smooth boundary $\partial\Omega$. Let $Q_T = \Omega \times (0,T)$, $c > 0$, and S be a nonvoid open subset of $\partial\Omega$. If we consider the air pressure $p = p_0 + p_1$, where p_0 is a constant and p_1 describes small vibrations, we arrive at the wave equation

$$\frac{\partial^2 p_1}{\partial t^2}(x,t) - c^2\Delta p_1(x,t) = 0, \ (x,t) \in Q_T. \tag{7.3.28}$$

Some part S of the boundary is assumed to be occupied by a sound source where $p_1(x, t) = f(t)$ is given. On the rest of the boundary the zero Neumann boundary condition is satisfied. Let $g(x, t) = -f''(t)$ and consider the wave equation

$$y_{tt}(x, t) - c^2 \Delta y(x, t) = g(x, t), \quad (x, t) \in Q_T, \qquad (7.3.29)$$

subject to the boundary conditions

$$y(x, t) = 0, \quad (x, t) \in S \times (0, T), \qquad (7.3.30)$$

$$\frac{\partial y}{\partial \nu}(x, t) = 0, \quad (x, t) \in (\partial\Omega \setminus S) \times (0, T), \qquad (7.3.31)$$

and the initial conditions

$$y(0, x) = y_0(x) \text{ and } y_t(0, x) = y_1(x), \quad x \in \Omega, \qquad (7.3.32)$$

where $y_0 \in H_E$, $y_1 \in L^2(\Omega)$, and $\frac{\partial}{\partial \nu}$ denotes the outward normal derivative. Clearly, $y + f$ is the solution of (7.3.28) subject to the original initial and boundary conditions.

Let $H = L^2(\Omega)$, $D(B) = \{u \in C_0^\infty(\Omega) \mid u = 0 \text{ on } S\}$, and $Bu = -\Delta u$. The energetic space H_E is now $H_E = \{u \in H_0^1(\Omega) \mid u = 0 \text{ on } S\}$ and the Friedrichs extension A of B is given by $D(A) = H_E \cap H^2(\Omega)$ and $Au = -\Delta u$. Let $g \in L^2(Q_T)$. (7.3.29)-(7.3.31) is a special case of (7.0.2) with $q(t) \equiv 0$ and $r(t) \equiv c^2$. The solutions of

$$y''(t) + c^2 B_E\, y(t) = g(t), \quad t \in (0, T), \qquad (7.3.33)$$

found in the theorems of this section, are generalized solutions of (7.3.29)-(7.3.31) in the sense of Sobolev. Thus (7.3.2) is a natural extension of (7.0.2).

If $g(x, t) = d \sin \nu t$, where $d \in \mathbb{R}$ and $\nu \approx \sqrt{\lambda_n}$ for some $n \in \mathbb{N}^*$, then a resonance phenomenon occurs: the n-th term in $\sum_j b_j(t) e_j$ is very large. Indeed,

$$b_n(t) = y_{0n} \cos c\sqrt{\lambda_n}\, t + \frac{1}{c\sqrt{\lambda_n}}\left(y_{1n} + \frac{\lambda_n (d, e_n)_H \nu}{c\lambda_n - c\nu^2}\right) \sin c\sqrt{\lambda_n}\, t +$$

$$+ \frac{\lambda_n(d, e_n)_H}{c^2\lambda_n - c^2\nu^2} \sin c\nu t \approx \frac{2\lambda_n(d, e_n)_H}{c^2\lambda_n - c^2\nu^2} \cos \frac{c(\sqrt{\lambda_n} + \nu)t}{2} \sin \frac{c(\nu - \sqrt{\lambda_n})t}{2}.$$

This formula allows one to develop a method to measure the frequencies of proper oscillation of sound in the room Ω, by tuning the frequency of sound in the source S. ☐

7.4   Semilinear second order equations

Let us next consider the equation

$$y''(t) + q(t)y'(t) + r(t)By(t) = g(y)(t), \ t \in (0,T), \tag{7.4.1}$$

where $q, r: [0,T] \to \mathbb{R}$, B satisfies (B.1)-(B.4), $y \mapsto g(y)$ is locally Lipschitzian and dominated by y'' and By. In order to be more precise, we define for each $k = 0,1,2\ldots$ and $t \in (0,T]$ a Banach space $Y_{k,t}^2$ by:

$$Y_{k,t}^2 = H^1(0,t;V_{k+1}) \cap H^2(0,t;V_{k-1}), \tag{7.4.2}$$

$$\|y\|_{Y_{k,t}^2}^2 = \|y(0)\|_{k+1}^2 + \|y'\|_{L^2(0,t;V_{k+1})}^2 + \|y''\|_{L^2(0,t;V_{k-1})}^2. \tag{7.4.3}$$

Let $\eta \in L^1(0,T)$, $k \in \mathbb{N}$, $l = 0,1,\ldots,k$, $y_0 \in V_{k+1}$, $K \geq 0$ and denote $Z_M^2 = \{y \in Y_{k,T}^2 \mid y(0) = y_0, \ \|y\|_{Y_{k,T}^2} \leq M\}$, for all $M > 0$. Now we state the following hypothesis on the perturbation $g(y)$:

(H_g) The function g maps $Y_{k,T}^2$ into $L^2(0,T;V_k)$. Whenever $M > 0$, there exists $\eta_M \in L^1(0,T)$ such that for a.a. $t \in (0,T)$, for each $\alpha, \beta \in Z_M^2$ and $\gamma \in Y_{k,T}^2$,

$$\|g(\alpha)(t) - g(\beta)(t)\|_k^2 \leq \eta_M(t)\|\alpha - \beta\|_{Y_{k,t}^2} +$$

$$+K\left(\|\alpha'(t) - \beta'(t)\|_{l+1}^2 + \|\alpha''(t) - \beta''(t)\|_{l-1}^2\right),$$

$$\|g(\gamma)(t)\|_k^2 \leq \eta(t)\left(\|\gamma\|_{Y_{k,t}^2}^2 + 1\right) + K\left(\|\gamma'(t)\|_{l+1}^2 + \|\gamma''(t)\|_{l-1}^2\right).$$

THEOREM 7.4.1

Let $k \in \mathbb{N}$ and assume (H_g) and all the conditions of Theorem 7.3.1 on B, q, and r. If $y_0 \in V_{k+1}$ and $y_1 \in V_k$, and C_6, given by Theorem 7.3.3, and λ_1, the lowest eigenvalue of A, satisfy $C_6 K \lambda_1^{l-k} < 1$, then there exists a unique $y \in Y_{k,T}^2 \cap C^1\big([0,T];V_k\big)$ such that

$$y''(t) + q(t)y'(t) + r(t)B_E\, y(t) = g(y)(t) \ \text{ for a.a. } t \in (0,T), \tag{7.4.4}$$

$$y(0) = y_0 \ \text{and} \ y'(0) = y_1. \tag{7.4.5}$$

PROOF   Let $N > 0$, $t \in (0,T]$, and $\xi \in L^1(0,T)$ be positive. We denote $\sigma = C_6 K \lambda_1^{l-k} < 1$ and define an equivalent norm $\|\cdot\|_{\xi,t}$ to $\|\cdot\|_{Y_{k,t}^2}$ by:

$$\|u\|_{\xi,t}^2 = \sup_{0 \leq s \leq t} e^{-E_\xi(s)}\|u\|_{Y_{k,s}^1}^2, \ \text{ where } E_\xi(t) = \frac{2C_6}{1-\sigma}\int_0^t \xi(s)\, ds. \tag{7.4.6}$$

Let $\tilde{Z}_N = \{y \in Y^2_{k,T} \mid y(0) = y_0, \ \|y\|_{\eta,T} \leq N\}$ and $\alpha, \beta \in \tilde{Z}_N$. By Theorem 7.3.3 there exist unique $y_\alpha, y_\beta \in \tilde{Z}_N \cap C^1\big([0,T]; V_k\big)$ such that

$$y''_\alpha(t) + q(t)y'(t) + r(t)B_H y_\alpha(t) = f(\alpha)(t) \ \text{ for a.a. } t \in (0,T), \quad (7.4.7)$$

$$y''_\beta(t) + q(t)y'(t) + r(t)B_H y_\beta(t) = f(\beta)(t) \ \text{ for a.a. } t \in (0,T), \quad (7.4.8)$$

$$y_\alpha(0) = y_0 \ \text{ and } \ y_\beta(0) = y_0. \quad (7.4.9)$$

The rest of the proof is similar to that of Theorem 7.2.1, except that $Y^1_{k,t}$ is replaced by $Y^2_{k,t}$. ∎

References

[Brézis1] H. Brézis, *Opérateurs maximaux monotones et semi-groupes de contractions dans les espaces de Hilbert*, North-Holland, Amsterdam, 1973.

[GurPip] M.E. Gurtin & A.C. Pipkin, A general theory of heat conduction with finite wave speed, *Arch. Rat. Mech. Anal.*, 31, 113–126.

[HMJLB] V.-M. Hokkanen, Gh. Moroşanu, P. Joulain, V. Locket & G. Bourceanu, An application of the Fourier method in acoustics, *Adv. Math. Sc. Appl.*, 7 (1997), no. 1, 69–77.

[MoGeG] Gh. Moroşanu, P. Georgescu & V. Grădinaru, The Fourier method for abstract differential equations and applications, *Comm. Appl. Anal.*, 3 (1999), no. 2, 173–188.

[MorSb] Gh. Moroşanu & S. Sburlan, Multiple orthogonal sequence method and applications, *An. Şt. Univ. "Ovidius" Constanţa*, 1 (1994), 188–200.

[Zeidler] E. Zeidler, *Applied Functional Analysis*, Appl. Math. Sci., 108, Springer-Verlag, 1995.

Chapter 8

The semigroup approach for abstract differential equations

In this chapter we investigate the existence and regularity of solutions for abstract semilinear first order differential equations using the notion of linear semigroup, whose definition and some basic properties are recalled in Section 1.4. The most important tool will be the variation of constants formula

$$y(t) = S(t)x + \int_0^t S(t-s)f(s)\,ds, \ t > 0,$$

which gives the solution of the nonhomogeneous initial value problem

$$y'(t) + Ay(t) = f(t), \ t > 0, \ y(0) = x, \tag{8.0.1}$$

in terms of the solution of the corresponding homogeneous problem.

The semigroup approach can be applied to a wide range of semilinear problems. As a special application we discuss here the regularity question for a class of linear hyperbolic partial differential systems with nonlinear boundary conditions. More precisely, we shall rewrite them in the abstract form (8.1.1) (see below) in a suitable Banach space X, where A is an unbounded linear operator in X, and $f \colon C\big([0,T];X\big) \to L^1(0,T;X)$.

8.1 Semilinear first order equations

Let X be a Banach space, $A \colon D(A) \subset X \to X$, $T > 0$ and f a mapping from $C\big([0,T];X\big)$ into $L^1(0,T;X)$, and consider the equation

$$u'(t) + Au(t) = f(u)(t), \ t \in (0,T). \tag{8.1.1}$$

A function $u \in W^{1,1}(0,T;X)$ is called a *strong solution* of (8.1.1), if $u(t) \in D(A)$ and $f(u)(t) - u'(t) = Au(t)$ a.e. on $(0,T)$. By a *weak solution* of (8.1.1) we mean a function $u \in C\big([0,T];X\big)$ such that there exist

$f_n\colon C\big([0,T];X\big) \to L^1(0,T;X)$ and strong solutions $u_n \in W^{1,1}(0,T;X)$ of

$$u_n'(t) + Au_n(t) \ni f_n(u_n)(t) \ \text{ for a.a. } t \in (0,T),\ n \in \mathbb{N}^*, \qquad (8.1.2)$$

satisfying

$$\lim_{n\to\infty} \|u_n - u\|_{C([0,T];X)} = \lim_{n\to\infty} \|f_n(u_n) - f(u)\|_{L^1(0,T;X)} = 0. \qquad (8.1.3)$$

Existence and uniqueness of solutions

We shall begin with the *existence* of weak and strong solutions for (8.1.1). We require the following two conditions:

($\mathbf{H}_A$) The operator A is linear and $-A$ generates a C_0-semigroup of bounded linear operators on X, denoted by $\{S(t)\colon X \to X \mid t \geq 0\}$.

($\mathbf{H}_f$) There exist u_0 in X and η in $L^1(0,T)$ with the following properties. If $\alpha \in Y_0 = \{\gamma \in C\big([0,T];X\big) \mid \gamma(0) = u_0\}$, then $f(\alpha) \in L^1(0,T;X)$ and

$$\|f(\alpha)(t)\|_X \leq \eta(t)\big(1 + \|\alpha\|_{L^\infty(0,t;X)}\big) \ \text{ for a.a. } t \in (0,T). \qquad (8.1.4)$$

For each $K > 0$, there exists $\eta_K \in L^1(0,T)$ such that

$$\|f(\alpha)(t) - f(\tilde\alpha)(t)\|_X \leq \eta_K(t)\|\alpha - \tilde\alpha\|_{L^\infty(0,t;X)} \qquad (8.1.5)$$

for a.a. $t \in (0,T)$, whenever $\alpha, \tilde\alpha \in \{\gamma \in Y_0 \mid \|\gamma\|_{L^\infty(0,T;X)} \leq K\}$.

THEOREM 8.1.1
Assume (H_A) and (H_f). Then problem (8.1.1) has a unique weak solution $u \in C\big([0,T];X\big)$ satisfying $u(0) = u_0$.

PROOF   By Theorem 1.4.2, $\overline{D(A)} = X$. Thus there are $u_{0n} \in D(A)$, $n \in \mathbb{N}^*$, converging in X toward u_0. Denote by χ_B the characteristic function of any set B, let $\alpha \in Y_0$, and consider the problems

$$u_n'(t) + Au_n(t) = f_n(\alpha)(t) \ \text{ for a.a. } t \in (0,T),\ u_n(0) = u_{0n}, \qquad (8.1.6)$$

$$f_n(\alpha)(t) = \int_{\mathbf{R}} \rho_{1/n}(t - s)\chi_{[0,T]}(s)f(\alpha)(s)\,ds, \qquad (8.1.7)$$

where $\rho_{1/n} \in C_0^\infty(\mathbf{R})$ is the usual mollifier satisfying $\int_{\mathbf{R}} \rho_{1/n}(s)\,ds = 1$, $\rho_{1/n} \geq 0$, and $\operatorname{supp}\rho_{1/n} \subset [-1/n, 1/n]$. Since $f_n(\alpha) \in C^\infty\big([0,T];X\big)$, then by Theorem 1.4.2, (8.1.6) has a unique strong solution $u_n \in W^{1,1}(0,T;X)$, with

$$u_n(t) = S(t)u_{0n} + \int_0^t S(t - s)f_n(\alpha)(s)\,ds \ \text{ for all } \in [0,T]. \qquad (8.1.8)$$

By Theorem 1.4.1, there are constants $M \geq 1$, $\omega \geq 0$ satisfying

$$\|S(t)\|_{L(X;X)} \leq M\mathrm{e}^{\omega t} \text{ for all } t \geq 0. \tag{8.1.9}$$

By (8.1.8) and (8.1.9), for each $t \in [0, T]$ and $m, n \in \mathbb{N}^*$,

$$\|u_n(t) - u_m(t)\|_X \leq M\mathrm{e}^{\omega T}\Big(\|u_{0m} - u_{0n}\|_X + \|f_n(\alpha) - f_m(\alpha)\|_{L^1(0,T;X)}\Big).$$

Since $f_n(\alpha) \to f(\alpha)$ in $L^1(0, T; X)$ as $n \to \infty$, then (u_n) converges toward some u_α in $C([0, T]; X)$. Hence, the equation

$$u'_\alpha(t) + Au_\alpha(t) = f(\alpha)(t), \ t \in (0, T), \ u_\alpha(0) = u_0, \tag{8.1.10}$$

has a weak solution $u_\alpha \in C([0, T]; X)$ given by

$$u_\alpha(t) = S(t)u_0 + \int_0^t S(t - s)f(\alpha)(s)\, ds \text{ for all } t \in [0, T]. \tag{8.1.11}$$

Let $\tilde{u}_\alpha \in C([0, T]; X)$ be another weak solution of (8.1.10). Then there are $\tilde{f}_n(\alpha) \in L^1(0, T; X)$ and strong solutions $\tilde{u}_n$ of (8.1.10) with $\tilde{f}_n(\alpha)$ instead of $f(\alpha)$ such that $\tilde{f}_n(\alpha) \to f(\alpha)$ in $L^1(0, T; X)$ and $\tilde{u}_n \to \tilde{u}_\alpha$ in $C([0, T]; X)$, as $n \to \infty$. By (8.1.8) and (8.1.9), for all $t \in [0, T]$,

$$\|u_\alpha(t) - \tilde{u}_\alpha(t)\|_X \leq \limsup_{n \to \infty} M\mathrm{e}^{\omega T}\|f_n(\alpha) - \tilde{f}_n(\alpha)\|_{L^1(0,T;X)} = 0.$$

Thus u_α is unique. So $P\colon Y_0 \to Y_0$, $P\alpha = u_\alpha$, is well defined.

Let $\xi \in L^1(0, T)$ be a positive function. Define, for each $\alpha \in L^\infty(0, T; X)$,

$$\|\alpha\|_{\xi,T} = \operatorname*{ess\,sup}_{0 \leq s \leq T} \mathrm{e}^{-\int_0^T \xi(\sigma)\, d\sigma}\|\alpha(s)\|_X. \tag{8.1.12}$$

Using (8.1.9), (8.1.11), and $\xi = M\mathrm{e}^{\omega T}\eta$ we obtain, for any $\alpha \in C([0, T]; X)$,

$$\|P\alpha\|_{\xi,T} \leq \left(1 - \mathrm{e}^{-\int_0^T \xi(\sigma)\, d\sigma}\right)\|\alpha\|_{\xi,T} +$$

$$+\mathrm{e}^{-\int_0^T \xi(\sigma)\, d\sigma} M\mathrm{e}^{\omega T}\Big(\|u_0\|_X + \|\eta\|_{L^1(0,T)}\Big).$$

Denote $Z_Q = \{\gamma \in Y_0 \mid \|\gamma\|_{\xi,T} \leq Q\}$. We choose $Q > M\mathrm{e}^{\omega T}(\|u_0\|_X + \|\eta\|_{L^1(0,T)})$. Then $P\colon Z_Q \to Z_Q$. Let $K = Q\exp\int_0^T \xi(\sigma)\, d\sigma$. So $K\|\gamma\|_{\xi,T} \geq Q\|\gamma\|_{L^\infty(0,T;X)}$. We use in Z_Q the metric induced by $\|\cdot\|_{\xi_Q,T}$, where $\xi_Q = M\mathrm{e}^{\omega T}\eta_K$. Now, Z_Q is a complete metric space. Using (8.1.5), (8.1.9), and (8.1.11) we can see that

$$\|P\alpha - P\beta\|_{\xi_Q,T} \leq \left(1 - \mathrm{e}^{-\int_0^T \xi_Q(\sigma)\, d\sigma}\right)\|\alpha - \beta\|_{\xi_Q,T} \tag{8.1.13}$$

for all $\alpha, \beta \in Z_Q$. Thus P is a strict contraction. By Banach's fixed point theorem it follows that P has a unique fixed point $u \in Z_Q$, which is the desired weak solution of (8.1.1). Indeed, if there were another weak solution $\tilde{u}$ of (8.1.1), we could choose Q above so big that $\tilde{u} \in Z_Q$.

Theorem 8.1.1 is proved. ∎

THEOREM 8.1.2

Assume (H_A), (H_f), $R(f) \subset W^{1,1}(0,T;X)$, and $u_0 \in D(A)$. Then, there is a unique strong solution u in $W^{1,1}(0,T;X)$ of (8.1.1) satisfying $u(0) = u_0$.

PROOF　　Let $\alpha \in Y_0$. By Theorem 1.4.2, the problem (8.1.10) has a unique strong solution $u_\alpha \in W^{1,1}(0,T;X)$, satisfying (8.1.11). Hence, the reasoning in the proof of Theorem 8.1.1 yields the existence of a unique strong solution of (8.1.1). ∎

Regularity of solutions

We proceed next to the study of *regularity* of solutions to (8.1.1). We shall need additional hypotheses, namely:

(H_f') There exist u_0 in $D(A)$ and η in $L^1(0,T)$ with the following properties. If

$$Y_1 = \{\gamma \in C^1([0,T];X) \mid \gamma(0) = u_0, \ \gamma'(0) = f(u_0)(0) - Au_0\},$$

then for each $\alpha \in Y_1$, (8.1.4) is satisfied and $f(\alpha) \in W^{1,1}(0,T;X)$. In addition, for each $K, Q > K_1 := \|u_0\|_X + \|f(u_0)(0) - Au_0\|_X$, there are $\eta_K, \tilde{\eta}_Q \in L^1(0,1)$ such that, for a.a. $t \in (0,T)$ and for each $\alpha, \beta, \gamma \in Y_1$,

$$\|f(\gamma)'(t)\|_X \le \eta_K(t)\big(1 + \|\gamma'\|_{L^\infty(0,t;X)}\big), \tag{8.1.14}$$

$$\sum_{i=0}^{1} \|f(\alpha)^{(i)}(t) - f(\beta)^{(i)}(t)\|_X \le \tilde{\eta}_L(t)\|\alpha - \beta\|_{W^{1,\infty}(0,t;X)}, \tag{8.1.15}$$

whenever $\|\gamma\|_{L^\infty(0,T;X)} \le K$, $\|\alpha\|_{W^{1,\infty}(0,T;X)} \le Q$, $\|\beta\|_{W^{1,\infty}(0,T;X)} \le Q$.

(H_f'') There exist u_0 in $D(A)$ and η in $L^1(0,1)$ such that $f(u_0)'(0)$ exists and $f(u_0)(0) - Au_0 \in D(A)$. Moreover, if

$$Y_2 = \Big\{\gamma \in C^2([0,T];X) \cap Y_1 \ \Big|$$

$$\gamma''(0) = f(u_0)'(0) - A\big(f(u_0)(0) - Au_0\big)\Big\},$$

then for each $\alpha \in Y_2$, $f(\alpha) \in W^{2,1}(0,T;X)$ and (8.1.4) is satisfied. Also, for each $K > K_1$ there is $\eta_K \in L^1(0,T)$ satisfying (8.1.14), whenever $\gamma \in Y_2$ and $\|\gamma\|_{L^\infty(0,T;X)} \leq K$. Finally, for each $Q, N > K_1$ there are $\tilde{\eta}_Q, \hat{\eta}_N \in L^1(0,T)$ such that for a.a. $t \in (0,T)$,

$$\|f(\gamma)''(t)\|_X \leq \tilde{\eta}_Q(t)\big(1 + \|\gamma''\|_{L^\infty(0,T;X)}\big) \text{ for all } \gamma \in Y_2, \quad (8.1.16)$$

$$\sum_{i=0}^{2} \|f(\alpha)^{(i)}(t) - f(\beta)^{(i)}(t)\|_X \leq \hat{\eta}_N(t)\|\alpha - \beta\|_{W^{2,\infty}(0,t;X)}, \quad (8.1.17)$$

whenever $\|\gamma'\|_{L^\infty(0,t;X)} \leq Q$ and

$$\alpha, \beta \in \big\{\gamma \in Y_2 \,\big|\, \|\gamma\|_{W^{2,\infty}(0,T;X)} \leq N\big\}.$$

THEOREM 8.1.3

Assume (H_A) and (H'_f). Then (8.1.1) together with the initial condition $u(0) = u_0$ has a unique (classical) solution $u \in C^1([0,T];X)$ such that $Au \in C([0,T];X)$.

PROOF   Let $\alpha \in Y_1$, $n \in \mathbb{N}^*$, and u_n be the strong solution of (8.1.6) with $u_n(0) = u_0$. Then

$$u_n(t) = S(t)u_0 + \int_0^t S(t-s)f_n(\alpha)(s)\,ds \text{ for all } t \in [0,T]. \quad (8.1.18)$$

Since $f'_n \in C([0,T];X)$ and $u_0 \in D(A)$, we can differentiate (8.1.18). Thus

$$u'_n(t) = S(t)\big(f(\alpha)(0) - Au_0\big) + \int_0^t S(s)f_n(\alpha)'(t-s)\,ds \quad (8.1.19)$$

for all $t \in [0,T]$. Thus $u_n \in C^1([0,T];X)$. Since $f_n(\alpha)$ converges to $f(\alpha)$ in $W^{1,1}(0,T;X)$ and (8.1.9) holds, there is $u_\alpha \in C^1([0,T];X)$, satisfying

$$u_n \to u_\alpha \text{ in } C^1([0,T];X), \text{ as } n \to \infty, \quad (8.1.20)$$

$$u_\alpha(t) = S(t)u_0 + \int_0^t S(t-s)f(\alpha)(s)\,ds, \quad (8.1.21)$$

$$u'_\alpha(t) = S(t)\big(f(\alpha)(0) - Au_0\big) + \int_0^t S(t-s)f(\alpha)'(s)\,ds \quad (8.1.22)$$

for all $t \in [0,T]$.

Since A is closed, (8.1.20) implies that u_α is a classical solution of (8.1.10). Since the weak solution of (8.1.10) is unique, so is the classical one. Thus the mapping $P: Y_1 \to Y_1, P\alpha = u_\alpha$, is well defined. (Note that by (8.1.5)

and by the continuity of $f(\alpha)$ and α', it follows that $f(\alpha)(0) = f(u_0)(0)$.)
Let $\xi \in L^1(0,T)$ be positive and, for any $\beta \in Y_1$, set

$$\|\beta\|'_{\xi,T} = \operatorname*{ess\,sup}_{0 \le s \le T} e^{-\int_0^s \xi(\sigma)\,d\sigma}\left(\|\beta(s)\|_X + \|\beta'(s)\|_X\right). \tag{8.1.23}$$

A calculation using (8.1.4), (8.1.14), (8.1.20), and (8.1.21) reveals that we can choose first K and then N such that $P\colon \hat{Z}_N \to \hat{Z}_N$, where $\hat{Z}_N = \{\gamma \in Y_1 \mid \|\gamma\|_{\xi,T} \le Q\}$ with $\xi = Me^{\omega T}(\eta + \eta_K)$. The set $\hat{Z}_N$ is a complete metric space if its metric is induced by the norm $\|\cdot\|'_{\tilde{\xi},T}$, where $\tilde{\xi} = Me^{\omega T}\tilde{\eta}_{N'}$ and $N' = N\exp\int_0^T \xi(\sigma)\,d\sigma$.

Let $\alpha, \beta \in \hat{Z}_N$. Then, by (8.1.9), (8.1.15), (8.1.21), and (8.1.22)

$$\|P\alpha - P\beta\|'_{\tilde{\xi},T} \le \left(1 - e^{-\int_0^T \tilde{\xi}(\sigma)\,d\sigma}\right)\|\alpha - \beta\|'_{\tilde{\xi},T}. \tag{8.1.24}$$

Thus P has a unique fixed point $u \in \hat{Z}_N$, which is the desired solution.
Theorem 8.1.3 is proved.   ■

THEOREM 8.1.4

Assume (H_A) and (H_f''). Then (8.1.1) with $u(0) = u_0$ has a unique (classical) solution $u \in C^2([0,T];X)$, $R(u') \subset D(A)$, and $Au' = (Au)' \in C([0,T];X)$. If, in addition, $R(f(\alpha)) \subset D(A)$ and $Af(\alpha) \in C([0,T];X)$ for any $\alpha \in Y_2$, then $R(u) \subset D(A^2)$ and $A^2 u \in C([0,T];X)$.

PROOF   By Theorems 1.4.2 and 1.4.3,

$$\|R(\lambda\colon A)\|_{L(X;X)} \le \frac{M}{\lambda - \omega} \quad \text{and} \quad \lim_{\lambda \to \infty} \lambda R(\lambda\colon A)x = x, \tag{8.1.25}$$

for all $\lambda > \omega$ and $x \in X$, with $M \ge 1$, $\omega \ge 0$ from (8.1.9). Let $n > \omega$, $\alpha \in Y_2$, $\tilde{f}_n(\alpha) = nR(n\colon A)f_{n^2}(\alpha)$, where $f_n(\alpha)$ are given by (8.1.7). Now $\tilde{f}_n(\alpha) \in C^\infty([0,T];X)$. By the properties of the convolution approximation,

$$\lim_{n \to \infty} \|f_n(\alpha) - f(\alpha)\|_{W^{2,1}(0,T;X)} = 0, \tag{8.1.26}$$

$$\|f_n(\alpha)(t) - f(\alpha)(t)\|_X \le \frac{1}{n}\|f(\alpha)'\|_{L^\infty(0,T;X)} \tag{8.1.27}$$

for all $t \in [0,T]$ and $n \in \mathbb{N}^*$.

By Theorem 1.4.2, the problem

$$u_n'(t) + Au_n(t) = \tilde{f}_n(\alpha)(t), \ t \in (0,T), \tag{8.1.28}$$

$$u_n(0) = nR(n\colon A)u_0, \tag{8.1.29}$$

has a unique strong solution u_n given by

$$u_n(t) = S(t)nR(n{:}A)u_0 + \int_0^t S(s)\tilde{f}_n(\alpha)(t-s)\,ds \tag{8.1.30}$$

on $[0,T]$. Since A and $R(n{:}A)$ commute and $\tilde{f}_n(\alpha)' \in C([0,T];X)$, we have

$$u_n'(t) = S(t)nR(n{:}A)\big(f_{n^2}(\alpha)(0) - Au_0\big) +$$

$$+ \int_0^t S(s)\tilde{f}_n(\alpha)'(t-s)\,ds \text{ for all } t \in [0,T]. \tag{8.1.31}$$

We can differentiate (8.1.31) again, since $\tilde{f}_n(\alpha)'' \in C([0,T];X)$. Thus

$$u_n''(t) = -S(t)AnR(n{:}A)\big(f_{n^2}(\alpha)(0) - Au_0\big) + S(t)\tilde{f}_n(\alpha)'(0) +$$

$$+ \int_0^t S(s)\tilde{f}_n(\alpha)''(t-s)\,ds \text{ for all } t \in [0,T]. \tag{8.1.32}$$

Calculations using (8.1.9) and (8.1.27)-(8.1.26) reveal that the right hand sides of (8.1.30)-(8.1.32) converge uniformly. Hence there exists a function $u \in C^2([0,T];X)$ such that, for all $t \in [0,T]$,

$$u(t) = S(t)u_0 + \int_0^t S(s)f(\alpha)(t-s)\,ds, \tag{8.1.33}$$

$$u'(t) = S(t)\big(f(u_0)(0) - Au_0\big) + \int_0^t S(s)f(\alpha)'(t-s)\,ds, \tag{8.1.34}$$

$$u''(t) = S(t)A\big(Au_0 - f(u_0)(0)\big) + S(t)f(\alpha)'(0) +$$

$$+ \int_0^t S(s)f(\alpha)''(t-s)\,ds, \tag{8.1.35}$$

$$u_n \to u \text{ in } C^2([0,T];X), \text{ as } n \to \infty. \tag{8.1.36}$$

Since A is closed, (8.1.27), (8.1.28)-(8.1.29), and (8.1.36) imply:

$$u(t) \in D(A) \text{ and } u'(t) + Au(t) = f(\alpha)(t) \text{ for all } t \in [0,T]. \tag{8.1.37}$$

Using the linearity of A and (8.1.28)-(8.1.29), we obtain

$$\frac{u_n'(t+h) - u_n'(t)}{h} + A\frac{u_n(t+h) - u_n(t)}{h} = \frac{\tilde{f}_n(\alpha)(t+h) - \tilde{f}_n(\alpha)(t)}{h}$$

for all $t \in [0,T]$ and $h \in [-t, T-t] \setminus \{0\}$. Letting $h \to 0$ and $n \to \infty$, successively, and employing the closedness of A, $u_n, \tilde{f}_n(\alpha) \in C^2([0,T];X)$, (8.1.26) and (8.1.36), we conclude that

$$u'(t) \in D(A) \text{ and } u''(t) + Au'(t) = f(\alpha)'(t) \tag{8.1.38}$$

for all $t \in [0, T]$. Since $f(\alpha)'$ and u'' are continuous, so is Au'. Since A is closed, $Au' = (Au)'$. Let $t \in [0, T]$. If $f(\alpha)(t) \in D(A)$, then by (8.1.37)-(8.1.38), $Au(t) = f(\alpha)(t) - u'(t) \in D(A)$. Thus $u(t) \in D(A^2)$. If, in addition, $Af(\alpha) \in C([0, T]; X)$, then by (8.1.37) $A^2 u = Af(\alpha) - Au' \in C([0, T]; X)$.

The mapping $P: Y_2 \to Y_2$, $P\alpha = u_\alpha$, the unique solution of (8.1.37) with $u(0) = u_0$, is well defined. Let $\xi \in L^1(0, T)$ be positive and define

$$\|\beta\|_{\xi, T}^* = \operatorname*{ess\,sup}_{0 \le s \le t} e^{-\int_0^s \xi(\sigma)\, d\sigma} \left(\|\beta(s)\|_X + \|\beta'(s)\|_X + \|\beta''(s)\|_X \right)$$

for all $\beta \in W^{2, \infty}(0, T; X)$. A calculation, using (8.1.33)-(8.1.35), (8.1.4), (8.1.14), and (8.1.16), shows that we can choose first K, then J, and finally N such that $P: Z_N^* \to Z_N^*$, where $Z_N^* = \{\gamma \in Y_2 \mid \|\gamma\|_{\xi, T}^* \le N\}$ and $\xi = Me^{\omega T}(\eta + \eta_K + \tilde{\eta}_J)$. We choose $N' = N \exp \int_0^T \xi(\sigma)\, d\sigma$ and use in Z_N^* the metric induced by $\|\cdot\|_{\xi^*, T}^*$, where $\xi^* = Me^{\omega T}\hat{\eta}_{N'}$. Then by (8.1.9), (8.1.16), and (8.1.33)-(8.1.35),

$$\|P\alpha - P\beta\|_{\xi^*, T}^* \le \left(1 - e^{-\int_0^T \hat{\xi}(\sigma)\, d\sigma}\right) \|\alpha - \beta\|_{\xi^*, T}^* \text{ for all } \alpha, \beta \in Z_N^*.$$

Hence P has a unique fixed point $u \in Y_N^*$.

Theorem 8.1.4 is proved.   ■

The assumption that f is locally Lipschitz can partly be avoided by using stronger differentiability conditions.

THEOREM 8.1.5

Assume the conditions of Theorem 8.1.3 and, in addition, that there exists a mapping f^* from $C^1([0, T]; X) \times C([0, T]; X)$ into $L^1(0, T; X)$ such that $f(\alpha) \in C^1([0, T]; X)$, $f(\alpha)'(t) = f^*(\alpha, \alpha')(t)$, $0 \le t \le T$, and the couple $\big(f(u_0)(0) - Au_0, f^*(\alpha, \cdot)\big)$ satisfies (H_f') for any given $\alpha \in C^1([0, T]; X)$ such that $\alpha(0) = u_0$ and $\alpha'(0) = f(u_0) - Au_0$. Then (8.1.1) has a unique (classical) solution $u \in C^2([0, T]; X)$ satisfying $u(0) = u_0$, $R(u') \subset D(A)$, and $Au' = (Au)' \in C([0, T]; X)$.

PROOF   By Theorem 8.1.3, Eq. (8.1.1) with $u(0) = 0$ has a unique solution $u \in C^1([0, T]; X)$, given by (8.1.33) with $\alpha = u$. By Theorem 8.1.3 there exists a unique $v \in C^1([0, T]; X)$, satisfying

$$v'(t) + Av(t) = f^*(u, v)(t), \ 0 < t < T, \ v(0) = f(u_0)(0) - Au_0. \quad (8.1.39)$$

On the other hand, by differentiating u,

$$u'(t) = S(t)\Big(f(u_0)(0) - Au_0\Big) + \int_0^t S(s)f^*(u, u')(t - s)\, ds$$

for all $t \in [0, T]$. Thus u' is the mild solution of (8.1.39). Since the mild solution is unique, $u' = v$. Hence $u \in C^2([0, T]; X)$, $Ru' \subset D(A)$, and $Au' \in C([0, T]; X)$. By the closedness of A, $(Au)' = Au'$.

Theorem 8.1.5 is proved. ∎

REMARK 8.1.1 One could similarly formulate and prove higher order regularity results as in Theorems 8.1.4 and 8.1.5 ∎

8.2 Hyperbolic partial differential systems with nonlinear boundary conditions

Consider the following hyperbolic boundary value problem:

$$u_t(r, t) + v_r(r, t) + Ru(r, t) = f_1(r, t), \tag{8.2.1}$$

$$v_t(r, t) + u_r(r, t) + Gv(r, t) = f_2(r, t), \quad (r, t) \in D_T, \tag{8.2.2}$$

$$\big(- u(0, t), u(1, t)\big) \in L\big(v(0, t), v(1, t)\big), \quad t \in (0, T), \tag{8.2.3}$$

$$u(r, 0) = u_0(r), \quad v(r, 0) = v_0(r), \quad r \in (0, 1), \tag{8.2.4}$$

where $L \subset \mathbb{R}^2 \times \mathbb{R}^2$ is a maximal monotone operator, $u_0, v_0 \colon [0, 1] \to \mathbb{R}$, $R, G \in \mathbb{R}$, and $f_1, f_2 \colon [0, 1] \times [0, T] \to \mathbb{R}$. Here $T > 0$ is fixed, $D_T = (0, 1) \times (0, T)$, and u_r, u_t denote the partial derivatives of $u = u(r, t)$. Clearly, (8.2.1)-(8.2.2) is the well known telegraph system and (8.2.3) includes as particular cases various classical boundary conditions. For details and an existence theory for (8.2.1)-(8.2.4), see Chapter 5, where the theory of evolution equations associated with monotone operators is used. Theorem 5.1.1 and its Comment 2 imply:

PROPOSITION 8.2.1

Let $f_1, f_2 \in L^1\big(0, T; L^2(0, 1)\big)$ and $u_0, v_0 \in L^2(0, 1)$. Then (8.2.1)-(8.2.4) has a unique weak solution $(u, v) \in C\big([0, T]; L^2(0, 1)\big)^2$. If, in addition, $f_1, f_2 \in W^{1,1}\big(0, T; L^2(0, 1)\big)$ and $u_0, v_0 \in H^1(0, 1)$ satisfy

$$\big(- u_0(0), u_0(1)\big) \in L\big(v_0(0), v_0(1)\big), \tag{8.2.5}$$

then (8.1.39)-(8.2.4) has a unique strong solution (u, v) such that

$$u, v \in W^{1,\infty}\big(0, T; L^2(0, 1)\big) \quad \text{and} \quad u_r, v_r \in L^\infty\big(0, T; L^2(0, 1)\big).$$

Without any loss of generality we may assume that $T < 1$. We transform (8.2.1)-(8.2.4) into a boundary value problem with homogeneous boundary conditions, namely

$$\tilde{u}_t(r,t) + \tilde{v}_r(r,t) = \tilde{f}_1(\tilde{u},\tilde{v})(r,t), \tag{8.2.6}$$

$$\tilde{v}_t(r,t) + \tilde{u}_r(r,t) = \tilde{f}_2(\tilde{u},\tilde{v})(r,t), \ (r,t) \in D_T, \tag{8.2.7}$$

$$\tilde{u}(0,t) = \tilde{u}(1,t) = 0, \ t \in (0,T), \tag{8.2.8}$$

$$\tilde{u}(r,0) = \tilde{u}_0(r), \ \tilde{v}(r,0) = \tilde{v}_0(r), \ r \in (0,1). \tag{8.2.9}$$

For each given $\tilde{u}, \tilde{v} \in C([0,T]; C[0,1])$, we define $a,c,d\colon [-T,1] \to \mathbb{R}$, $b\colon [0,T+1] \to \mathbb{R}$, and $u,v,\tilde{f}_1(\tilde{u},\tilde{v}), \tilde{f}_2(\tilde{u},\tilde{v})\colon \overline{D_T} \to \mathbb{R}$. First we write

$$u(r,t) = e^{-Rt}\Big(\tilde{u}(r,t) + a(r-t) + b(r+t) +$$

$$+ r(1-r)^2 c(r-t) - r^2(1-r)d(r-t)\Big), \tag{8.2.10}$$

$$v(r,t) = e^{-Rt}\Big(\tilde{v}(r,t) + a(r-t) - b(r+t) +$$

$$+ r(1-r)^2 c(r-t) - r^2(1-r)d(r-t)\Big), \tag{8.2.11}$$

for all $(r,t) \in \overline{D_T}$. Next we denote

$$\tilde{f}_1(\tilde{u},\tilde{v})(r,t) = e^{Rt}f_1(r,t) - (1-r)(1-3r)c(r-t) -$$

$$- r(3r-2)d(r-t), \tag{8.2.12}$$

$$\tilde{f}_2(\tilde{u},\tilde{v})(r,t) = e^{-Rt}\Big(f_2(r,t) - (G-R)v(r,t)\Big) -$$

$$- (1-r)(1-3r)c(r-t) - r(3r-2)d(r-t), \tag{8.2.13}$$

$$c(-t) = e^{Rt}f_1(0,t), \ d(1-t) = e^{Rt}f_1(1,t) \tag{8.2.14}$$

for all $(r,t) \in \overline{D_T}$. Then, c and d are continued smoothly to $[-T,1]$. We set $\tilde{L}(t) = I - \left(I + e^{Rt}Le^{-Rt}\right)^{-1}$ and write, for all $0 \le r \le 1$ and $0 < t \le T$,

$$a(r) = \frac{1}{2}\big(u_0(r) + v_0(r)\big) - r(1-r)^2 c(r) + r^2(1-r)d(r), \tag{8.2.15}$$

$$b(r) = \frac{1}{2}\big(u_0(r) - v_0(r)\big), \tag{8.2.16}$$

$$\begin{pmatrix} -a(-t) \\ b(1+t) \end{pmatrix} = \tilde{L}(t)\begin{pmatrix} \tilde{v}(0,t) - 2b(t) \\ \tilde{v}(1,t) + 2a(1-t) \end{pmatrix} - \begin{pmatrix} -b(t) \\ a(1-t) \end{pmatrix}. \tag{8.2.17}$$

If $\tilde{u}$ and $\tilde{v}$ are C^1-functions satisfying (8.2.6)-(8.2.9) and a,b,c,d are also C^1-functions, then u,v satisfy (8.2.1)-(8.2.4). Indeed, (8.2.17) is equivalent to (8.2.3) under (8.2.8)-(8.2.11). Moreover, by choosing $\tilde{u}_0 = \tilde{v}_0 = 0$, we obtain the solution of (8.2.1)-(8.2.4) from the solution of (8.2.6)-(8.2.9).

The *general solution* $(\tilde{u},\tilde{v})$ of (8.2.6)-(8.2.7) can be expressed by D'Alembert's type formulae (see Chapter 5). However, we need them only in the homogeneous case $\tilde{f}_1 = \tilde{f}_2 = 0$. Then, they read as

$$\tilde{u}(r,t) = \frac{1}{2}\big(\phi(r-t) + \psi(r+t)\big), \tag{8.2.18}$$

$$\tilde{v}(r,t) = \frac{1}{2}\big(\phi(r-t) - \psi(r+t)\big) \tag{8.2.19}$$

for all $t \in [0,T]$ and $r \in [0,1]$, where $\phi : [-T,1] \to \mathbb{R}$ and $\psi : [0,1+T] \to \mathbb{R}$ are arbitrary smooth functions.

Assume (8.2.18)-(8.2.19). Then (8.2.8) and (8.2.9) are equivalent to

$$\phi(r) = \tilde{u}_0(r) + \tilde{v}_0(r), \quad \psi(r) = \tilde{u}_0(r) - \tilde{v}_0(r), \ 0 < r < 1, \tag{8.2.20}$$

$$\phi(-t) = -\psi(t), \quad \psi(1+t) = -\phi(1-t) \ \text{ for all } t \in (0,T]. \tag{8.2.21}$$

If $\tilde{u}_0$ and $\tilde{v}_0$ are differentiable and satisfy the compatibility conditions $\tilde{u}_0(0) = \tilde{u}_0(1) = 0$, then (8.2.18)-(8.2.21) yield the classical solution of (8.2.6)-(8.2.9) with $\tilde{f}_1 = \tilde{f}_2 = 0$. However, (8.2.18)-(8.2.21) make sense under weaker assumptions. Thus $(\tilde{u}, \tilde{v})$, given by (8.2.18)-(8.2.21), is called the *generalized solution* of (8.2.6)-(8.2.9) with $\tilde{f}_1 = \tilde{f}_2 = 0$.

PROPOSITION 8.2.2
Assume that $u_0 \in C_0[0,1]$ and $v_0 \in C[0,1]$. Then, (8.2.6)-(8.2.9) with $\tilde{f}_1 = \tilde{f}_2 = 0$ has a unique generalized solution $(\tilde{u}, \tilde{v}) \in C\big(\overline{D_T}\big)^2$ that coincides with its weak solution.

PROOF   See Proposition 8.2.1. The mapping $(\tilde{u}_0, \tilde{v}_0) \mapsto (\tilde{u}, \tilde{v})$ from $L^2(0,1)^2$ into $C\big([0,T]; L^2(0,1)\big)^2$ is continuous by (8.2.18)-(8.2.21). Since the weak solution is a limit of classical solutions, given by D'Alembert's type formulae, the weak solution coincides with the generalized one. Clearly, $(\tilde{u}, \tilde{v}) \in C\big(\overline{D_T}\big)^2$. ∎

We return to the study of (8.2.1)-(8.2.4). Let $0 < T < 1$. By Proposition 8.2.2, the problem (8.2.6)-(8.2.9) with $\tilde{f}_1 = \tilde{f}_2 = 0$ and $T = \infty$ has a unique generalized solution $(\tilde{u}, \tilde{v}) \in C\big([0,\infty) \times [0,1]\big)^2$, if $\tilde{u}_0 \in C_0[0,1]$ and $\tilde{v}_0 \in C[0,1]$. We introduce the Banach space

$$X = C_0[0,1] \times C[0,1], \quad \|(y,z)\|_X = \|y\|_{C[0,1]} + \|z\|_{C[0,1]}. \tag{8.2.22}$$

Then $S(t) \colon X \to X$, $S(t)(\tilde{u}_0, \tilde{v}_0) = \big(\tilde{u}(t), \tilde{v}(t)\big)$ is well defined for any $t \geq 0$.

LEMMA 8.2.1
The family $\{S(t) \colon X \to X \mid t \geq 0\}$ is a C_0-semigroup of bounded linear operators in X, whose infinitesimal generator is $-A$,

$$D(A) = \{(y,z) \in C^1[0,1]^2 \mid y(0) = y(1) = z'(0) = z'(1) = 0\}, \tag{8.2.23}$$

$$A(y, z) = (z', y'). \tag{8.2.24}$$

PROOF   Let us consider the Hilbert space $H = L^2(0,1)^2$, and the operator $B \subset H \times H$, given by

$$((y_1, y_2), (z_1, z_2))_H = (y_1, z_1)_{L^2(0,1)} + (y_2, z_2)_{L^2(0,1)},$$

$$D(B) = \{(y, z) \in H^1(0,1)^2 \mid y(0) = y(1) = 0\}, \ B(y, z) = (z', y').$$

Since B is linear and maximal monotone, $-B$ generates a C_0-semigroup $\{T(t): H \to H \mid t \geq 0\}$ of linear contractions on H. On the other hand, $T(\cdot)(\tilde{u}_0, \tilde{v}_0) \in C\big([0, \infty); H\big)$ and it is the weak solution of (8.2.6)-(8.2.9) with $\tilde{f}_1 = \tilde{f}_2 = 0$ (see Proposition 8.2.2). Hence it coincides with $(\tilde{u}, \tilde{v})$ and

$$\big\|T(t)(\tilde{u}_0, \tilde{v}_0) - S(t)(\tilde{u}_0, \tilde{v}_0)\big\|_H = 0 \ \text{ for all } t \geq 0. \tag{8.2.25}$$

So also $\{S(t): X \to X \mid t \geq 0\}$ is a semigroup of linear operators, since $S(\cdot)(y, z) \in C\big([0,1] \times [0, \infty)\big)$ whenever $(y, z) \in X$. Direct calculations with (8.2.18)-(8.2.21) using the Weierstrass Theorem and the Mean Value Theorem show that

$$\|S(t)\|_{L(X;X)} \leq 2 \ \text{ for all } t \in [0, 1], \tag{8.2.26}$$

$$\lim_{t \to 0+} \|S(t)(y, z) - (y, z)\|_X = 0 \ \text{ for all } (y, z) \in X, \tag{8.2.27}$$

$$\lim_{t \to 0+} \Big\| \frac{1}{t}\big(S(t)(y, z) - (y, z)\big) + A(y, z) \Big\|_X = 0, \tag{8.2.28}$$

for all $(y, z) \in D(A)$. By (8.2.26)-(8.2.28), $\{S(t) \mid t \geq 0\}$ is a C_0-semigroup of bounded linear operators. By (8.2.27), the generator of $\{S(t) \mid t \geq 0\}$ is $-A$.   ∎

REMARK 8.2.1   Since A is not accretive in X, the Lumer-Phillips Theorem could not be used.   ∎

The proofs of Lemmas 8.2.2-8.2.4 below are straightforward and so we shall omit them. Observe that the compatibility conditions are needed to prove that a and b are smooth at 0 and at 1, respectively.

LEMMA 8.2.2

Assume that f_1 and f_2 are continuous. Let $u_0, v_0 \in C[0, 1]$ and satisfy (8.2.5). Then $a, c, d \in C[-T, 1]$ and $b \in C[0, 1 + T]$, for all $\tilde{u}, \tilde{v} \in C\big([0, T]; X\big)$ such that $\tilde{u}(0, \cdot) = \tilde{v}(0, \cdot) = 0$. Moreover, $\tilde{f} = (\tilde{f}_1, \tilde{f}_2)$ satisfies (H_f) with zero as the initial value.

LEMMA 8.2.3

Assume that $f_1, f_2 \in C^1\big([0,T]; C[0,1]\big)$, $u_0, v_0 \in C^1[0,1]$, and $L \in C^1(\mathbb{R}^2)^2$ and satisfy the compatibility conditions (8.2.5) as well as

$$\begin{pmatrix} -f_1(0,0) + Ru_0(0) + v_0'(0) \\ f_1(1,0) - Ru_0(1) - v_0'(1) \end{pmatrix} =$$

$$= L' \begin{pmatrix} v_0(0) \\ v_0(1) \end{pmatrix} \begin{pmatrix} f_2(0,0) - Gv_0(0) - u_0'(0) \\ f_2(1,0) - Gv_0(1) - u_0'(1) \end{pmatrix}. \tag{8.2.29}$$

Then $a, c, d \in C^1[-T, 1]$ and $b \in C^1[0, 1+T]$, for all $\tilde{u}, \tilde{v} \in C^1\big([0,T]; X\big)$ such that

$$\tilde{u}(0,\cdot) = \tilde{v}(0,\cdot) = 0, \ \tilde{u}_t(0,\cdot) = \tilde{f}_1(0,0)(0,\cdot), \ and \ \tilde{v}_t(0,\cdot) = \tilde{f}_2(0,0)(0,\cdot).$$

If, in addition, $L'(\cdot)\mathbf{x}$ is locally Lipschitzian for all $\mathbf{x} \in \mathbb{R}^2$, then $\tilde{f} = (\tilde{f}_1, \tilde{f}_2)$ satisfies (H_f') with zero as the initial value.

LEMMA 8.2.4

Let $f_1, f_2 \in C^2\big([0,T]; C[0,1]\big)$, $u_0, v_0 \in C^2[0,1]$, and let $L \in C^2(\mathbb{R}^2)^2$ satisfy (8.2.5), (8.2.29), and

$$\begin{pmatrix} -f_{1t}(0,0) + f_{2r}(0,0) - Gv_0'(0) - u_0''(0) \\ f_{1t}(1,0) - f_{2r}(1,0) + Gv_0'(0) + u_0''(1) \end{pmatrix} =$$

$$= L'' \begin{pmatrix} v_0(0) \\ v_0(1) \end{pmatrix} \begin{pmatrix} f_2(0,0) - Gv_0(0) - u_0'(0) \\ f_2(1,0) - Gv_0(1) - u_0'(1) \end{pmatrix} +$$

$$+ L' \begin{pmatrix} v_0(0) \\ v_0(1) \end{pmatrix} \begin{pmatrix} f_{2t}(0,0) - f_{1r}(0,0) - v_0''(0) + \\ + (R-G)\big(f_2(0,0) - Gv_0(0) - u_0'(0)\big) \\ f_{2t}(1,0) - f_{1r}(1,0) - v_0''(1) + \\ + (R-G)\big(f_2(1,0) - Gv_0(1) - u_0'(1)\big) \end{pmatrix},$$

where L'' is the second order differential of L. Then $a, c, d \in C^2[-T, 1]$ and $b \in C^2[0, 1+T]$ for all $\tilde{u}, \tilde{v} \in C^2\big([0,T]; X\big)$ such that

$$\tilde{u}(0,\cdot) = 0, \ \tilde{v}(0,\cdot) = 0,$$
$$\tilde{u}_t(0,\cdot) = \tilde{f}_1(0,0)(0,\cdot), \ \tilde{v}_t(0,\cdot) = \tilde{f}_2(0,0)(0,\cdot),$$
$$\tilde{u}_{tt}(0,\cdot) = \tilde{f}_1(0,0)_t(0,\cdot) - \tilde{f}_2(0,0)_r(0,\cdot),$$
$$\tilde{v}_{tt}(0,\cdot) = \tilde{f}_2(0,0)_t(0,\cdot) - \tilde{f}_1(0,0)_r(0,\cdot).$$

If, in addition, $L'(\cdot)\mathbf{x}$ and $L''(\cdot)\mathbf{x}$ are locally Lipschitzian for all $\mathbf{x} \in \mathbb{R}^2$, $f_1, f_2 \in C^1\big(\overline{D_T}\big)$, and

$$2c(1) = f_{2r}(1,0) - (G-R)v_0'(1) + Rf_1(1,0) + f_{1t}(1,0) - 4f_1(1,0),$$
$$2d(0) = -f_{2r}(1,0) + (G-R)v_0'(0) - Rf_1(0,0) - f_{1t}(0,0) - 4f_1(0,0),$$

then $\tilde{f} = (\tilde{f}_1, \tilde{f}_2)$ satisfies (H''_f) with zero as the initial value.

THEOREM 8.2.1

Assume all the conditions of Lemma 8.2.3. Then the problem (8.2.1)-(8.2.4) has a unique classical solution $(u, v) \in C^1\left(\overline{D_T}\right)^2$.

PROOF   By Theorem 8.1.3 the problem (8.2.6)-(8.2.9) has a unique classical solution $(\tilde{u}, \tilde{v}) \in C^1(0, T; X)^2$. Moreover, $t \mapsto A(\tilde{u}, \tilde{v})$ is continuous on $[0, T]$. Thus $(\tilde{u}, \tilde{v}) \in C^1\left(\overline{D_T}\right)^2$. Since a, b, c, d are continuously differentiable, u and v are C^1-functions, as well. $\blacksquare$

THEOREM 8.2.2

Assume all the conditions of Lemma 8.2.4. Then the problem (8.2.1)-(8.2.4) has a unique classical solution $(u, v) \in C^2\left(\overline{D_T}\right)^2$.

PROOF   By Theorem 8.1.4, the problem (8.2.6)-(8.2.9) has a unique classical solution $(\tilde{u}, \tilde{v}) \in C^2(0, T; X)^2$, and $t \mapsto A(\tilde{u}, \tilde{v})$ is a C^1-function. Hence $\tilde{u}_{tr}$ and $\tilde{v}_{tr}$ exist and they are continuous on $\overline{D_T}$. We do not know whether $R\tilde{f}(\tilde{u}, \tilde{v}) \subset D(A)$. However, by (8.1.38) $\tilde{u}_{tr}$ and $\tilde{v}_{tr}$ exist on $\overline{D_T}$. Hence $\tilde{v}_{tr} = \tilde{v}_{rt}$ and $\tilde{u}_{tr} = \tilde{u}_{rt}$, and we obtain from (8.2.1) that $\tilde{u}_{rr}, \tilde{v}_{rr} \in C\left(\overline{D_T}\right)$. Thus $(\tilde{u}, \tilde{v}) \in C^2\left(\overline{D_T}\right)^2$. Since a, b, c, d are twice continuously differentiable, u and v are C^2-functions, as well. $\blacksquare$

References

[HokMo2]   V.-M. Hokkanen & Gh. Moroşanu, Existence and regularity for a class of nonlinear hyperbolic boundary value problems, *J. Math. Anal. Appl.*, 266, (2002), 432–450.

Chapter 9

Nonlinear nonautonomous abstract differential equations

Many boundary value problems arising from the concrete applications have a nonautonomous character and require a more advanced treatment. For example, let us consider the following problem:

$$u_t(r,t) - w_r(r,t) + \hat{w}(r,t) = f(r,t), \ (r,t) \in Q_T, \tag{9.0.1}$$

$$w(r,t) \in G(r,t)u_r(r,t), \ (r,t) \in Q_T, \tag{9.0.2}$$

$$\hat{w}(r,t) \in K(r,t)u(r,t), \ (r,t) \in Q_T, \tag{9.0.3}$$

$$\big(w(0,t), -w(1,r)\big) \in \beta(t)\big(u(0,t), u(1,t)\big), \ t \in (0,T), \tag{9.0.4}$$

$$u(r,0) = u_0(r), \ r \in (0,1), \tag{9.0.5}$$

where $T > 0$ is fixed, $Q_T = (0,1) \times (0,T)$, and u_r, u_t denotes the partial derivatives of $u = u(r,t)$. Here $G(r,t)$, $K(r,t) \subset \mathbb{R} \times \mathbb{R}$ and $\beta(t) \subset \mathbb{R}^2 \times \mathbb{R}^2$, $t \in [0,T]$, are maximal monotone operators, possibly multi-valued. This very general model describes heat conduction and diffusion phenomena. In Chapter 3 we studied the autonomous case with G and K single-valued. Here we investigate (9.0.1)-(9.0.5) by using the following abstract Cauchy problem in a real Hilbert space (with subdifferential operators $A(t)$, which are perturbed by maximal monotone operators $B(t)$ and locally Lipschitzian or strongly-weakly closed functionals $u \mapsto f(u)$):

$$u'(t) + A(t)u(t) + B(t)u(t) \ni f(u)(t), \ t > 0, \ u(0) = u_0. \tag{9.0.6}$$

We study the existence of a solution for (9.0.6) in the next section. Our condition (H.3) there allows $A(t)$ to be a subdifferential with a time dependent domain. (H.3) is close to that of H. Attouch et al. [AtBDP]. Our proofs are based on the methods used for the case of time independent $A(t)$ in [Brézis1]. The functional perturbation $f(u)$ is handled by fixed point theorems. In Section 9.2 our results will be applied to (9.0.1)-(9.0.5) with unbounded $\beta(t)$. Our abstract theory can be applied to a wide range of other practical problems containing nonlinear partial differential or integral systems with time-dependent coefficients. Their common character, presupposed by (H.3) below, is that they are perturbed *parabolic* problems.

Our results have similarities to those of H. Attouch & A. Damlamian, T. Kato, and D. Tătaru, etc. (see [AttDam], [Kato], [Tătaru]), but we point out that our proofs are quite natural, since they actually follow the same steps as the original proof for the autonomous case in [Brézis1]. This is new.

9.1   First order differential and functional equations containing subdifferentials

For our convenience, let us recall some concepts we shall use in this section. Let $\lambda > 0$, $\psi \colon H \to (-\infty, \infty]$ be convex and A maximal monotone in H, a real Hilbert space. The *subdifferential*, the *Moreau-Yosida regularization*, and the positive part of ψ are denoted by $\partial \psi$, ψ_λ, and ψ_+, respectively. More precisely, $\psi_+(x) = \max\{0, \psi(x)\}$. The *resolvent* and the *Yosida approximation* of A are denoted by J_λ and A_λ, respectively, i.e., $J_\lambda = (I + \lambda A)^{-1}$, $A_\lambda = \frac{1}{\lambda}(I - J_\lambda)$. $A^0 x$ is the element of Ax with the minimal norm. For further details, see Section 1.2.

We continue with our main hypotheses (H.1)-(H.4) below that involve positive numbers M, T, and λ_0, functions $\eta \in L^1(0,T)$, $\phi \colon [0,T] \times H \to (-\infty, \infty]$, and operators $B(t) \subset H \times H$, $t \in [0,T]$.

(H.1) For each $t \in [0,T]$, $\phi(t, \cdot)$ is a proper convex lower semicontinuous function and $B(t)$ is a maximal monotone operator in H.

(H.2) There is a $z \in L^2(0,T;H)$ such that $z(t) \in D\big(\partial\phi(t, \cdot)\big)$ for a.a. $t \in (0,T)$, and the functions $t \mapsto \phi\big(t, z(t)\big)$, $t \mapsto \|\partial\phi^0(t, \cdot)z(t)\|_H^2$, and $t \mapsto \|B^0(t)z(t)\|_H^2$ are integrable.

(H.3) For each $\lambda \in (0, \lambda_0)$ and $y \in H$, $\phi_\lambda(\cdot, y) \in W^{1,1}(0,T)$ and

$$\max\left\{ \frac{\partial \phi_\lambda}{\partial t}(t,y), \|B_\lambda(t)y\|_H^2 \right\} \leq \frac{1}{3}\|\partial\phi_\lambda(t, \cdot)y\|_H^2 + M\|y\|_H^2 +$$

$$+ \eta(t)\Big(1 + \phi_{\lambda+}(t,y)\Big) \text{ for a.a. } t \in (0,T).$$

(H.4) For each $y \in H$ and $\lambda \in (0, \lambda_0)$, the mappings $t \mapsto \big(I + \lambda\partial\phi(t, \cdot)\big)^{-1} y$ and $t \mapsto \big(I + \lambda B(t)\big)^{-1} y$ are measurable.

REMARK 9.1.1   Condition (H.3) allows the domain of $\partial\phi(t,\cdot)$ to depend on time. Indeed, let $H = \mathbb{R}$, $c_1, c_2 \in H^1(0,1)$ with $c_1 \leq c_2$ and

$$\phi: [0,1] \times \mathbb{R} \to \{0,\infty\}, \quad \phi(t,x) = \begin{cases} 0 & \text{if } c_1(t) \leq x \leq c_2(t) \\ \infty & \text{otherwise.} \end{cases}$$

Then, for all $x \in \mathbb{R}$ and a.a. $t \in (0,1)$, $\phi_\lambda(\cdot, x) \in W^{1,1}(0,1)$ and

$$\frac{\partial\phi_\lambda}{\partial t}(t,x) \leq \frac{1}{3}\left|\partial\phi_\lambda(t,\cdot)x\right|^2 + \frac{3}{4}\max\left\{c_1'(t)^2, c_2'(t)^2\right\}.$$

∎

REMARK 9.1.2   The coefficient $1/3$ in (H.3) can be slightly increased, depending on $B(t)$. If $B(t) \equiv 0$, then $1/3$ can be replaced by any $\alpha \in (0,1)$.
∎

LEMMA 9.1.1
Assume (H.1)-(H.4) and let $\delta \in [0,T)$. Then the realizations of $\partial\phi(t,\cdot)$ and of $B(t)$ in $L^2(\delta, T; H)$, i.e.,

$$\partial\phi = \left\{(x,y) \in L^2(\delta,T;H)^2 \mid y(t) \in \partial\phi(t,\cdot)x(t) \text{ for a.a. } t \in (\delta,T)\right\},$$
$$B = \left\{(x,y) \in L^2(\delta,T;H)^2 \mid y(t) \in B(t)x(t) \text{ for a.a. } t \in (\delta,T)\right\},$$

are maximal monotone in $L^2(\delta, T; H)$.

PROOF   Clearly, $\partial\phi$ and B are monotone. They are maximal because $R(I + B) = R(I + \partial\phi) = L^2(\delta, T; H)$. This holds since $t \mapsto \left(I + \partial\phi(t,\cdot)\right)^{-1}y(t)$ and $t \mapsto \left(I + B(t)\right)^{-1}y(t)$ are square integrable, for each $y \in L^2(\delta, T; H)$. ∎

First order differential equations perturbed only by maximal monotone operators

THEOREM 9.1.1
Assume (H.1)-(H.4). If $u_0 \in D(\phi(0,\cdot))$ and $f \in L^2(0,T;H)$, then there exist $v, w \in L^2(0,T,H)$, a unique $u \in H^1(0,T;H)$, and some constants $M_1, M_2 > 0$, independent of f and u_0, such that

$$u'(t) + v(t) + w(t) = f(t) \text{ for a.a. } t \in (0,T), \tag{9.1.1}$$
$$v(t) \in \partial\phi(t,\cdot)u(t), \; w(t) \in B(t)u(t) \text{ for a.a. } t \in (0,T), \tag{9.1.2}$$
$$u(0) = u_0; \tag{9.1.3}$$

$$\|u\|^2_{L^\infty(0,t;H)} + \int_0^t \left(\|u'(\tau)\|^2_H + \|v(\tau)\|^2_H + \|w(\tau)\|^2_H\right) d\tau +$$

$$+ \operatorname*{ess\,sup}_{\tau \in [0,t]} \phi_+\big(\tau, u(\tau)\big) \leq M_1 \int_0^t \|f(\tau)\|^2_H \, d\tau +$$

$$+ M_2\big(1 + \|u_0\|^2_H + \phi_+(0, u_0)\big) \ \textit{for all } t \in (0, T]. \qquad (9.1.4)$$

PROOF We are inspired by Brézis' well known proof of existence of solution for an autonomous equation (see [Brézis1, pp. 54-57, 72-78] or [Moro1, pp. 46-61]). We denote different constants, independent of f, λ, t, and u_0, by $C_1, C_2, \ldots$. Let $\lambda \in (0, \lambda_0)$. The mappings $x \mapsto \partial\phi_\lambda(t, \cdot)x + B_\lambda(t)x$, $H \to H$, $t \in [0, T]$, are Lipschitzian and $t \mapsto \partial\phi_\lambda(t, \cdot)x + B_\lambda(t)x$, $[0, T] \to H$, $x \in H$, are square integrable. Thus by Lemma 1.5.3, the following problem has a unique strong solution $u_\lambda \in H^1(0, T; H)$.

$$u'_\lambda(t) + \partial\phi_\lambda(t, \cdot)u_\lambda(t) + B_\lambda(t)u_\lambda(t) = f(t) \text{ a.e. on } (0, T), \quad (9.1.5)$$

$$u_\lambda(0) = u_0. \qquad\qquad\qquad\qquad\qquad\qquad\qquad\qquad\qquad (9.1.6)$$

We shall prove that there exist positive constants M_1 and M_2, independent of λ, f, and u_0, such that for all $t \in (0, T]$,

$$\|u_\lambda\|^2_{L^\infty(0,t;H)} + \sup_{\tau \in [0,t]} \phi_{\lambda+}\big(\tau, u_\lambda(\tau)\big) +$$

$$+ \int_0^t \left(\|u'_\lambda(\tau)\|^2_H + \|\partial\phi_\lambda(\tau \cdot)u_\lambda(\tau)\|^2_H + \|B_\lambda(\tau)u_\lambda(\tau)\|^2_H\right) d\tau \leq$$

$$\leq M_1 \int_0^t \|f(\tau)\|^2_H \, d\tau + M_2\big(1 + \|u_0\|^2_H + \phi_+(0, u_0)\big). \qquad (9.1.7)$$

Since ϕ_λ, u_λ, and $\partial\phi_\lambda u_\lambda$ satisfy the conditions of the chain rule in Theorem 1.2.18, then $\phi_\lambda\big(\cdot, u_\lambda(\cdot)\big) \in W^{1,1}(0, T)$ and, for a.a. $t \in (0, T)$,

$$\frac{d}{dt}\phi_\lambda\big(t, u_\lambda(t)\big) = \big(\partial\phi_\lambda(t, \cdot)u_\lambda(t), u'_\lambda(t)\big)_H + \frac{\partial\phi_\lambda}{\partial t}\big(t, u_\lambda(t)\big). \qquad (9.1.8)$$

We multiply (9.1.5) by $u'_\lambda(t)$, by $100\partial\phi_\lambda(t, \cdot)u_\lambda(t)$, and by $B_\lambda(t)u_\lambda(t)$, successively. By summing the resulting equations, by (9.1.8) and (H.3), we get

$$\frac{1}{303}\|u'_\lambda(t)\|^2_H + \frac{d}{dt}\phi_\lambda\big(t, u_\lambda(t)\big) + \frac{1}{22}\|\partial\phi_\lambda(t, \cdot)u_\lambda(t)\|^2_H +$$

$$+ \frac{1}{101}\|B_\lambda(t)u_\lambda(t)\|^2_H \leq 25\|f(t)\|^2_H + 3M\|u_\lambda(t)\|^2_H +$$

$$+ 3\eta(t)\big(1 + \phi_{\lambda+}\big(t, u_\lambda(t)\big)\big) \text{ for a.a. } t \in (0, T), \qquad (9.1.9)$$

whence by integrating and by $\phi_\lambda \leq \phi \leq \phi_+$ (see Theorem 1.2.16),

$$\int_0^t \left(\frac{1}{303} \|u_\lambda'(\tau)\|_H^2 + \frac{1}{22} \|\partial\phi_\lambda(\tau, \cdot)u_\lambda(\tau)\|_H^2 + \right.$$
$$\left. + \frac{1}{101} \|B_\lambda(\tau)u_\lambda(\tau)\|_H^2 \right) d\tau + \phi_\lambda\big(t, u_\lambda(t)\big) \leq \phi(0, u_0) +$$
$$+ 25\|f\|_{L^2(0,T;H)}^2 + 3\|\eta\|_{L^1(0,T)} + 3M \int_0^t \|u_\lambda(\tau)\|_H^2 \, d\tau +$$
$$+ \int_0^t 3\eta(\tau)\frac{1}{2}\Big(1 + \operatorname{sgn}\phi_\lambda\big(\tau, u_\lambda(\tau)\big)\Big)\phi_\lambda\big(\tau, u_\lambda(\tau)\big) \, d\tau. \qquad (9.1.10)$$

By Gronwall's inequality, there is a constant $C_1 > 0$ such that

$$\phi_\lambda\big(t, u_\lambda(t)\big) \leq C_1 \left(1 + \phi_+(0, u_0) + \|f\|_{L^2(0,T;H)}^2 + \int_0^t \|u_\lambda(\tau)\|_H^2 \, d\tau \right) \qquad (9.1.11)$$

for all $t \in [0, T]$. By (9.1.10) and by (9.1.11), for all $t \in [0, T]$,

$$\|u_\lambda(t)\|_H^2 \leq 2\|u_0\|_H^2 + 606T\frac{1}{303}\int_0^t \|u_\lambda'(\tau)\|_H^2 \, d\tau \leq$$
$$\leq 2\|u_0\|_H^2 - 606T\phi_\lambda\big(t, u_\lambda(t)\big) +$$
$$+ C_2\left(1 + \phi_+(0, u_0) + \|f\|_{L^2(0,T;H)}^2 + \int_0^t \|u_\lambda(\tau)\|_H^2 \, d\tau \right). \qquad (9.1.12)$$

By the definition of the subdifferential and on account of

$$\phi_\lambda(t, \cdot) \geq \phi\big(t, J_\lambda(t)\,\cdot\,\big), \quad \|\partial\phi_\lambda(t, \cdot)\cdot\|_H \leq \|\partial\phi^0(t, \cdot)\cdot\|_H$$

and $J_\lambda(t) - I = -\lambda\partial\phi_\lambda(t, \cdot)$, we obtain that

$$\phi_\lambda(t, x) \geq \phi\big(t, J_\lambda(t)z(t)\big) + \big(\partial\phi_\lambda(t, \cdot)z(t), x - z(t)\big)_H \geq$$
$$\geq \phi\big(t, z(t)\big) + \big(\partial\phi^0(t, \cdot)z(t), J_\lambda(t)z(t) - z(t)\big)_H -$$
$$- \|\partial\phi^0(t, \cdot)z(t)\|_H \big(\|x\|_H + \|z(t)\|_H\big) \geq \phi\big(t, z(t)\big) -$$
$$- \frac{1}{909T}\|x\|_H^2 - (228T + 2)\|\partial\phi^0(t, \cdot)z(t)\|_H^2 - \|z(t)\|_H^2 \qquad (9.1.13)$$

for all $t \in [0, T]$ and $x \in H$. By (9.1.12)-(9.1.13), integrating over $[0, t]$ and using Gronwall's inequality, we obtain that, for all $t \in (0, T]$,

$$\int_0^t \|u_\lambda(\tau)\|_H^2 \, d\tau \leq C_3\left(1 + \|u_0\|_H^2 + \phi_+(0, u_0) + \|f\|_{L^2(0,t;H)}^2 \right). \qquad (9.1.14)$$

Since $\phi(T, \cdot)$ is a proper convex lower semicontinuous function, it is bounded from below by an affine function (see Theorem 1.2.8). Since there is a $\tilde{z} \in D\big(\partial\phi(T, \cdot)\big)$ and $\phi\big(T, J_\lambda(T)\,\cdot\,\big) \leq \phi_\lambda(T, \cdot)$, we have, for any $x \in H$,

$$-\phi_\lambda(T, x) \leq C_4\|J_\lambda(T)x\|_H + C_4 \leq \frac{1}{909T}\|x\|_H^2 + C_5. \qquad (9.1.15)$$

We can now derive (9.1.7) from (9.1.10)-(9.1.12) and (9.1.14).

Next, we shall prove that there exist a subsequence of (λ) that converges to 0, and $v, w \in L^2(0, T; H)$, and $u \in H^1(0, T; H)$ such that, as $\lambda \to 0+$,

$$u_\lambda \to u \text{ in } C([0, T]; H), \tag{9.1.16}$$

$$u'_\lambda \to u', \ \partial\phi_\lambda u_\lambda \to v, \ B_\lambda u_\lambda \to w \text{ weakly in } L^2(0, T; H). \tag{9.1.17}$$

By (9.1.9), the sequences (u'_λ), $(\partial\phi_\lambda u_\lambda)$, and $(B_\lambda u_\lambda)$ are bounded in the Hilbert space $L^2(0, T; H)$. Thus there are $u^*, v, w \in L^2(0, T; H)$ and a subsequence such that, as $\lambda \to 0+$,

$$u'_\lambda \to u^*, \ \partial\phi_\lambda u_\lambda \to v, \text{ and } B_\lambda u_\lambda \to w \text{ weakly in } L^2(0, T; H). \tag{9.1.18}$$

Let $\lambda, \mu \in (0, \lambda_0)$ be arbitrarily chosen. By (9.1.5) and the monotonicity of $\partial\phi_\lambda(t, \cdot)$ and $B_\lambda(t)$, we have

$$\frac{1}{2}\frac{d}{dt}\left\|u_\lambda(t) - u_\mu(t)\right\|_H^2 = \left(u'_\lambda(t) - u'_\mu(t), u_\lambda(t) - u_\mu(t)\right)_H \le$$

$$\le (\lambda + \mu)\left(2\|\partial\phi_\lambda(t, \cdot)u_\lambda(t)\|_H^2 + 2\|\partial\phi_\mu(t, \cdot)u_\mu(t)\|_H^2\right) +$$

$$+ (\lambda + \mu)\left(2\|B_\lambda(t)u_\lambda(t)\|_H^2 + 2\|B_\mu(t)u_\mu(t)\|_H^2\right)$$

for a.a. $t \in (0, T)$, whence by integrating and by (9.1.6) and (9.1.8), we get

$$\|u_\lambda - u_\mu\|_{C([0,T];H)}^2 \le 8(\lambda + \mu)\left(M_1\|f\|_{L^2(0,T;H)}^2 + M_2\left(1 + \|u_0\|_H^2 + \phi(0, u_0)\right)\right).$$

So (u_λ) is a Cauchy sequence in the complete metric space $C([0, T]; H)$. Hence it converges toward some $u \in C([0, T]; H)$. Since the derivative d/dt is a strongly-weakly closed mapping in $L^2(0, T; H)$, $u' = u^*$. Thus $u \in H^1(0, T; H)$.

By (9.1.8), (9.1.16), and the definition of the Yosida approximation, we have

$$\|u - J_\lambda u_\lambda\|_{L^2(0,T;H)} \le \|u - u_\lambda\|_{L^2(0,T;H)} + \lambda\|\partial\phi_\lambda u_\lambda\|_{L^2(0,T;H)} \to 0,$$

as $\lambda \to 0+$. Since $\partial\phi_\lambda u_\lambda \in \partial\phi J_\lambda u_\lambda$, $\partial\phi_\lambda u_\lambda \to v$ weakly in $L^2(0, T; H)$, and $\partial\phi$ is demiclosed (see Theorem 1.2.3), then $v \in \partial\phi u$. Similarly,

$$(I + \lambda B)^{-1}u_\lambda \to u \text{ in } L^2(0, T; H), \text{ as } \lambda \to 0+,$$

and $B_\lambda u_\lambda \in B(I + \lambda B)^{-1}u_\lambda$, whence $w \in Bu$.

Let $x \in H$ and $t \in (0, T)$. Since $u'_\lambda \to u'$, $\partial\phi_\lambda u_\lambda \to v$, and $B_\lambda u_\lambda \to w$ weakly in $L^2(0, t; H)$, we have

$$\int_0^t \left(x, f(\tau) - u'_\lambda(\tau) - \partial\phi_\lambda(\tau, \cdot)u_\lambda(\tau) - B_\lambda(\tau)u_\lambda(\tau)\right)_H d\tau =$$

$$= \lim_{\lambda \to 0+} \int_0^t \left(x, f(\tau) - u'_\lambda(\tau) - \partial\phi_\lambda(\tau, \cdot)u_\lambda(\tau) - B_\lambda(\tau)u_\lambda(\tau)\right)_H d\tau = 0.$$

By differentiating we obtain (9.1.1). The initial condition (9.1.3) is satisfied due to the convergence $u_\lambda \to u$ in $C([0,T];H)$. Hence (u,v,w) satisfies (9.1.1)-(9.1.3).

The weak convergence in $L^2(0,T;H)$ implies the weak convergence in $L(0,t;H)$, for any $t \in (0,T]$. Thus the estimate (9.1.4) is implied by (9.1.7) and the weak lower semicontinuity of both the norm of $L^2(0,t;H)$ and $\phi(t,\cdot)$. Indeed, for a subsequence and for a.a. $t \in (0,T)$,

$$\phi\big(t,u(t)\big) \leq \liminf_{\lambda \to 0+} \phi\big(t,J_\lambda(t)u_\lambda(t)\big) \leq \liminf_{\lambda \to 0+} \phi_\lambda\big(t,u_\lambda(t)\big).$$

Let (u,v,w), $(\tilde{u},\tilde{v},\tilde{w}) \in H^1(0,T;H) \times L^2(0,T;H)^2$ satisfy (9.1.1)-(9.1.3). Then by (9.1.1) and the monotonicity of both $\partial\phi(t,\cdot)$ and $B(t)$,

$$\frac{1}{2}\frac{d}{dt}\big\|u(t) - \tilde{u}(t)\big\|_H^2 = \big(u(t) - \tilde{u}(t), -v(t) - w(t) + \tilde{v}(t) + \tilde{w}(t)\big)_H \leq 0,$$

for a.a. $t \in (0,T)$, whence by integrating over $[0,t] \subset [0,T]$ and by (9.1.3), we get that $u \equiv \tilde{u}$. Thus the solution is unique.

Theorem 9.1.1 is proved. $\blacksquare$

Next, we relax our assumptions on u_0 and f. A strong solution for (9.1.1)-(9.1.3) still exists, but a little bit less regular, similarly to the autonomous case (see [Brézis1, p. 76]).

THEOREM 9.1.2

Assume (H.1)-(H.4). Let $u_0 \in \overline{D\big(\phi(0,\cdot)\big)}$, $f \in L^1(0,T;H)$ and we further assume that

$$\int_0^t \tau\|f(\tau)\|_H^2\,d\tau + \|f\|_{L^1(0,t;H)}^2 < \infty \quad \text{and} \quad \int_0^t \frac{1}{\tau}\|z(\tau) - u_0\|_H^2\,d\tau + \|u_0\|_H^2 < \infty$$

for all $t \in [0,T]$. Then there exist measurable v, $w\colon [0,T] \to H$, a unique $u \in C\big([0,T];H\big)$, differentiable a.e. on $(0,T)$, and constants M_3, $M_4 > 0$, independent of u_0 and f. They satisfy (9.1.1)-(9.1.3) and

$$\|u\|_{L^\infty(0,t;H)}^2 + \operatorname*{ess\,sup}_{\tau \in [0,t]} \tau\phi_+\big(\tau,u(\tau)\big) +$$

$$+ \int_0^t \tau\Big(\|u'(\tau)\|_H^2 + \|v(\tau)\|_H^2 + \|w(\tau)\|_H^2\Big)\,d\tau \leq$$

$$\leq M_3 M_f(t) + M_4\Big(M_{u_0}(t) + 1\Big) \quad \text{for all } t \in (0,T], \quad (9.1.19)$$

where $M_f(t)$ and $M_{u_0}(t)$ are defined by

$$M_f(t) := \int_0^t \tau\|f(\tau)\|_H^2\,d\tau + \|f\|_{L^1(0,t;H)}^2 < \infty, \qquad (9.1.20)$$

$$M_{u_0}(t) := \int_0^t \frac{1}{\tau}\|z(\tau) - u_0\|_H^2\,d\tau + \|u_0\|_H^2 < \infty. \qquad (9.1.21)$$

PROOF   For each $n \in \mathbb{N}^*$, we choose $u_{0n} \in D\big(\phi(0, \cdot)\big)$ and $f_n \in L^2(0, T; H)$ such that

$$\|u_0 - u_{0n}\|_H \leq \frac{1}{n}, \ f_n(t) = \begin{cases} f(t) & \text{if } \|f(t)\|_H \leq n, \\ 0 & \text{otherwise.} \end{cases} \tag{9.1.22}$$

By Theorem 9.1.1, there exist $u_n \in H^1(0, T; H)$ and $v_n, w_n \in L^2(0, T; H)$ that satisfy:

$$u_n'(t) + v_n(t) + w_n(t) = f_n(t), \ \text{a.e. on } (0, T), \tag{9.1.23}$$
$$v_n(t) \in \partial\phi(t, \cdot)u_n(t), \ w_n(t) \in B(t)u_n(t), \ \text{a.e. on } (0, T), \tag{9.1.24}$$
$$u_n(0) = u_{0n}. \tag{9.1.25}$$

We shall show that there exist some constants $M_3, M_4 > 0$, which are independent of f, u_0, and n and satisfy, for every $n \in \mathbb{N}^*$,

$$\|u_n\|^2_{L^\infty(0,t;H)} + \operatorname*{ess\,sup}_{\tau \in [0,t]} \tau \phi_+\big(\tau, u_n(\tau)\big) +$$
$$+ \int_0^t \tau \Big(\|u_n'(\tau)\|^2_H + \|v_n(\tau)\|^2_H + \|w_n(\tau)\|^2_H\Big)\, d\tau \leq$$
$$\leq M_3 M_f(t) + M_4 M_{u_0}(t) + M_4 \ \text{for all } t \in (0, T]. \tag{9.1.26}$$

By the proof of Theorem 9.1.1, there are solutions $(u_{n\lambda})$, $\lambda \in (0, \lambda_0)$, of (9.1.5)-(9.1.6) with (u_{0n}, f_n) instead of (u_0, f). Moreover, they converge toward (u_n, v_n, w_n) in the following manner (see (9.1.16)-(9.1.17)):

$$u_\lambda' \to u', \ \partial\phi_\lambda u_\lambda \to v, \ B_\lambda u_\lambda \to w \text{ weakly in } L^2(0, T; H), \tag{9.1.27}$$
$$u_\lambda \to u \text{ in } C([0, T]; H), \text{ as } \lambda \to 0+. \tag{9.1.28}$$

Taking into account the definition of the subdifferential, the modified (9.1.5), $\phi_\lambda \leq \phi$, and the monotonicity of $B_\lambda(t)$, we obtain for a.a. $t \in (0, T)$,

$$\phi_\lambda\big(t, u_{n\lambda}(t)\big) \leq \big(f_n(t) - B_\lambda(t)u_{n\lambda}(t) - u_{n\lambda}'(t), u_{n\lambda}(t) - z(t)\big)_H +$$
$$+ \phi_\lambda\big(t, z(t)\big) \leq \Big(\|f(t)\|_H + \|B^0(t)z(t)\|_H\Big)\|u_{n\lambda}(t) - u_0\|_H -$$
$$- \frac{d}{dt}\frac{1}{2}\|u_{n\lambda}(t) - u_0\|^2_H + \frac{t}{2}\|f(t)\|^2_H + \frac{t}{2}\|B^0(t)z(t)\|^2_H +$$
$$+ \frac{t}{606}\|u_{n\lambda}'(t)\|^2_H + \frac{152}{t}\|u_0 - z(t)\|^2_H + \phi\big(t, z(t)\big).$$

We multiply the modified (9.1.9) by t and integrate it over $[0, t] \subset [0, T]$. Then

$$t\phi_\lambda(t, u_{n\lambda}(t)) + \frac{1}{2}\|u_{n\lambda}(t) - u_0\|^2_H +$$

$$+\frac{1}{606}\int_0^t \tau\Big(\|u'_{n\lambda}(\tau)\|_H^2 + \|\partial\phi_\lambda(\tau,\cdot)u_{n\lambda}(\tau)\|_H^2 + \|B_\lambda(\tau)u_{n\lambda}(\tau)\|_H^2\Big)\,d\tau \le$$

$$\le C_6\int_0^t \big(\|f(\tau)\|_H + \|u_{n\lambda}(\tau)\|_H\big)\|u_{n\lambda}(\tau) - u_0\|_H\,d\tau +$$

$$+C_6\big(1 + M_f(t) + M_{u_0}(t)\big) +$$

$$+\int_0^t 3\eta(\tau)\frac{1}{2}\Big(1 + \operatorname{sgn}\phi_\lambda\big(\tau, u_{n\lambda}(\tau)\big)\Big)\tau\phi_\lambda\big(\tau, u_{n\lambda}(\tau)\big)\,d\tau \qquad (9.1.29)$$

for all $t \in [0, T]$. By Gronwall's inequality, we have for all $t \in [0, T]$,

$$t\phi_\lambda(t, u_{n\lambda}(t)) \le C_7\big(1 + M_f(t) + M_{u_0}(t)\big) +$$

$$+C_7\int_0^t \Big(\|f(\tau)\|_H + \|u_{n\lambda}(\tau)\|_H\Big)\|u_{n\lambda}(\tau) - u_0\|_H\,d\tau. \qquad (9.1.30)$$

Since $B_\lambda(t)$ and $\partial\phi_\lambda(t, \cdot)$ are monotone, (9.1.5)-(9.1.6) yield

$$\frac{1}{2}\|u_{n\lambda}(t) - u_0\|_H^2 = \frac{1}{2}\|u_{0n} - u_0\|_H^2 + \int_0^t \big(u_{n\lambda}(\tau) - u_0, u'_{n\lambda}(\tau)\big)_H\,d\tau \le$$

$$\le \frac{1}{2} + \int_0^t \Big(\frac{\tau}{2}\|u'_{n\lambda}(\tau)\|_H^2 + \frac{1}{\tau}\|z(\tau) - u_0\|_H^2\Big)\,d\tau +$$

$$+\int_0^t \Big(-\big(u_{n\lambda}(\tau) - z(\tau), \partial\phi_\lambda(\tau,\cdot)z(\tau) + B_\lambda(\tau)z(\tau)\big)_H +$$

$$+\|f(\tau)\|_H\|u_{n\lambda}(\tau) - u_0\|_H + \frac{\tau}{2}\|f(\tau)\|_H^2\Big)\,d\tau \le C_8 + M_f(t) +$$

$$+M_{u_0}(t) + \int_0^t \frac{\tau}{2}\|u'_{n\lambda}(\tau)\|_H^2\,d\tau + \int_0^t \tilde{\eta}(\tau)\|u_{n\lambda}(\tau) - u_0\|_H\,d\tau,$$

where $\tilde{\eta} \in L^1(0, 1)$. By the Gronwall type inequality (Lemma 1.5.2), (9.1.29)-(9.1.30), and (9.1.13), we can easily show that

$$\|u_{n\lambda}(t)\|_H^2 \le C_9 + 4M_f + 4M_{u_0} + 2\int_0^t \tau\|u'_{n\lambda}(\tau)\|_H^2\,d\tau \le$$

$$\le C_{10} + 4M_f(t) + 4M_{u_0}(t) + \frac{1}{2}\|u_{n\lambda}(t)\|_H^2 \text{ for all } t \in [0, T].$$

By (9.1.29)-(9.1.30) and (9.1.13), $(u_{n\lambda}, \partial\phi_\lambda u_{n\lambda}, B_\lambda u_{n\lambda})$ satisfies (9.1.26). Using the weak lower semicontinuity of the norms, the inequality

$$\liminf_{\lambda \to 0+} \phi_\lambda\big(t, u_{n\lambda}(t)\big) \ge \phi\big(t, u_n(t)\big),$$

and (9.1.27)-(9.1.28), we finally obtain (9.1.26).

We denote $u_n^*(t) = \sqrt{t}\,u'_n(t)$, $\hat{v}_n(t) = \sqrt{t}\,v_n(t)$, and $\hat{w}_n(t) = \sqrt{t}\,w_n(t)$, for each $t \in [0, T]$. By (9.1.26) the sequences (u_n^*), $(\hat{v}_n)$, and $(\hat{w}_n)$ are bounded in the Hilbert space $L^2(0,T;H)$. Thus there are $u^*, \hat{v}, \hat{w} \in L^2(0,T;H)$ such that, on a subsequence,

$$u_n^* \to u^*, \ \hat{v}_n \to \hat{v}, \ \hat{w}_n \to \hat{w} \text{ weakly in } L^2(0,T;H), \text{ as } n \to \infty. \qquad (9.1.31)$$

By (9.1.5) and by the monotonicity of $\partial\phi(t,\cdot)$ and of $B(t)$, we have

$$\frac{1}{2}\frac{d}{dt}\|u_n(t) - u_m(t)\|_H^2 \leq \|u_n(t) - u_m(t)\|_H \|f_n(t) - f_m(t)\|_H,$$

for a.a. $t \in (0,T)$. Therefore,

$$\|u_n - u_m\|_{C([0,T];H)} \leq \|u_{0n} - u_{0m}\|_H + \int_0^T \|f_n(t) - f_m(t)\|_H \, dt,$$

i.e., (u_n) is a Cauchy sequence. Since $C([0,T];H)$ is complete, there exists $u \in C([0,T];H)$ such that

$$u_n \to u \text{ in } C([0,T];H), \text{ as } n \to \infty. \qquad (9.1.32)$$

Hence the initial condition (9.1.3) is satisfied, due to (9.1.22) and (9.1.25).

We denote $v(t) = \hat{v}(t)/\sqrt{t}$, $w(t) = \hat{w}(t)/\sqrt{t}$, for all $t \in (0,T]$. Let $\delta \in (0,T)$ and $\xi \in C_0^\infty(\delta,T)$. Then, as $n \to \infty$,

$$\int_\delta^T u^*(t)\frac{\xi(t)}{\sqrt{t}} \, dt \leftarrow \int_\delta^T u_n^*(t)\frac{\xi(t)}{\sqrt{t}} \, dt =$$

$$= -\int_\delta^T u_n(t)\xi'(t) \, dt \to -\int_\delta^T u(t)\xi'(t) \, dt.$$

Hence u belongs to $H^1(\delta,T;H)$, and $u'(t) = u^*(t)/\sqrt{t}$ for a.a. $t \in (\delta,T)$. Since $\delta \in (0,T)$ is arbitrary, $u'(t) = u^*(t)/\sqrt{t}$ a.e. on $(0,T)$. Let $x \in H$ and $t \in (0,T)$ be arbitrary. Now, using (9.1.31), we get

$$\int_\delta^t \left(x, f(\tau) - u'(\tau) - v(\tau) - w(\tau)\right)_H d\tau =$$

$$= \lim_{n\to\infty} \int_\delta^t \left(x, f(\tau) - u_n'(\tau) - v_n(\tau) - w_n(\tau)\right)_H d\tau = 0.$$

By differentiating, we see that (9.1.1) is satisfied.

Since $u_n \to u$ strongly and $v_n \to v$, $w_n \to w$ weakly in $L^2(\delta,T;H)$, the demiclosedness of maximal monotone operators and Lemma 9.1.1 give that $v \in \partial\phi u$ and $w \in Bu$, i.e., (9.1.2) is satisfied.

By the weak lower semicontinuity of the norms and of $\phi(t,\cdot)$ and by (9.1.31)-(9.1.32), we deduce that (u,v,w) satisfy (9.1.5). The uniqueness of u can be established as in the proof of Theorem 9.1.1.

Theorem 9.1.2 is proved. $\blacksquare$

First order differential equations perturbed by functionals

Let $V \subset H$ be a real Banach space, $a > 0$, and $j = 0, 1$. Let us define the sets X_j and the norms $\| \cdot \|_{j,t}$, $t \in (0, T]$, by

$$X_j = \left\{ y \in C([0, T]; H) \mid t \mapsto \sqrt{t}^j y(t) \text{ belongs to } L^2(0, T; V) \text{ and} \right.$$
$$\left. y(t) \in \overline{D(\phi(t, \cdot))} \text{ for all } t \in [0, T] \right\},$$

$$\|y\|_{j,t}^2 = \|y\|_{L^\infty(0,t;H)}^2 + \int_0^t s^j \|y(s)\|_V^2 \, ds, \text{ for all } y \in X_j.$$

Now, we state auxiliary hypotheses for the case of locally Lipschitzian perturbations.

(H.5) The set $V \subset H$ is separable and $V \subset H$ densely and continuously.

(H.6) For all $t \in [0, T]$ and $x \in D(\phi(t, \cdot))$,

$$D(\phi(t, \cdot)) \subset V, \text{ and } a\|x\|_V^2 \leq \phi_+(t, x) + \eta(t)\left(1 + \|x\|_H^2\right).$$

(H.7) The functional f maps X_j into $L^{2-j}(0, T; H)$.

(H.8) For all $K > 0$, there exist $\eta_K \in L^1(0, T)$ and $a_K > 0$ such that

$$\|f(x)(t) - f(y)(t)\|_H \leq \eta_K(t)\|x - y\|_{j,t},$$
$$a_K\|x(t) - y(t)\|_V^2 \leq \left(x(t) - y(t), \hat{x} - \hat{y}\right)_H + \eta_K(t)\|x(t) - y(t)\|_H^2$$

for a.a. $t \in (0, T)$ and all $\hat{x} \in \partial\phi(t, \cdot)x(t) + B(t)x(t)$, $\hat{y} \in \partial\phi(t, \cdot)y(t) + B(t)y(t)$, and

$$x, y \in X_{j,K} := \left\{ \tilde{x} \in X_j \,\Big|\, \|\tilde{x}\|_{j,T}^2 + \operatorname*{ess\,sup}_{0 \leq t \leq T} t^j \phi_+\left(t, \tilde{x}(t)\right) \leq K^2 \right\}.$$

(H.9) For all $y \in X_j$ and a.a. $t \in (0, T)$,

$$t^j \|f(y)(t)\|_H^2 \leq \eta(t)\left(1 + \|y\|_{j,t}^2 + \operatorname*{ess\,sup}_{0 \leq \tau \leq t} \tau^j \phi_+\left(\tau, y(\tau)\right)\right).$$

(H.10) For all $y \in X_1$ and a.a. $t \in (0, T)$,

$$\|f(y)(t)\|_H \leq \eta(t)\left(1 + \|y\|_{1,t} + \operatorname*{ess\,sup}_{0 \leq \tau \leq t} \sqrt{\tau \phi_+\left(\tau, y(\tau)\right)}\right).$$

REMARK 9.1.3   In typical applications $H = L^2(0, 1)$, $V = H^1(0, 1)$, and $\partial\phi(t, \cdot)u = -u_{rr}$. The space V is introduced in order to allow f to depend on u_r, not only on u. If this is not needed, one can simply remove all terms including V in (H.5)-(H.6), the second inequality of (H.8), and Theorem 9.1.3. ∎

THEOREM 9.1.3

Assume (H.1)-(H.9). If $j = 0$ and $u_0 \in D\big(\phi(0,\cdot)\big)$, then there exist $u \in H^1(0,T;H) \cap L^2(0,T;V)$ and $v,w \in L^2(0,T;H)$ that satisfy

$$u'(t) + v(t) + w(t) = f(u)(t) \text{ for a.a. } t \in (0,T), \tag{9.1.33}$$

$$v(t) \in \partial\phi(t,\cdot)u(t),\ w(t) \in B(t)u(t) \text{ for a.a. } t \in (0,T), \tag{9.1.34}$$

$$u(0) = u_0. \tag{9.1.35}$$

If $j = 1$, $u_0 \in \overline{D\big(\phi(0,\cdot)\big)}$, (H.10) is satisfied, and

$$\int_0^T \frac{1}{\tau}\|z(\tau) - u_0\|_H^2\, d\tau < \infty,$$

then there exist a unique $u \in C\big([0,T];H\big)$, which is differentiable a.e. on $(0,T)$ and V-valued measurable on $[0,T]$, and measurable functions $v,w\colon [0,T] \to H$ that satisfy (9.1.33)-(9.1.35) and

$$\int_0^T \tau\Big(\|u(\tau)\|_V^2 + \|u'(\tau)\|_H^2 + \|v(\tau)\|_H^2 + \|w(\tau)\|_H^2\Big)\, d\tau < \infty. \tag{9.1.36}$$

PROOF   Let us first show that X_0 and X_1 are *nonempty*. Since $\phi(0,\cdot)$ is proper, there exists a $\hat{u}_0 \in D\big(\phi(0,\cdot)\big)$ and thus by Theorem 9.1.1 there is $\tilde{u} \in H^1(0,T;H)$, which satisfies (9.1.1)-(9.1.2) with $\tilde{u}(0) = \hat{u}_0$ such that $\tilde{u}(t) \in D\big(\phi(t,\cdot)\big)$ for a.a. $t \in (0,T)$. Since $\tilde{u}$ is continuous, $\tilde{u}(t) \in \overline{D\big(\phi(t,\cdot)\big)}$, for all $t \in [0,T]$.

We identify H^*, the dual of H, with H. Since V is dense in H, then $H = H^*$ is dense in V^*. Thus for any $x \in V^*$, there is a sequence (x_n) of elements from H such that

$$\big\langle \tilde{u}(t), x \big\rangle = \lim_{n\to\infty} \langle \tilde{u}(t), x_n \rangle_V = \lim_{n\to\infty} \big(\tilde{u}(t), x_n\big)_H,$$

for a.a $t \in (0,T)$. Hence $t \mapsto \tilde{u}(t)$ is weakly measurable in V. By the separability of V and Pettis' theorem, $t \mapsto \tilde{u}(t)$ is also measurable in V. Using (H.6) and (9.1.4), we get

$$a\|\tilde{u}(t)\|_V^2 \le \phi_+\big(t, \tilde{u}(t)\big) + \eta(t)\big(1 + \|\tilde{u}(t)\|_H^2\big) \le$$
$$\le \big(1 + \eta(t)\big)M_2\big(1 + \|\hat{u}_0\|_H^2 + \phi_+(0,\hat{u}_0)\big) + \eta(t)$$

for a.a. $t \in (0,T)$. Hence $u \in L^2(0,T;H)$ and so $\tilde{u} \in X_0 \cap X_1$.

Next, we define the norms $\|\cdot\|_{j,\xi}$ by

$$\|x\|_{j,\xi}^2 = \sup_{0\le t\le T} e^{-2E_\xi(t)} \left(\|x(t)\|_H^2 + \int_0^t s^j \|x(s)\|_V^2\, ds \right),$$

$$E_\xi(t) = \max\left\{1, c_0, c_1, \sqrt{2c_1}\right\} \int_0^t \xi(s)\, ds,$$

$$c_j = 4M_{2j+1}\left(1 + \frac{T^j}{a}\|\eta\|_{L^1(0,T)} + \frac{T}{a} \right),\ j = 0,1,$$

where $\xi \in L^1(0,T)$ is nonnegative. Observe that

$$\|x\|_{j,t}^2 \le 2 \operatorname*{ess\,sup}_{0\le s\le t} \left(\|x(s)\|_H^2 + \int_0^s \sigma^j \|x(\sigma)\|_V^2\, d\sigma \right) \le$$

$$\le 2e^{2E_\xi(t)} \|x\|_{j,\xi}^2 \tag{9.1.37}$$

for all $t \in (0,T]$. We denote, for $N > 0$ and $j = 0,1$,

$$\tilde{X}_{j,N} = \left\{ y \in X_j \,\middle|\, \|y\|_{j,\eta}^2 + \operatorname*{ess\,sup}_{0\le t\le T} e^{-2E_\eta(t)} t^j \phi_+\big(t, u(t)\big) \le N^2 \right\}.$$

Consider the mapping $P\colon X_j \mapsto C([0,T]; H)$, $Px = u$, where u satisfies

$$u'(t) + v(t) + w(t) = f(x)(t) \text{ for a.a. } t \in (0,T) \tag{9.1.38}$$

$$v(t) \in \partial\phi(t,\cdot)u(t),\ w(t) \in B(t)u(t) \text{ for a.a. } t \in (0,T), \tag{9.1.39}$$

$$u(0) = u_0. \tag{9.1.40}$$

By (H.7) and Theorem 9.1.1 or 9.1.2 such u exists and is unique.

Let $x \in X_j$, $j = 0,1$, and $u = Px$. As $t \mapsto \tilde{u}(t)$ above, the function $t \mapsto u(t)$ turns out to be measurable in V. By (H.6), we have

$$\|u(t)\|_H^2 + \int_0^t s^j \|u(s)\|_V^2\, ds + \operatorname*{ess\,sup}_{0\le s\le t} s^j \phi_+\big(s, u(s)\big) \le$$

$$\le \|u(t)\|_H^2 + \frac{1}{a}\int_0^t s^j \Big(\phi_+\big(s, u(s)\big) + \eta(s)\big(1 + \|u(s)\|_H^2\big) \Big)\, ds +$$

$$+ \operatorname*{ess\,sup}_{0\le s\le t} s^j \phi_+\big(s, u(s)\big) \le \Big(1 + \frac{T^j}{a}\|\eta\|_{L^1(0,T)}\Big)\|u\|_{L^\infty(0,t;H)}^2 +$$

$$+ \Big(1 + \frac{T}{a}\Big) \operatorname*{ess\,sup}_{0\le s\le t} s^j \phi_+\big(s, u(s)\big) + \frac{T^j}{a}\|\eta\|_{L^1(0,T)} \tag{9.1.41}$$

for all $t \in [0,T]$. By (H.9) and (9.1.37), we get for all $t \in [0,T]$,

$$\int_0^t s^j \|f(x)(s)\|_H^2\, ds \le \int_0^t \eta(s)\Big(\operatorname*{ess\,sup}_{0\le \tau\le s} \tau^j \phi_+\big(\tau, x(\tau)\big) +$$

$$+ (1 + \|x\|_{j,s}^2 + d\sigma) \, ds \leq \frac{e^{2E_\eta(t)} - 1}{2 \max\{1, c_0, c_1, \sqrt{2c_1}\}} \times$$

$$\times \left(\|x\|_{j,\eta}^2 + \operatorname*{ess\,sup}_{0 \leq s \leq T} e^{-2E_\eta(s)} s^j \phi_+(s, x(s)) \right) + \|\eta\|_{L^1(0,T)}. \quad (9.1.42)$$

Let $x, y \in X_{j,N}$, $j = 0, 1$ and $N, K > 0$. We subtract the corresponding equations (9.1.38) and multiply the result by $Px(t) - Py(t)$. By (H.8), we have

$$\frac{d}{dt} \frac{1}{2} \|Px(t) - Py(t)\|_H^2 + a_K \|Px(t) - Py(t)\|_V^2 \leq \qquad\qquad (9.1.43)$$

$$\leq \eta_K(t) \|Px(t) - Py(t)\|_H^2 + \big(f(x)(t) - f(y)(t), Px(t) - Py(t)\big)_H$$

for a.a. $t \in (0, T)$. We drop the second term, integrate over $[0, t] \subset [0, T]$, and use the initial condition. Then

$$\frac{1}{2} \|Px(t) - Py(t)\|_H^2 \leq \int_0^t \eta_K(s) \|Px(s) - Py(s)\|_H^2 \, ds +$$

$$+ \int_0^t \|f(x)(s) - f(y)(s)\|_H \|Px(s) - Py(s)\|_H \, ds$$

for all $t \in [0, T]$. Now, Gronwall's inequality implies

$$\frac{1}{2} \|Px(t) - Py(t)\|_H^2 \leq$$

$$\leq e^{2\|\eta_K\|_{L^1(0,T)}} \int_0^t \|f(x)(s) - f(y)(s)\|_H \|Px(s) - Py(s)\|_H \, ds$$

for all $t \in [0, T]$. Using the Gronwall type inequality, we arrive at

$$\|Px(t) - Py(t)\|_H \leq e^{2\|\eta_K\|_{L^1(0,T)}} \int_0^t \|f(x)(s) - f(y)(s)\|_H \, ds \quad (9.1.44)$$

for all $t \in [0, T]$. We multiply (9.1.43) by t^j, integrate over $[0, t] \subset [0, T]$, and use the initial condition. Then

$$\frac{1}{2} t^j \|Px(t) - Py(t)\|_H^2 + a_K \int_0^t s^j \|Px(s) - Py(s)\|_V^2 \, ds \leq$$

$$\leq \int_0^t \eta_K(s) \Big(s + \frac{j}{2}\Big) \|Px(s) - Py(s)\|_H^2 \, ds +$$

$$+ \int_0^t s^j \|f(x)(s) - f(y)(s)\|_H \|Px(s) - Py(s)\|_H \, ds$$

for all $t \in [0, T]$. In view of (9.1.44) this yields

$$\int_0^t \|Px(s) - Py(s)\|_V^2 \, ds \leq \frac{1}{a_K} \left(\int_0^t \|f(x)(s) - f(y)(s)\|_H \, ds \right)^2 \times$$

$$\times \left(\left(T + \frac{j}{2} \right) \|\eta_K\|_{L^1(0,T)} e^{4\|\eta_K\|_{L^1(0,T)}} + T^j e^{2\|\eta_K\|_{L^1(0,T)}} \right) =$$

$$=: c_K \left(\int_0^t \|f(x)(s) - f(y)(s)\|_H \, ds \right)^2 \tag{9.1.45}$$

for all $t \in [0,T]$. We denote

$$b_K := \max\{ e^{4\|\eta_K\|_{L^1(0,T)}}, c_K \}.$$

By (9.1.44), (H.8), and (9.1.37), we obtain

$$\|Px(t) - Py(t)\|_H^2 + \int_0^t s^j \|Px(s) - Py(s)\|_V^2 \, ds \leq$$

$$\leq b_K \left(\int_0^t \|f(x)(s) - f(y)(s)\|_H \, ds \right)^2$$

$$\leq b_K \left(\int_0^t \eta_K(s) \sqrt{2} e^{E_{\tilde{\eta}}(s)} \|x - y\|_{j,\tilde{\eta}} \, ds \right)^2$$

$$\leq b_K \left(\int_0^t \eta_K(s) e^{E_{\tilde{\eta}}(s)} \, ds \right)^2 2 \|x - y\|_{j,\tilde{\eta}}^2$$

$$\leq \frac{1}{2} e^{2 E_{\tilde{\eta}}(t)} \|x - y\|_{j,\tilde{\eta}}^2,$$

where $\tilde{\eta}(t) = 2\sqrt{b_K} \eta_K(t)$. Hence,

$$\|Px - Py\|_{j,\tilde{\eta}} \leq \frac{1}{\sqrt{2}} \|x - y\|_{j,\tilde{\eta}}, \tag{9.1.46}$$

i.e., $P: X_{j,K} \mapsto X_j$ is a strict contraction with respect to the norm $\| \cdot \|_{j,\tilde{\eta}}$.

Now, assume (H.1)-(H.9) with $j = 0$ and $u_0 \in D(\phi(0,\cdot))$. By (9.1.41), (9.1.4), (9.1.42), and (9.1.37), we can show that

$$\|u(t)\|_H^2 + \int_0^t \|u(s)\|_V^2 \, ds + \operatorname*{ess\,sup}_{0 \leq s \leq t} \phi_+\big(s, u(s)\big) \leq \frac{1}{a} \|\eta\|_{L^1(0,T)} +$$

$$+ \frac{c_0}{4M_1} \left(M_1 \int_0^t \|f(x)(s)\|_H^2 \, ds + M_2 \big(1 + \|u_0\|_H^2 + \phi_+(0, u_0)\big) \right) \leq$$

$$\leq C_1 + \frac{c_0}{4} \int_0^t \eta(s) \left(1 + 2 e^{2 E_\eta(s)} \|x\|_{0,\eta}^2 + \operatorname*{ess\,sup}_{0 \leq \tau \leq s} \phi_+\big(\tau, x(\tau)\big) \right) ds \leq$$

$$\leq C_2 + \frac{1}{4} \big(e^{2 E_\eta(t)} - 1 \big) \left(\|x\|_{0,\eta}^2 + \operatorname*{ess\,sup}_{0 \leq s \leq T} e^{-2 E_\eta(s)} \phi_+\big(s, x(s)\big) \right)$$

for all $t \in [0,T]$, where C_1 and C_2 are some positive constants. Hence

$$\|Px\|_{0,\eta}^2 + \operatorname*{ess\,sup}_{0 \leq t \leq T} e^{-2 E_\eta(t)} \phi_+\big(t, Px(t)\big) \leq$$

$$\leq \frac{1}{2} \left(\|x\|_{0,\eta}^2 + \operatorname*{ess\,sup}_{0 \leq t \leq T} e^{-2 E_\eta(t)} \phi_+\big(t, x(t)\big) \right) + 2 C_2.$$

We define positive numbers N, K and redefine the function $\tilde{\eta} \in L^1(0,T)$ by

$$N^2 = \max\left\{4C_2, \|\tilde{u}\|_{0,\eta}^2 + \operatorname*{ess\,sup}_{0 \le t \le T} e^{-2E_\eta(t)}\phi_+\left(t, \tilde{u}(t)\right)\right\},$$

$$K = N e^{E_\eta(T)}, \quad \tilde{\eta}(t) = 2\sqrt{b_K}\,\eta_K(t). \tag{9.1.47}$$

Then $\tilde{X}_{0,N} \ne \emptyset$ and P *maps* $\tilde{X}_{0,N}$ *into itself.* We endow $\tilde{X}_{0,N}$ with the norm $\| \cdot \|_{0,\tilde{\eta}}$, under which $X_{0,N}$ remains a complete metric space. By (9.1.46), $P: \tilde{X}_{0,N} \mapsto \tilde{X}_{0,N}$ is a strict contraction. By the Banach Fixed Point Theorem it has a unique fixed point in $\tilde{X}_{0,K}$. The desired solution is found. We have only to show its uniqueness in the whole X_0. So, let $x, y \in X_0$ be two fixed points of P. Then we can choose N above to be sufficiently large so that they both belong to $\tilde{X}_{0,N}$. In that case (9.1.46) is still valid and so $x = y$.

Now, assume (H.1)-(H.10) with $j = 1$, $M_{u_0} < \infty$ (see (9.1.19)), and $u_0 \in \overline{D\left(\phi(0, \cdot)\right)}$. Let $x \in X_1$. By (H.10) and (9.1.44), we have for all $t \in [0,T]$

$$\int_0^t \|f(x)(s)\|_H \, ds \le$$

$$\le \int_0^t \eta(s)\left(\operatorname*{ess\,sup}_{0 \le \tau \le s} \sqrt{\tau\phi_+\left(\tau, x(\tau)\right)} + \sqrt{2}e^{E_\eta(s)}\|x\|_{1,\eta} + 1\right) ds \le$$

$$\le \|\eta\|_{L^1(0,T)} + \int_0^t \sqrt{2}\eta(s)e^{E_\eta(s)} \, ds \times$$

$$\times\left(\|x\|_{1,\eta} + \operatorname*{ess\,sup}_{0 \le s \le T} \sqrt{e^{-2E_\eta(s)}s\phi_+\left(s, x(s)\right)}\right) \le \|\eta\|_{L^1(0,T)} +$$

$$+\frac{e^{E_\eta(t)} - 1}{\sqrt{c_1}}\sqrt{\|x\|_{1,\eta}^2 + \operatorname*{ess\,sup}_{0 \le s \le T} e^{-2E_\eta(s)}s\phi_+\left(s, x(s)\right)}. \tag{9.1.48}$$

On account of (9.1.41), (9.1.19), (9.1.48), and (9.1.44), we get for all $t \in [0,T]$,

$$\|(Px)(t)\|_H^2 + \int_0^t s\|Px(s)\|_V^2 \, ds + \operatorname*{ess\,sup}_{0 \le s \le t} s\phi_+\left(s, Px(s)\right) \le$$

$$\le \frac{c_1}{4M_3}\left(M_3 \int_0^t \|f(x)(s)\|_H^2 \, ds + M_3\left(\int_0^t \|f(x)(s)\|_H \, ds\right)^2 +\right.$$

$$\left.+M_4 M_{u_0}(T) + M_4\right) + \frac{T}{a}\|\eta\|_{L^1(0,T)} \le C_3 +$$

$$+\left(\frac{1}{8} + \frac{1}{4}\right)e^{2E_\eta(t)}\left(\|x\|_{1,\eta}^2 + \operatorname*{ess\,sup}_{0 \le s \le T} e^{-2E_\eta(s)}s\phi_+\left(s, x(s)\right)\right) + C_3,$$

where C_3 is a positive constant. Thus

$$\|Px\|_{1,\eta}^2 + \operatorname*{ess\,sup}_{0 \le s \le T} e^{-2E_\eta(s)}s\phi_+\left(s, Px(s)\right) \le$$

$$\leq \frac{3}{4}\left(\|x\|_{1,\eta}^2 + \operatorname*{ess\,sup}_{0\leq s\leq T} e^{-2E_\eta(s)} s\phi_+\big(s,x(s)\big)\right) + 2C_3.$$

Again we can choose N to be sufficiently large so that $\tilde{X}_{1,N}$ *is nonempty and P maps it into itself.* Moreover, $\tilde{X}_{1,N}$ is a complete metric space with respect to the norm $\|\cdot\|_{1,\tilde{\eta}}$, where K and $\tilde{\eta}$ are given by (9.1.47). By (9.1.46), $P\colon \tilde{X}_{1,N} \to \tilde{X}_{1,N}$ is a strict contraction and hence it has a fixed point $u \in \tilde{X}_{1,N}$. That is the desired solution. By (9.1.19) and Theorem 9.1.2, (9.1.36) is satisfied. The uniqueness of the solution can be proved similarly as in the first case.

Theorem 9.1.3 is now completely proved. ∎

In the next and last theorem of this section we shall replace the local Lipschitzianity of f by a compactness assumption. In order to formulate further hypotheses we denote

$$Y = \big\{y \in C([0,T];H) \mid y(t) \in D\big(\partial\phi(t,\cdot)\big) \text{ for a.a. } t \in (0,T)\big\}.$$

The set Y is nonempty under (H.1)-(H.4), as we can see by applying Theorem 9.1.1 with $f = 0$ and some $u_0 \in D\big(\phi(0,\cdot)\big)$. Let $j = 0,1$.

(H.11) The functional f maps Y into $L^1(0,T;H)$. The functional $y \mapsto \tilde{f}(y)$, $\tilde{f}(y)(t) = t^{j/2}f(y)(t)$ maps Y into $L^2(0,T;H)$ and it is strongly-weakly closed as a mapping $C\big([0,T];H\big) \to L^2(0,T;H)$.

(H.12) For each $\delta > 0$, the closure of the following set is compact in H:

$$\big\{y \in H \mid \|y\|_H^2 + \phi(t,y) \leq \delta \text{ for a.a. } t \in [0,T]\big\}.$$

(H.13) The function $\psi\colon [0,T] \times H \to [0,\infty]$ is measurable with respect to the σ-field generated by the products of Lebesgue sets in $[0,T]$ and Borel sets in H. For a.a. $t \in (0,T)$, $\psi(t,\cdot)$ is proper, convex, and lower semicontinuous.

(H.14) For each $y \in Y$ and a.a. $t \in (0,T)$, $y(t) \in D\big(\partial\psi(t,\cdot)\big)$ and

$$\psi(t,y(t)) \leq \big\|\partial\phi^0(t,\cdot)y(t)\big\|_H^2 + \big\|B^0(t)y(t)\big\|_H^2 + t^{-\alpha}\eta(t).$$

(H.15) For each $y \in Y$ and a.a. $t \in (0,T)$,

$$t^j\|f(y)(t)\|_H^2 \leq \eta(t)\left(1 + \|y\|_{C([0,t];H)}^2 + \int_0^t s^j\psi\big(s,y(s)\big)\,ds\right),$$

$$\|f(y)(t)\|_H^2 \leq \eta(t)^2\left(1 + \|y\|_{C([0,t];H)}^2 + \int_0^t s^j\psi\big(s,y(s)\big)\,ds\right).$$

THEOREM 9.1.4
Assume (H.1)-(H.4) and (H.11)-(H.15). If $u_0 \in D(\phi(0, \cdot))$ and $j = 0$, then there exist $u \in H^1(0, T; H)$ and $v, w \in L^2(0, T; H)$ satisfying (9.1.33)-(9.1.35). If $u_0 \in \overline{D(\phi(0, \cdot))}$, $j = 1$ and

$$M_{u_0}(T) := \int_0^T \frac{1}{\tau} \|z(\tau) - u_0\|_H^2 < \infty,$$

then there exist $u \in C([0, T]; H)$, differentiable a.a. on $(0, T)$, and measurable $v, w: [0, T] \to H$ that satisfy (9.1.33)-(9.1.35) and

$$\int_0^T \tau \left(\|u'(\tau)\|_H^2 + \|v(\tau)\|_H^2 + \|w(\tau)\|_H^2 \right) d\tau < \infty.$$

PROOF Denote for $j = 0, 1$,

$$Y_j = \left\{ y \in C([0, T]; H) \mid F_j(y) \leq N_j \right\},$$

$$F_j(y) = \sup_{0 \leq t \leq T} e^{-E_j(t)} \|y(t)\|_H^2 + \sup_{0 \leq t \leq T} e^{-E_j(t)} \int_0^t s^j \psi(s, y(s)) \, ds,$$

$$E_0(t) = 4M_1 \int_0^t \eta(s) \, ds, \ E_1(t) = 8 \max \left\{ M_3, 4\sqrt{M_3} \right\} \int_0^t \eta(s) \, ds,$$

$$N_0 = 4(1 + 2M_1)\|\eta\|_{L^1(0,T)} + 4M_2 \left(1 + \|u_0\|_H^2 + \phi(0, u_0) \right),$$

$$N_1 = 4(1 + M_3)\|\eta\|_{L^1(0,1)} + 4M_4 M_{u_0}(T) + 4M_4 + 8M_3\|\eta\|_{L^1(0,T)}^2.$$

Let $u_0 \in D(\phi(0, \cdot))$ and $j = 0$. By Theorem 9.1.1, problem (9.1.1)-(9.1.3) with $f = 0$ has a solution $\hat{u}$. By (9.1.4) and (H.14), we have

$$\max \left\{ \|\hat{u}(t)\|_H^2, \int_0^t \psi(s, \hat{u}(s)) \, ds \right\} \leq$$

$$\leq \|\eta\|_{L^1(0,T)} + M_2 \left(1 + \|u_0\|_H^2 + \phi_+(0, u_0) \right),$$

whence $F_0(\hat{u}) \leq N_0$ and thus $\hat{u} \in Y_0$, i.e., Y_0 is *nonempty*. Clearly, Y_0 is *convex*. It is also *closed* in $C([0, T]; H)$, by Fatou's lemma and the lower semicontinuity of $\psi(t, \cdot)$.

Consider the mapping $P: Y_0 \to C([0, T]; H)$, $Px = u$, where u satisfies (9.1.38)-(9.1.40). By Theorem 9.1.1, such $u \in H^1(0, T; H)$ exists and is unique. Let $x \in Y_0$. By (H.14), (9.1.4), and (H.15), we get

$$\|Px\|_{C([0,t];H)}^2 + \int_0^t \psi\left(s, (Px)(s)\right) ds \leq \int_0^t \eta(s) \, ds +$$

$$+ M_1 \int_0^t \|f(x)(s)\|_H^2 \, ds + M_2 \left(1 + \|u_0\|_H^2 + \phi_+(0, u_0) \right) \leq$$

$$\leq M_1 \int_0^t \eta(s)\left(\|x\|_{L^\infty(0,s;H)} + \int_0^s \psi\big(\tau, x(\tau)\big)\, d\tau + 1\right) ds +$$

$$+ \frac{1}{4}N_0 - 2M_1\|\eta\|_{L^1(0,T)} \leq \frac{1}{4}\int_0^t 4M_1\eta(s)\mathrm{e}^{E_0(s)}F_0(x)\, ds +$$

$$+ \frac{1}{4}N_0 = \frac{1}{4}\big(\mathrm{e}^{E_0(t)} - 1\big)N_0 + \frac{1}{4}N_0 = \frac{1}{4}\mathrm{e}^{E_0(t)}N_0 \qquad (9.1.49)$$

for all $t \in [0,T]$. Hence $F_0(Px) \leq \frac{1}{2}N_0 \leq N_0$ and so P *maps Y_0 into itself.*

Let $x, x_n \in Y_0$, $n = 1, 2\ldots$, be such that $x_n \to x$ in $C([0,T]; H)$, as $n \to \infty$. Denote $u_n = Px_n$. By Theorem 9.1.1 there exist v_n and w_n such that (u_n, v_n, w_n) satisfies (9.1.38)-(9.1.40) with x_n instead of x. Moreover, using (9.1.4) and (H.15) as in (9.1.49), we get

$$\int_0^T \left(\|u_n'(\tau)\|_H^2 + \|v_n(\tau)\|_H^2 + \|w_n(\tau)\|_H^2\right) d\tau +$$

$$+ \operatorname*{ess\,sup}_{0 \leq t \leq T} \phi_+\big(t, u(t)\big) \leq \frac{1}{4}\mathrm{e}^{E_0(T)}N_0 < \infty. \qquad (9.1.50)$$

Thus there are $u^*, v, w, c \in L^2(0,T; H)$ such that, on a subsequence,

$$u_n' \to u^*, \quad v_n \to v, \quad w_n \to w, \quad \text{and } f(x_n) \to c \text{ weakly in } L^2(0,T; H),$$

as $n \to \infty$. By (H.12) and (9.1.50), there is $S \subset [0,T]$ of zero measure such that $\{u_n(t) \mid t \in [0,T]\backslash S,\ n \in \mathbb{N}^*\}$ is relatively compact in H. By a general variant of Ascoli's theorem (see Theorem 1.1.4) and the equicontinuity of $\{u_n\}$ (see (9.1.50)), there exists a $\tilde{u} \in C([0,T]\backslash S; H)$ such that $u_n \to \tilde{u}$ in $C([0,T]\backslash S; H)$, as $n \to \infty$, on a subsequence. Thus $u_n \to \tilde{u}$ in $L^2(0,T; H)$, as $n \to \infty$. As in the proof of Theorem 9.1.1, $\tilde{u}' = u^*$, $v \in \partial\phi\tilde{u}$, and $w \in B\tilde{u}$. We extend $\tilde{u}$ to a function u from $C([0,T]; H)$. Then

$$\|u_n - u\|_{C([0,T];H)} = \|u_n - u\|_{C([0,T]\backslash S;H)} \to 0, \text{ as } n \to \infty.$$

By (H.11), $c = f(u)$. In view of the uniqueness result of Theorem 9.1.1, $\tilde{u} = u = Px$. Then,

$$\|Px_n - Px\|_{C([0,T];H)} = \|u_n - \tilde{u}\|_{C([0,T]\backslash S;H)} \to 0, \text{ as } n \to \infty.$$

This limit holds for the whole sequence, since the solution of (9.1.38)-(9.1.40) is unique. Hence $P: Y_0 \to Y_0$ is *continuous.*

Let $y_n \in P(Y_0)$, $n \in \mathbb{N}^*$. Then y_n satisfy (9.1.49). Again, (y_n) has a subsequence converging in $C([0,T]; H)$. Since $P(Y_0) \subset Y_0$ and Y_0 is closed, then $\overline{P(Y_0)}$ is sequentially compact. Hence it is compact (see [KanAki, § I.5.1]).

We have shown so far, that $P: Y_0 \to Y_0$ satisfies all the conditions of Schauder's fixed point theorem (see, e.g., [KanAki, § XVI.4.2]). Thus it has a fixed point $u \in Y_0$, which is the desired solution of (9.1.33)-(9.1.35).

Now, assume that $u_0 \in \overline{D(\phi(0,\cdot))}$, $M_{u_0}(T) < \infty$, and $j = 1$. By Theorem 9.1.2, problem (9.1.33)-(9.1.35) with $f = 0$ has a solution $\hat{u}$. Using (9.1.19) and (H.14), we get

$$\|\hat{u}(t)\|_H + \int_0^t s\psi\big(s, \hat{u}(s)\big)\, ds \leq \|\eta\|_{L^1(0,T)} + M_4 + M_{u_0}(T) + M_4 \leq \frac{1}{4}N_1.$$

Hence $F_1(\hat{u}) \leq N_1$ and so $\hat{u} \in Y_1$, i.e., Y_1 is *nonempty*. Clearly, it is *convex* and *closed*.

Let $x \in Y_1$. By (9.1.19) and (H.15), one gets

$$\|Px\|_{L^\infty(0,t;H)}^2 + \int_0^t s\psi\big(s,(Px)(s)\big)\, ds \leq \int_0^t \eta(s)\, ds +$$

$$+ M_3 M_f(t) + M_4 M_{u_0}(t) + M_4 \leq \|\eta\|_{L^1(0,T)} + M_4 M_{u_0}(T) +$$

$$+ M_4 + M_3 \int_0^t s\|f(x)(s)\|_H^2\, ds + M_3 \left(\int_0^t \|f(x)(s)\|_H\, ds\right)^2 \leq$$

$$\leq M_3 \int_0^t \eta(s)\left(\|x\|_{L^\infty(0,s;H)} + \int_0^s \tau\psi\big(\tau, y(\tau)\big)\, d\tau\right) ds + \frac{1}{4}N_0 -$$

$$- 2M_3\|\eta\|_{L^1(0,T)}^2 + M_3\left\{\int_0^t \eta(s)\left(1 + \|x\|_{L^\infty(0,s;H)} + \right.\right.$$

$$+ \left.\left.\left(\int_0^s \tau\psi\big(\tau, x(\tau)\big)\, d\tau\right)^{1/2}\right) ds\right\}^2 \leq \frac{1}{4}N_1 - 2M_3\|\eta\|_{L^1(0,T)}^2 +$$

$$+ M_3\left(\|\eta\|_{L^1(0,T)} + M_3 \int_0^t \eta(s)e^{E_1(s)} F_1(x)\, ds + \right.$$

$$+ \left.\int_0^t \eta(s)e^{\frac{1}{2}E_1(s)} F_1(x)^{1/2} s\, ds\right)^2 \leq \frac{1}{8}N_1\left(e^{E_1(t)} - 1\right) +$$

$$+ \frac{1}{4}N_1 + 2M_3\left(\frac{1}{4\sqrt{M_3}}\left(e^{\frac{1}{2}E_1(t)} - 1\right)\right)^2 \leq \frac{1}{4}N_1 e^{E_1(t)} \qquad (9.1.51)$$

for all $t \in [0,T]$. Hence $Px \in Y_1$, i.e., P *maps* Y_1 *into itself*.

Let $x_n, x \in Y_1$, $n = 1, 2, \cdots$, with $x_n \to x$ in $C([0,T]; H)$, as $n \to \infty$. Denote $u_n = Px_n$. By virtue of Theorem 9.1.3 there are v_n and w_n satisfying (9.1.33)-(9.1.35) with $f(x_n)$ instead of $f(x)$, and

$$\|u_n\|_{C([0,T];H)}^2 + \operatorname{ess\,sup}\big\{t\phi_+\big(t, u_n(t)\big) \mid t \in [0,T]\big\} +$$

$$+ \int_0^T t\left(\|u_n'(t)\|_H^2 + \|v_n(t)\|_H^2 + \|w_n(t)\|_H^2\right) dt \leq M_1', \qquad (9.1.52)$$

where M_1' is some positive constant. Next, we exploit the weighted space $L^2(0,T;H,t)$, which is a Hilbert space with the inner product

$$(p, q)_{L^2(0,T;H,t)} = \int_0^T \big(p(t), q(t)\big)_H\, t\, dt.$$

By (9.1.52) and (H.12) there exist $t' \in (0, T)$, $u^*, v, w, c \in L^2(0, T; H, t)$, $u \in L^2(0, T; H)$, and $u^0 \in H$ such that, on a subsequence, as $n \to \infty$,

$$u'_n \to u^*, \ v_n \to v, \ w_n \to w,$$
$$f(x_n) \to c \text{ weakly in } L^2(0, T; H, t), \tag{9.1.53}$$
$$u_n \to u \text{ weakly in } L^2(0, T; H), \text{ and } u_n(t') \to u^0 \text{ in } H. \tag{9.1.54}$$

Let $\epsilon \in (0, T)$ be arbitrary. Again $u^* = u'$ a.a. on (ϵ, T) and thus we may redefine u on a set of zero measure such that $u \in H^1(\epsilon, T; H)$. Since ϵ is arbitrary, $u \in C((0, T]; H)$. Indeed, for all $t \in (0, T]$,

$$u(t) = u^0 + \int_{t'}^{t} u'(\tau) \, d\tau \text{ and } u_n(t) \to u(t) \text{ weakly in } H, \text{ as } n \to \infty. \tag{9.1.55}$$

Let $\delta \in (0, T)$. By (H.12) there is a set $S \subset (\delta, T]$ of zero measure such that $\{u_n\}$ is an equicontinuous family of mappings from $[\delta, T] \setminus S$ to a compact metric space. By the general variant of Ascoli's theorem (see Theorem 1.1.4), there are a further subsequence (n_k) and a $\hat{u}$ such that

$$\|u_{n_k} - \hat{u}\|_{C([\delta, T] \setminus S; H)} \to 0, \text{ as } k \to \infty. \tag{9.1.56}$$

However, by (9.1.55), $\hat{u}(t) = u(t)$, for each $t \in [\delta, T] \setminus S$. Thus (9.1.56) holds for the whole original subsequence. Since u and u_n are continuous on $[\delta, T]$, we have

$$\|u_n - u\|_{C([\delta, T]; H)} = \|u_n - u\|_{C([\delta, T] \setminus S; H)} \to 0, \text{ as } n \to \infty. \tag{9.1.57}$$

By (9.1.33)-(9.1.35) and the monotonicity of $\partial \phi(t, \cdot)$ and of $B(t)$,

$$\frac{1}{2} \frac{d}{dt} \|u_n(t) - u_0\|_H^2 =$$
$$= \left(u_n(t) - u_0, f(x_n)(t) - \partial \phi^0(t, \cdot) z(t) - B^0(t) z(t)\right)_H +$$
$$+ \left(u_n(t) - u_0, \partial \phi^0(t, \cdot) z(t) + B^0(t) z(t) - v_n(t) - w_n(t)\right)_H \leq$$
$$\leq \|u_n(t) - u_0\|_H \|f(x_n)(t) + \partial \phi^0(t, \cdot) z(t) - B^0(t) z(t)\|_H +$$
$$+ \left(z(t) - u_0, \partial \phi^0(t, \cdot) z(t) + B^0(t) z(t) - v_n(t) - w_n(t)\right)_H \leq$$
$$\leq \|u_n(t) - u_0\|_H \left(\|f(x_n)(t)\|_H + \eta_1(t)\right) +$$
$$+ \eta_2(t) + 4\epsilon t \|v_n(t) + w_n(t)\|_H^2 + \frac{1}{\epsilon t} \|z(t) - u_0\|_H^2$$

for a.a. $t \in (0, T)$ and for any $\epsilon > 0$. Here the functions $\eta_1, \eta_2 \in L^1(0, T)$ are independent of ϵ and n. The last term is $\eta_3(t)/\epsilon$, where $\eta_3 \in L^1(0, T)$. We integrate over $[0, t]$, and then use (9.1.52) and the Gronwall type inequality (Lemma 1.5.2). So, we obtain that, for each $t \in [0, T]$ and $\epsilon \in (0, 1)$,

$$\|u_n(t) - u_0\|_H \leq \left(8\epsilon M_1' + \int_0^t \eta_2(s) \, ds + \frac{1}{\epsilon} \int_0^t \eta_3(s) \, ds\right)^{\frac{1}{2}} +$$

$$+ \int_0^t \eta_1(\tau)\, d\tau + \frac{1}{\sqrt{M_3}}\left(e^{\frac{1}{2}E_1(t)} - 1 \right), \qquad (9.1.58)$$

where the last term is the upper bound for $\int_0^t \|f(x_n)(s)\|_H\, ds$ and it has previously been derived in (9.1.51). Define $u(0) = u_0$. By (9.1.55) and the weak lower semicontinuity of the norm, one has

$$\|u_n - u\|_{C([0,T];H)} \le \sup_{0 < t \le \delta} \|u_n(t) - u(t)\|_H + \sup_{\delta < t \le T} \|u_n(t) - u(t)\|_H \le$$

$$\le \sup_{0 < t \le \delta} \left(\|u_n(t) - u_0\|_H + \liminf_{m \to \infty} \|u_0 - u_m(t)\|_H \right) + \|u_n - u\|_{C([\delta,T];H)}$$

$$\le \sup_{0 < t \le \delta} \; \sup_{m \in \mathbb{N}^*} 2\|u_m(t) - u_0\|_H + \|u_n - u\|_{C([\delta,T];H)}.$$

Then, by (9.1.57) and (9.1.58), one gets

$$\limsup_{n \to \infty} \|u_n - u\|_{C([0,T];H)} \le \left(8\epsilon M_1' + \int_0^\delta \eta_2(s)\, ds + \frac{1}{\epsilon}\int_0^\delta \eta_3(s)\, ds \right)^{\frac{1}{2}} +$$

$$+ \int_0^\delta \eta_1(\tau)\, d\tau + \frac{1}{\sqrt{M_3}}\left(e^{\frac{1}{2}E_1(\delta)} - 1 \right)^{\frac{1}{2}} \to \sqrt{8\epsilon M_1'} \to 0, \qquad (9.1.59)$$

as $\delta \to$ and $\epsilon \to 0$, successively. Hence $u_n \to u$ in $C([0,T];H)$, as $n \to \infty$, and thus $u \in C([0,T];H)$.

By the demiclosedness of maximal monotone operators, Lemma 9.1.1 and (H.11), it is easily seen that

$$v(t) \in \partial\phi(t,\cdot)u(t), \; w(t) \in B(t)u(t), \; c(t) = fu(t) \; \text{ for a.a. } t \in (\delta, T)$$

and for each $\delta \in (0, T)$. Thus u, v, w satisfy (9.1.33)-(9.1.35). Since the solution u of (9.1.33)-(9.1.35) is unique, it follows that the whole sequence (u_n) converges toward $u = Px$ in $C([0,T];H)$, as $n \to \infty$. Thus P is *continuous*.

Let $u_n \in P(Y_1)$, $n \in \mathbb{N}^*$. As above, there exist a $u \in C([0,T];H)$ such that $u_n \to u$ in $C([0,T];H)$, as $n \to \infty$, on a subsequence. Since Y_1 is closed and $P(Y_1) \subset Y_1$, then $\overline{P(Y_1)}$ is sequentially compact. Thus $\overline{P(Y_1)}$ is compact.

By Schauder's fixed point theorem, $P\colon Y_1 \to Y_1$ has a fixed point $u \in Y_1$, which is the desired solution. Theorem 9.1.4 is now completely proved.

9.2   An application

Let us apply our previous results to problem (9.0.1)-(9.0.5). For the sake of simplicity, we consider only the case where $G(r,t) \equiv G$ and $K(r,t) \equiv 0$.

For a more general application with r- and t-dependent G and K, see [HokMo1]. Let δ, $M, T > 0$, $H = L^2(0,1)$, and $V = H^1(0,1)$. We state the following hypotheses on G and $\beta(t)$, $t \in [0,T]$:

(J.1) There exist functions $g: \mathbb{R} \to (-\infty, \infty]$ and $j: [0,T] \times \mathbb{R}^2 \to (-\infty, \infty]$ such that g and $j(t, \cdot)$ are proper convex lower semicontinuous and $G = \partial g$, $\beta(t) = \partial j(t, \cdot)$ for all $t \in [0,T]$.

(J.2) The operator $G \subset \mathbb{R} \times \mathbb{R}$ satisfies

$$(\tilde{y}_1 - \tilde{y}_2)(y_1 - y_2) \geq \delta(y_1 - y_2)^2 \text{ for all } (y_1, \tilde{y}_1), (y_2, \tilde{y}_2) \in G,$$

i.e., it is strongly monotone.

(J.3) There exists a $z : [0,T] \to V$ such that $z \in L^2(0,T;H)$, $G^0 z_r(t, \cdot) \in H$, $\big(z(t,0), z(t,1)\big) \in D(\beta(t))$ and $g\big(z_r(t, \cdot)\big) \in L^1(0,1)$ for all $t \in [0,T]$.

(J.4) Either $\beta(t)$ is bounded for all $t \in [0,T]$ or G is bounded.

We define function $\phi: [0,T] \times H \to (-\infty, \infty]$ and operators $A(t) \subset H \times H$, $t \in [0,T]$, by

$$\phi(t,y) = \begin{cases} \int_0^1 g\big(y'(r)\big)\, dr + j\big(t, y(0), y(1)\big) & \text{if } y \in H^1(0,1), \\ \infty & \text{otherwise,} \end{cases} \tag{9.2.1}$$

$$A(t) = \big\{(u, -w') \mid u, w \in V, w(r) \in Gu'(r) \text{ for a.a. } r \in (0,1),$$
$$\big(w(0), -w(1)\big) \in \beta(t)\big(u(0), u(1)\big)\big\}. \tag{9.2.2}$$

LEMMA 9.2.1

Assume (J.1)-(J.3). Then, for every $t \in [0,T]$, $\phi(t, \cdot)$ is a proper convex lower semicontinuous function. If, in addition, (J.4) is satisfied, then $\partial\phi(t, \cdot) = A(t)$ and there exists a constant $M_\delta > 0$, independent of t, such that $u_\lambda(t) = \big(I + \lambda A(t)\big)^{-1} y$ satisfies, for any $y \in H$, and $\lambda \in (0,1]$,

$$\|u_\lambda(t)\|_H^2 + \lambda\|u_\lambda(t)\|_V^2 \leq M_\delta \Big(\|y\|_H^2 + \|z(t)\|_H^2 +$$

$$+ \lambda\|G^0 z_x(t, \cdot)\|_H^2 + \lambda\|\beta^0(t)\big(z(t,0), z(t,1)\big)\|_{\mathbb{R}^2}^2\Big). \tag{9.2.3}$$

PROOF Let $t \in [0,T]$ be arbitrary but fixed. By the lower semicontinuity of g, the integrand in (9.2.1) is measurable. Since G is a subdifferential, one has

$$g(y) \geq g\big(z_r(r,0)\big) + G^0 z_r(r,0)\big(y - z_r(r,0)\big)$$

for all $y \in \mathbb{R}$ and a.a. $r \in (0,1)$. Hence the integral in (9.2.1) is well defined, taking values from $(-\infty, \infty]$. The convexity of $\phi(t, \cdot)$ is clear. In order to prove its lower semicontinuity, let $\lambda > 0$ and consider the level sets $S_\lambda = \{y \in H \mid \phi(t, y) \leq \lambda\}$. Let $y_n \in S_\lambda$ and $y \in H$, $n \in \mathbb{N}^*$, be such that $y_n \to y$ in H, as $n \to \infty$. By (J.1) and the definition of the subdifferential, we obtain for a.a. $r \in (0,1)$,

$$j\big(t, y_n(0), y_n(1)\big) \geq j\big(t, z(0,t), z(1,t)\big) +$$
$$+ \Big(\beta^0(t)\big(z(0,t), z(1,t)\big), \big(y_n(0), y_n(1)\big) - \big(z(0,t), z(1,t)\big)\Big)_{\mathbb{R}^2} \geq$$
$$\geq j\big(t, z(0,t), z(1,t)\big) - \frac{\delta}{8}\|y_n\|_V^2 - c_t, \tag{9.2.4}$$

where $c_t > 0$ does not depend on n. Let $\mu > 0$ and $\xi \in \mathbb{R}$. Since G_μ is the Fréchet differential of g_μ, we have

$$g_\mu(\xi) = g_\mu\big(z_r(r,t)\big) + \int_{z_r(r,t)}^{\xi} G_\mu \tau \, d\tau = g_\mu\big(z_r(r,t)\big) +$$
$$+ \int_0^1 G_\mu\big(s\xi + (1-s)z_r(r,t)\big)\big(\xi - z_r(r,t)\big)\, ds = g_\mu\big(z_r(r,t)\big) +$$
$$+ \int_0^1 \frac{1}{s}\Big(G_\mu\big(s\xi + (1-s)z_r(r,t)\big) - G_\mu z_r(r,t)\Big) \times$$
$$\times \big(s\xi + (1-s)z_r(r,t) - z_r(r,t)\big)\, ds + G_\mu z_r(r,t)\big(\xi - z_r(r,t)\big).$$

Passing to the limit as $\mu \to 0+$ and using Fatou's lemma, Theorems 1.2.4 and 1.2.16, and (J.2), we get

$$g(\xi) \geq g\big(z_r(r,t)\big) + \int_0^1 \frac{1}{s}\Big(G^0\big(s\xi + (1-s)z_r(r,t)\big) - G^0 z_r(r,t)\Big) \times$$
$$\times \big(s\xi + (1-s)z_r(r,t) - z_r(r,t)\big)\, ds +$$
$$+ G^0 z_r(r,t)\big(\xi - z_r(r,t)\big) \geq g\big(z_r(r,t)\big) + \frac{\delta}{2}\big(\xi - z_r(r,t)\big)^2 -$$
$$- \frac{\delta}{8}\big(\xi - z_r(r,t)\big)^2 - \frac{2}{\delta}\big(G^0 z_r(r,t)\big)^2 \geq$$
$$\geq g\big(z_r(r,t)\big) + \frac{3\delta}{16}\xi^2 - \frac{3\delta}{8}z_r(r,t)^2 - \frac{2}{\delta}\big(G^0 z_r(r,t)\big)^2. \tag{9.2.5}$$

Taking into account (9.2.4), (J.3), (9.2.1) and substituting $\xi = y_n'(r)$, we obtain that

$$\frac{\delta}{16}\|y_n'\|_H^2 \leq \int_0^1 g\big(y_n'(r)\big)\, dr + j\big(t, y_n(0), y_n(1)\big) + \tilde{c}_t\big(1 + \|y_n\|_H^2\big) =$$
$$= \phi(t, y_n) + \tilde{c}_t\big(1 + \|y_n\|_H^2\big) \leq \lambda + \tilde{c}_t\big(1 + \|y_n\|_H^2\big),$$

where $\tilde{c}_t > 0$ does not depend on n. Since $y_n \to y$ in H, (y_n) is bounded in V. So, on a subsequence, $y_n \to y$ weakly in V, as $n \to \infty$. By Mazur's lemma we can form convex combinations z_n from y_n's such that $z_n \to y$ strongly in V. Hence, on a subsequence,

$$z_n(r) \to y(r) \text{ for all } r \in [0,1], \text{ and } z'_n(r) \to y'(r) \text{ for a.a. } r \in (0,1).$$

Since S_λ is convex, all $z_n \in S_\lambda$. Using Fatou's lemma and the lower semicontinuity of g and $j(t,\cdot)$, we obtain that

$$\lambda \geq \liminf_{n\to\infty} \int_0^1 g\big(z'_n(r)\big)\,dr + \liminf_{n\to\infty} j\big(t, z_n(0), z_n(1)\big) \geq$$

$$\geq \int_0^1 \liminf_{n\to\infty} g\big(z_n(r)\big)\,dr + j\big(t, y(0), y(1)\big) \geq \phi(t,y).$$

Thus $y \in S_\lambda$ and so $\phi(t,\cdot)$ is lower semicontinuous. Therefore, the subdifferential $\partial\phi(t,\cdot)$ exists. Evidently, it contains $A(t)$. Since $A(t) \subset H \times H$ is monotone, it is maximal if $R\big(I + A(t)\big) = H$.

We shall show that indeed $R\big(I + A(t)\big) = H$. Our proof is essentially the same as that of Proposition 2.1.1. As an exercise, we do not assume that $(0,0) \in G$ and $(0,0) \in \beta(t)$, since in nonautonomous cases that would not be achieved by redefining the corresponding operators without any loss of generality. By Lemma 2.1.1, the operator $F \subset H \times H$, defined by $Fw(r) = -w''(r)$ and

$$D(F) = \big\{ w \in H^2(0,1) \mid \big(w'(0), w'(1)\big) \in \gamma\big(w(0), -w(1)\big) \big\}, \qquad (9.2.6)$$

is maximal monotone in H, whenever $\gamma \subset \mathbb{R}^2 \times \mathbb{R}^2$ is maximal monotone.

Let $y \in H$ and $y_n \in C_0^\infty\big((0,1)\big)$, $n \in \mathbb{N}^*$, be such that $y_n \to y$ in H, as $n \to \infty$. Consider the problems

$$-w''_n(r) + G^{-1} w_n(r) + \frac{1}{n} w_n(r) = y'_n(r) \text{ for a.a. } r \in (0,1), \quad (9.2.7)$$

$$\big(w'_n(0), w'_n(1)\big) \in \beta(t)^{-1}\big(w_n(0), -w_n(1)\big). \qquad (9.2.8)$$

Choosing in (9.2.6) $\gamma = \beta(t)^{-1}$ and denoting by $\hat{F}$ the realization of $G^{-1} + 1/n$ in H, we see that (9.2.7)-(9.2.8) is equivalent to $Fw_n + \hat{F}w_n = y'_n$. Since G is strongly monotone, its inverse G^{-1} is Lipschitzian and monotone. Therefore, $\hat{F}$ is maximal monotone and coercive. Thus $F + \hat{F}$ is surjective (cf. Theorems 1.2.6 and 1.2.7). Hence there exist $w_n \in H^2(0,1)$ that satisfy (9.2.7)-(9.2.8). Define

$$u_n(r) = w'_n(0) + \int_0^r G^{-1} w_n(\sigma)\,d\sigma, \quad v_n(r) = u_n(r) + \frac{1}{n} \int_0^r w_n(\sigma)\,d\sigma \quad (9.2.9)$$

for all $r \in [0,1]$. Then $u_n, v_n \in H^1(0,1)$ and

$$v_n(r) - w'_n(r) = y_n(r) \text{ for all } r \in [0,1], \tag{9.2.10}$$

$$w_n(r) \in G u'_n(r) \text{ for a.a. } r \in (0,1), \tag{9.2.11}$$

$$\big(w_n(0), -w_n(1)\big) \in \beta(t)\big(v_n(0), v_n(1)\big). \tag{9.2.12}$$

We multiply (9.2.10) by $v_n - z(t)$ and then integrate over $[0,1]$. Therefore,

$$\big(v_n, v_n - z(t)\big)_H - \big(w'_n, v_n - z(t)\big)_H = \big(y_n, v_n - z(t)\big)_H. \tag{9.2.13}$$

We integrate the second term by parts, use the strong monotonicity of G, the monotonicity of $\beta(t)$, and (9.2.9). Then

$$-\big(w'_n, v_n - z(t)\big)_H = \big(w_n, v'_n - z_r(\cdot, t)\big)_H + \int_0^1 w_n(r)\big(v_n(r) - z(r,t)\big) =$$

$$= (w_n, \frac{1}{n} w_n)_H + \big(w_n - G^0 z_r(\cdot, t) + G^0 z_r(\cdot, t), u'_n - z_r(\cdot, t)\big)_H +$$

$$+ \left(\begin{pmatrix} w_n(0) \\ -w_n(1) \end{pmatrix}, \begin{pmatrix} v_n(0) \\ v_n(1) \end{pmatrix} - \begin{pmatrix} z(0,t) \\ z(1,t) \end{pmatrix} \right)_{\mathbb{R}^2} \geq \frac{1}{n}\|w_n\|_H^2 +$$

$$+ \frac{\delta}{2}\|u'_n - z_r(\cdot, t)\|_H^2 - m_t - m_t \big\|\big(v_n(0), v_n(1)\big)\big\|_{\mathbb{R}^2} \geq \frac{1}{n}\|w_n\|_H^2 +$$

$$+ \frac{\delta}{4}\|u'_n\|_H^2 - \frac{\delta}{2}\|z_r(\cdot, t)\|_H^2 - m_t - m_t\sqrt{2}\big(\|v_n\|_H + \|v'_n\|_H\big),$$

where $m_t > 0$ does not depend on n. By virtue of (9.2.13) and $y_n \to y$ in H, it follows that

$$\frac{1}{2}\|v_n\|_H^2 + \frac{\delta}{4}\|u'_n\|_H^2 + \frac{1}{n}\|w_n\|_H^2 \leq m_t^* + m_t\sqrt{2}\|v'_n\|_H,$$

where $m_t^* > 0$ does not depend on n. By (9.2.9)-(9.2.10) and $y_n \to y$ in H, we finally obtain the following estimate

$$\|u_n\|_{H^1(0,1)}^2 + \|v_n\|_{H^1(0,1)}^2 + \|w'_n\|_H^2 + \frac{1}{n}\|w_n\|_H^2 \leq M_t^* < \infty, \tag{9.2.14}$$

where $M_t^* > 0$ does not depend on n.

If $\beta(t)$ is bounded, then $\big(w_n(0), -w_n(1)\big)$ is also bounded. Assume that G is bounded and $\big(w_n(0), -w_n(1)\big)$ is unbounded. Then, by (9.2.14), $w_n(r)$ is unbounded, for each $r \in [0,1]$. By Fatou's lemma,

$$\liminf_{n \to \infty} \int_0^1 u'_n(r)^2 \, dr \geq \int_0^1 \liminf_{n \to \infty} \big(G^{-1} w_n(r)\big)^2 \, dr = \infty,$$

which contradicts (9.2.14). Hence in any case $\big(w_n(0), -w_n(1)\big)$ is bounded. Thus there exist $u, w \in H^1(0,1)$ such that, as $n \to \infty$,

$$v_n \to u, \ w_n \to w \text{ in } H, \text{ and } u'_n \to u', \ w'_n \to w', \text{ weakly in } H,$$

$$v_n(0) \to u(0), \ v_n(1) \to u(1), \ w_n(0) \to w(0), \ w_n(1) \to w(1),$$

on a subsequence.

Since $\beta(t)$ and the realization of G in H are maximal monotone, it is easily seen that

$$u(r) - w'(r) = y(r) \text{ for a.a. } r \in (0,1), \tag{9.2.15}$$

$$w(r) \in Gu'(r) \text{ for a.a. } r \in (0,1), \tag{9.2.16}$$

$$\big(w(0), -w(1)\big) = \beta(t)\big(u(0), u(1)\big). \tag{9.2.17}$$

Hence $A(t)$ given by (9.2.2) is maximal monotone in H.

It remains to prove (9.2.3). Let $y \in H$, $t \in [0,T]$, and $\lambda \in (0,1]$. The function u_λ satisfies $u_\lambda(t) + \lambda A(t)u_\lambda(t) \ni y$. Hence there exists a $w_\lambda(\cdot, t) \in H^1(0,1)$ such that

$$u_\lambda(r,t) - \lambda w_{\lambda,r}(r,t) = y(r) \text{ for a.a. } r \in (0,1), \tag{9.2.18}$$

$$w_\lambda(r,t) \in Gu_{\lambda,r}(r,t) \text{ for a.a. } r \in (0,1), \tag{9.2.19}$$

$$\big(w_\lambda(0,t), -w_\lambda(1,t)\big) \in \beta(t)\big(u_\lambda(0,t), u_\lambda(1,t)\big). \tag{9.2.20}$$

We now multiply (9.2.18) by $u_\lambda(r,t) - z(r,t)$ and then proceed as when we derived (9.2.14). In such a manner we arrive at (9.2.3).

Lemma 9.2.1 is proved. $\blacksquare$

REMARK 9.2.1 The operator $A(t)$ defined by (9.2.2) is still maximal monotone in H if $\beta(t)$ is only maximal monotone, not necessarily a subdifferential. Indeed, in proving the fact that $R\big(I + A(t)\big) = H$ we only needed that $\beta(t)$ is maximal monotone. $\blacksquare$

Let us now introduce further hypotheses.

(J.5) There exists an $h_0 \in (0,T)$ such that

$$(\hat{\mathbf{y}} - \hat{\mathbf{x}}, \mathbf{y} - \mathbf{x})_{\mathbb{R}^2} \geq -\frac{\delta}{4}\|\mathbf{y} - \mathbf{x}\|_{\mathbb{R}^2}^2 - h^2 M\big(1 + \|\mathbf{x}\|_{\mathbb{R}^2}^2 + \|\mathbf{y}\|_{\mathbb{R}^2}^2\big)$$

for all $h \in (0, h_0)$, $t \in [0, T - h]$, $(\mathbf{x}, \hat{\mathbf{x}}) \in \beta(t)$, and $(\mathbf{y}, \hat{\mathbf{y}}) \in \beta(t + h)$.

(J.6) For a.a. $t \in (0,T)$,

$$\big\|z(t)\big\|_H + \big|j\big(t, z(0,t), z(1,t)\big)\big| +$$
$$+ \big\|\beta^0(t)\big(z(0,t), z(1,t)\big)\big\|_{\mathbb{R}^2} + \big\|G^0 z_r(\cdot, t)\big\|_H \leq M.$$

(J.7) For each $h \in (0, h_0)$, $t \in [0, T - h]$, $\mathbf{y} \in \mathbb{R}^2$, and $\lambda \in (0,1]$,

$$j_\lambda(t + h, \mathbf{y}) - j_\lambda(t, \mathbf{y}) \leq$$
$$\leq Mh\big(1 + \|\mathbf{y}\|_{\mathbb{R}^2}\|\beta_\lambda(t+h)\mathbf{y}\|_{\mathbb{R}^2} + \|\mathbf{y}\|_{\mathbb{R}^2}^2 + j_{\lambda+}(t + h, \mathbf{y})\big).$$

(J.8) For each $(y, \xi) \in G$, it holds $|\xi| \leq M(1 + |y|)$.

LEMMA 9.2.2
Assume (J.1)-(J.6) and let $y \in H$, $0 < \lambda \leq \min\{1, \delta^{-1}\}$. Then $t \mapsto u_\lambda(t) = (I + \lambda A(t))^{-1} y$ belongs to $H^1(0, T; V)$, $t \mapsto \partial\phi_\lambda(t, \cdot)y = A_\lambda(t)y$ belongs to $H^1(0, T; H)$, and there exists a constant $M_5 > 0$, which satisfies, for a.a. $t \in (0, T)$,

$$\|u_\lambda'(t)\|_H^2 + \lambda\|u_\lambda'(t)\|_V^2 + \lambda^2\left\|\frac{d}{dt}A_\lambda(t)y\right\|_H^2 \leq \lambda M_5\left(1 + \|u_\lambda(t)\|_V^2\right). \quad (9.2.21)$$

PROOF   We subtract (9.2.18) for the values $t + h$ and t of the time variable, multiply the result by $u_\lambda(r, t + h) - u_\lambda(r, t)$, and then integrate with respect to r over $[0, 1]$. Hence, after integrating by parts, we get

$$\lambda\left(w_\lambda(t + h) - w_\lambda(t), u_{\lambda,r}(\cdot, t + h) - u_{\lambda,r}(\cdot, t)\right)_H +$$

$$+\|u_\lambda(t) - u_\lambda(t)\|_H^2 + \lambda\left(\begin{pmatrix} w_\lambda(0, t + h) \\ -w_\lambda(1, t + h) \end{pmatrix} - \right.$$

$$\left. - \begin{pmatrix} w_\lambda(0, t) \\ -w_\lambda(1, t) \end{pmatrix}, \begin{pmatrix} u_\lambda(0, t + h) \\ u_\lambda(1, t + h) \end{pmatrix} - \begin{pmatrix} u_\lambda(0, t) \\ u_\lambda(1, t) \end{pmatrix}\right)_{\mathbb{R}^2} = 0.$$

By the strong monotonicity of G and (J.5), it follows

$$\|u_\lambda(t + h) - u_\lambda(t)\|_H^2 + \lambda\delta\|u_{\lambda,r}(\cdot, t + h) - u_{\lambda,r}(\cdot, t)\|_H^2 \leq$$

$$\leq \lambda\frac{\delta}{4}\left\|\begin{pmatrix} u_\lambda(0, t + h) \\ u_\lambda(1, t + h) \end{pmatrix} - \begin{pmatrix} u_\lambda(0, t) \\ u_\lambda(1, t) \end{pmatrix}\right\|_{\mathbb{R}^2}^2 +$$

$$+\lambda h^2 M\left(1 + \left\|\left(u_\lambda(0, t + h), u_\lambda(1, t + h)\right)\right\|_{\mathbb{R}^2}^2\right) +$$

$$+\lambda h^2 M\left\|\left(u_\lambda(0, t), u_\lambda(1, t)\right)\right\|_{\mathbb{R}^2}^2. \quad (9.2.22)$$

We subtract (9.2.18) for the values $t + h$ and t of the time variable, multiply the result by $w_{\lambda,r}(r, t + h) - u_{\lambda,r}(r, t)$, and then integrate with respect to r over $[0, 1]$. So, after integrating by parts and by the strong monotonicity of G and (J.5), we obtain the following estimate

$$\delta\|u_{\lambda,r}(t + h) - u_{\lambda,r}(t)\|_H^2 + \lambda\|w_{\lambda,r}(\cdot, t + h) - w_{\lambda,r}(t)\|_V^2 \leq$$

$$\leq \frac{\delta}{4}\left\|\begin{pmatrix} u_\lambda(0, t + h) \\ u_\lambda(1, t + h) \end{pmatrix} - \begin{pmatrix} u_\lambda(0, t) \\ u_\lambda(1, t) \end{pmatrix}\right\|_{\mathbb{R}^2}^2 +$$

$$+h^2 M\left(1 + \left\|\left(u_\lambda(0, t + h), u_\lambda(1, t + h)\right)\right\|_{\mathbb{R}^2}^2\right) +$$

$$+h^2 M\left\|\left(u_\lambda(0, t), u_\lambda(1, t)\right)\right\|_{\mathbb{R}^2}^2. \quad (9.2.23)$$

Using the inequality $\|(x(0), x(1))\|_{\mathbb{R}^2} \leq \sqrt{2}\|x\|_V$ for all $x \in V$, we obtain from (9.2.22)-(9.2.23), (J.6), and (9.2.3) that there exists a $\hat{c}_\lambda > 0$, which does not depend on h, such that

$$\int_0^{T-h} \|u_\lambda(t+h) - u_\lambda(t)\|_V^2 \, dt \le \hat{c}_\lambda h^2,$$

$$\int_0^{T-h} \|w_{\lambda,r}(\cdot, t+h) - w_{\lambda,r}(\cdot,t)\|_H^2 \, dt \le \hat{c}_\lambda h^2.$$

Therefore, $t \mapsto u_\lambda(t)$ belongs to $H^1(0,T;V)$ and $t \mapsto w_{\lambda,r}(\cdot, t)$ belongs to $H^1(0,T;H)$. Thus $u'_\lambda(t)$ and $\frac{d}{dt} w_{\lambda,r}(\cdot,t)$ exist for a.a. $t \in (0,T)$. Hence (9.2.22) and (9.2.23) yield

$$\max\left\{\frac{1}{2}\|u'_\lambda(t)\|_H^2, \frac{\lambda\delta}{2}\|u'_\lambda(t)\|_V^2\right\} \le \lambda M\left(1 + 2\|(u_\lambda(0,t), u_\lambda(1,t))\|_{\mathbb{R}^2}^2\right),$$

$$\lambda \left\|\frac{d}{dt} w_{\lambda,r}(\cdot,t)\right\|_H^2 \le \frac{\delta}{2}\|u'_\lambda(t)\|_H^2 + M\left(1 + 2\|(u_\lambda(0,t), u_\lambda(1,t))\|_{\mathbb{R}^2}^2\right).$$

These inequalities and (9.2.18) imply (9.2.21). ∎

LEMMA 9.2.3
Assume (J.1)-(J.8). Let ϕ be given by (9.2.1), and $B(t) \equiv 0$. Then (H.1)-(H.4) are satisfied.

PROOF　The conditions (H.1) and (H.2) are guaranteed by Lemma 9.2.1 and (J.4). We are now going to prove (H.3). Let $y \in H$, $\lambda_0 = \min\{1, \delta^{-1}\}$, $\lambda \in (0, \lambda_0)$, $h \in (0,T)$, $t \in (0, T-h)$, and $u_\lambda(t) = (I + \lambda A(t))^{-1} y$. By the definition of the subdifferential and by Theorems 1.2.4 and 1.2.16, we obtain that

$$\phi_\lambda(t+h, y) - \phi_\lambda(t, y) = \phi(t+h, u_\lambda(t+h)) - \phi(t, u_\lambda(t)) +$$

$$+ \frac{1}{2\lambda}\|u_\lambda(t+h) - y\|_H^2 - \frac{1}{2\lambda}\|u_\lambda(t) - y\|_H^2 =$$

$$= j(t+h, u_\lambda(0, t+h), u_\lambda(1, t+h)) - j(t, u_\lambda(0,t), u_\lambda(1,t)) +$$

$$+ \int_0^1 \Big(g(u_{\lambda,r}(r, t+h)) - g(u_{\lambda,r}(r,t))\Big) \, dr +$$

$$+ \frac{1}{2\lambda}\big(u_\lambda(t+h) + u_\lambda(t) - 2y, u_\lambda(t+h) - u_\lambda(t)\big)_H \le$$

$$\le \liminf_{\mu \to 0+}\Big\{j\Big(t+h, (I + \mu\beta(t+h))^{-1}(u_\lambda(0, t+h), u_\lambda(1, t+h))\Big) -$$

$$- j_\mu(t, u_\lambda(0,t), u_\lambda(1,t))\Big\} +$$

$$+ \int_0^1 w_\lambda(r,t)(u_{\lambda,r}(r, t+h) - u_{\lambda,r}(r,t)) \, dr +$$

$$+ \frac{1}{2\lambda}\big(\lambda w_{\lambda,r}(\cdot, t+h) + \lambda w_{\lambda,r}(\cdot,t), u_\lambda(t+h) - u_\lambda(t)\big)_H. \tag{9.2.24}$$

By integrating by parts, we see that the sum of the last two terms in (9.2.24) is equal to

$$-\left(\begin{pmatrix} w_\lambda(0,t) \\ w_\lambda(1,t) \end{pmatrix}, \begin{pmatrix} u_\lambda(0,t+h) \\ u_\lambda(1,t+h) \end{pmatrix} - \begin{pmatrix} u_\lambda(0,t) \\ u_\lambda(1,t) \end{pmatrix}\right)_{\mathbb{R}^2} +$$

$$+\left(w_{\lambda,r}(\cdot,t+h) - w_{\lambda,r}(\cdot,t), u_\lambda(t+h), u_\lambda(t)\right)_H.$$

By Theorem 1.2.16, (J.7), and the definition of the subdifferential, the term whose limit was taken in (9.2.24) is less or equal to

$$j_\mu\big(t+h, u_\lambda(0,t+h), u_\lambda(1,t+h)\big) - j_\mu\big(t, u_\lambda(0,t+h), u_\lambda(1,t+h)\big) +$$

$$+j_\mu\big(t, u_\lambda(0,t+h), u_\lambda(1,t+h)\big) - j_\mu\big(t, u_\lambda(0,t), u_\lambda(1,t)\big) \le$$

$$\le Mh\Big(1 + \big\|\beta_\mu(t+h)\big(u_\lambda(0,t+h), u_\lambda(1,t+h)\big)\big\|_{\mathbb{R}^2} \times$$

$$\times \big\|\big(u_\lambda(0,t+h), u_\lambda(1,t+h)\big)\big\|_{\mathbb{R}^2} +$$

$$+j_{\mu+}\big(t+h, u_\lambda(0,t+h), u_\lambda(1,t+h)\big)\Big) +$$

$$+\left(\beta_\mu(t)\begin{pmatrix} u_\lambda(0,t) \\ u_\lambda(1,t) \end{pmatrix}, \begin{pmatrix} u_\lambda(0,t+h) \\ u_\lambda(1,t+h) \end{pmatrix} - \begin{pmatrix} u_\lambda(0,t) \\ u_\lambda(1,t) \end{pmatrix}\right)_{\mathbb{R}^2}.$$

We have used the inequalities

$$\|\beta_\mu \mathbf{y}\|_{\mathbb{R}^2} \le \|\beta^0(t)\mathbf{y}\|_{\mathbb{R}^2}, \quad \|(x(0), y(0))\|_{\mathbb{R}^2} \le \sqrt{2}\|x\|_V$$

and $w_\lambda(t)$ given by (9.2.18)-(9.2.20). Then (9.2.24) yields

$$\phi_\lambda(t,y) - \phi_\lambda(t,y) \le \frac{h}{9}\|w_\lambda(t+h)\|_V^2 + 4\|w_\lambda(t)\|_V\|u_\lambda(t+h) - u_\lambda(t)\|_V +$$

$$+hM_6\Big(1 + \|u_\lambda(t+h)\|_V^2 + j_+\big(t+h, u_\lambda(0,t+h), u_\lambda(1,t+h)\big)\Big) +$$

$$+\|w_{\lambda,r}(\cdot,t+h) - w_{\lambda,r}(\cdot,t)\|_H\|u_\lambda(t+h) - u_\lambda(t)\|_H, \qquad (9.2.25)$$

where $M_6 > 0$ is a constant. Similarly, we can obtain a lower estimate for $\phi_\lambda(t+h,y) - \phi_\lambda(t,y)$. By the definition of the subdifferential and (J.6), it is easily seen that

$$j_+\big(t+h, u_\lambda(0,t+h), u_\lambda(1,t+h)\big) \le M\big(1 + \sqrt{2}\|u_\lambda(t+h)\|_V\big). \qquad (9.2.26)$$

Since $u_\lambda \in H^1(0,T;V)$, we have

$$\int_0^{T-h} |\phi_\lambda(t+h,y) - \phi_\lambda(t,y)|^2 \, dt \le c_\lambda h^2,$$

for some $c_\lambda > 0$, which does not depend on h. Thus $\phi_\lambda(\cdot,y) \in H^1(0,T)$. On the other hand, (9.2.25)-(9.2.26), (J.8), and (9.2.21) imply that

$$\frac{\partial \phi_\lambda}{\partial t}(t,y) \le \limsup_{h\to 0+} \frac{1}{9}\Big(\|w_\lambda(t+h)\|_H^2 + \|w_{\lambda,r}(\cdot,t+h)\|_H^2\Big) + M_6 +$$

$$+M_6\|u_\lambda(t)\|_V^2 + M_6 M\left(1 + \sqrt{2}\|u_\lambda(t)\|_V\right) + \frac{2}{9}\|w_\lambda(t)\|_V^2 +$$

$$+\frac{18}{2}\|u_\lambda'(t)\|_V^2 \le \limsup_{h\to 0+} \frac{1}{9} M^2 \left(\|u_\lambda(t+h)\|_V + 1\right)^2 +$$

$$+\frac{1}{3}\|w_{\lambda,r}(\cdot,t)\|_H^2 + M_7\left(1 + \|u_\lambda(t)\|_V^2\right) + \frac{2}{9} M^2 \left(\|u_\lambda(t)\|_V + 1\right)^2 +$$

$$+\frac{18}{2} M_5\left(1 + 2\|u_\lambda(t)\|_V^2\right) \le \frac{1}{3}\|w_{\lambda,r}(\cdot,t)\|_H^2 + M_8\left(1 + \|u_\lambda(t)\|_V^2\right),$$

where $M_7, M_8 > 0$ are constants. By (9.2.3), (J.6), and (9.2.5), we get

$$\|u_\lambda(t)\|_V^2 \le \|u_\lambda(t)\|_H^2 + \|u_{\lambda,r}(\cdot,t)\|_H^2 \le M_9\left(1 + \|y\|_H^2 + \int_0^1 g\big(u_{\lambda,r}(r,t)\big)\,dr\right),$$

where $M_9 > 0$ is a constant. Using (9.2.4) with $u_\lambda(t)$ instead of y_n, we see that

$$j\big(t, u_\lambda(0,t), u_\lambda(1,t)\big) \ge \frac{\delta}{8}\|u_\lambda(t)\|_H^2 - M_{10},$$

where $M_{10} > 0$ is a constant. Hence

$$\frac{\partial\phi_\lambda}{\partial t}(t,y) \le \frac{1}{3}\|\partial\phi_\lambda(t,\cdot)y\|_H^2 + M_{10}\left(1 + \|y\|_H^2 + \phi_+\big(t, u_\lambda(t)\big)\right).$$

Thus (H.3) is satisfied.

The mapping $t \mapsto \left(I + \lambda\partial\phi(t,\cdot)\right)^{-1} y = u(t)$ has turned out to belong even to $H^1(0,T;V) \subset H^1(0,T;H)$. Thus it is measurable. Trivially, $t \mapsto \left(I + B(t)\right)^{-1} y = y$ is also measurable. Hence (H.4) is satisfied. ∎

We conclude this chapter by the following theorem, which is a direct consequence of Theorem 9.1.2, and Lemma 9.2.3.

THEOREM 9.2.1

Assume (J.1)-(J.8). Let $u_0 \in L^2(0,1)$ and $f \in L^1\big(0,T;L^2(0,1)\big)$ such that

$$\int_0^T \frac{1}{t}\int_0^1 |f(r,t)|^2 \, dr\, dt < \infty.$$

Then there exist $u \in C\big([0,T];L^2(0,1)\big)$, differentiable a.e. on $(0,T)$, and measurable $w:[0,T] \to H^1(0,1)$ that satisfy (9.0.1)-(9.0.5) with $K(t,x) \equiv 0$ and

$$\operatorname*{ess\,sup}_{0\le t\le T} \int_0^1 t\big(u_r(r,t)^2 + w(r,t)^2\big)\, dr +$$

$$+ \int_0^T \int_0^1 t\big(u_t(r,t)^2 + w_r(r,t)^2\big)\, dr\, dt < \infty.$$

References

[AtBDP]   H. Attouch, P. Bénilan, A. Damlamian & P. Picard, Equations d'évolution avec condition unilaterale, *C.R. Acad. Sci.*, A279 (1974), 606–609

[AttDam]   H. Attouch & A. Damlamian, Strong solutions for parabolic variational inequalities, *Nonlin. Anal.*, 2 (1978), no. 3, 329–353.

[Brézis1]   H. Brézis, *Opérateurs maximaux monotones et semi-groupes de contractions dans les espaces de Hilbert*, North-Holland, Amsterdam, 1973.

[HokMo1]   V.-M. Hokkanen & Gh. Moroşanu, A class of nonlinear parabolic boundary value problems, *Math. Sci. Res. Hot-Line*, 3 (4) (1999), 1–22.

[KanAki]   L.V. Kantorovitch & G.P. Akilov, *Functional Analysis*, 2nd edition, Nauka, Moscow, 1977. (In Russian).

[Kato]   T. Kato, Nonlinear semigroups and evolution equations, *J. Math. Soc. Japan.*, 19 (4) (1967), 508–520.

[Moro1]   Gh. Moroşanu, *Nonlinear Evolution Equations and Applications*, Editura Academiei and Reidel, Bucharest and Dordrecht, 1988.

[Tătaru]   D. Tătaru, Time dependent m-accretive operators generating differentiable evolutions, *Diff. Int. Eq.*, 4 (1991), 137–150.

Chapter 10

Implicit nonlinear abstract differential equations

We are motivated by the two-phase Stefan problem with convection and nonlinear Robin-Steklov type boundary conditions. This problem has arisen recently in modeling the continuous casting of steel. Let $T > 0$ be fixed and $\Omega \subset \mathbb{R}^n$, $n \in \mathbb{N}$, $n \geq 2$, be an open connected bounded domain such that the Gauss theorem, the Rellich theorem, [Agmon, p. 30], and the trace theorem, [Nečas, p. 84], are valid and, moreover, let the boundary measure of $\partial\Omega$ be finite.

Suppose that Ω is filled with some material in liquid or solid phase. Denote by $v(x,t)$ the energy density, by $u(x,t)$ the temperature, by $j(x,t)$ the density of the energy flow, and by $f(x,t)$ the density of the energy production rate at the point x and at the time t. This situation can be described by the system

$$\frac{\partial v}{\partial t}(x,t) + \nabla \cdot \mathbf{j}(x,t) = f(x,t), \tag{10.0.1}$$

$$v(x,t) \in Eu(x,t), \ (x,t) \in \Omega \times (0,T). \tag{10.0.2}$$

We assume that

$$Ey = \begin{cases} c_2 y + c_0 & \text{if } y > 0, \\ [0, c_0] & \text{if } y = 0, \\ c_1 y & \text{if } y < 0, \end{cases} \tag{10.0.3}$$

where c_1 and c_2 are some positive constants and c_0 is a nonnegative constant. Assume that the energy transfer is conductive or convective. Then

$$\mathbf{j}(x,t) = -k\nabla u(x,t) + g\big(x,t,u(x,t)\big), \ (x,t) \in \Omega \times (0,T), \tag{10.0.4}$$

where k is a positive constant and $g(x,t,y) = 0$ if $y < 0$ or $x \in \partial\Omega$. We *assume* that g is *given*, e.g., we know the mass flows. As boundary conditions and initial conditions we have

$$\hat{n}(x) \cdot \mathbf{j}(x,t) \in h(x,t)u(x,t), \ (x,t) \in \partial\Omega \times (0,T), \tag{10.0.5}$$

$$v(x,0) = v_0(x), \ x \in \Omega, \tag{10.0.6}$$

where $\hat{n}(x)$ is the outward unit normal to $\partial\Omega$ at $x \in \partial\Omega$ and $h(x,t) \subset \mathbb{R} \times \mathbb{R}$ is a monotone and affinely bounded operator.

From (10.0.1)-(10.0.2) and (10.0.4), we obtain

$$\int_\Omega y(x)v_t(x,t)\,dx + \int_\Omega y(x)\nabla \cdot \big(-\nabla u(x,t) + g(x,t)\big)\,dx =$$

$$= \int_\Omega yf\,dx \ \text{ for all } y \in H^1(\Omega). \tag{10.0.7}$$

Let $V = H^1(\Omega)$, $W = L^2(\Omega)$, and $i: V \to W$ be the canonical injection. Define $\phi: W \to \mathbb{R}$,

$$\phi(y) = \int_\Omega \left(\frac{1}{2}c_2 y_+(x)^2 + c_0 y_+(x) + \frac{1}{2}c_1 y_-(x)^2\right) d^n x,$$

where

$$y_- = \begin{cases} -y & \text{if } y < 0, \\ 0 & \text{otherwise,} \end{cases} \quad \text{and} \quad y_+ = \begin{cases} y & \text{if } y > 0, \\ 0 & \text{otherwise.} \end{cases}$$

By Theorem 1.2.19 we see that ϕ is a lower semicontinuous proper convex function and

$$v(x,t) \in Eu(x,t) \ \text{ for a.a. } x \in \Omega \iff v(\cdot,t) \in \partial\phi\, i\, u(t,\cdot).$$

Define, for each $t \in [0,T]$, $\mathcal{B}(t) \subset V \times V^*$ by $w \in \mathcal{B}(t)z$ if there exists $w_1 \in L^2(\partial\Omega)$ satisfying $w_1(x) \in h(x,t)z(x)$ a.a. on $\partial\Omega$ such that

$$\langle y, w \rangle = \int_\Omega \nabla y(x) \cdot \nabla z(x)\,dx + \int_{\partial\Omega} y(x)\,w_1(x). \tag{10.0.8}$$

Moreover, set $\tilde{f}(t) \in V^*$ and $\mathcal{C}(t): V \to V^*$,

$$\langle y, \mathcal{C}(t)z \rangle = \int_\Omega y(x)\nabla \cdot g\big(x,t,z(x)\big)\,dx, \tag{10.0.9}$$

$$\langle y, \tilde{f}(t) \rangle = \int_\Omega y(x)f(x,t)\,dx \ \text{ for all } y \in V, \tag{10.0.10}$$

where $\langle \cdot, \cdot \rangle$ is the pairing between V and its dual V^*. By Gauss' theorem and by the boundary conditions, we obtain from (10.0.1)-(10.0.2) that

$$v(t) \in \partial\phi(\cdot)iu(t), \ w(t) \in \mathcal{B}(t)u(t),$$
$$\langle iy, v_t \rangle_{W,W^*} + \langle y, w(t) \rangle + \langle y, \mathcal{C}(t)u(t) \rangle = \langle y, \tilde{f}(t) \rangle$$

for all $y \in V$ and $t \in [0,T]$. We notice that

$$\langle iy, \partial_t v(t) \rangle_{W,W^*} = \left\langle y, \frac{d}{dt}i^* v(t) \right\rangle,$$

where $\langle \cdot, \cdot \rangle_{W,W^*}$ is the pairing between W and W^*. Thus we have reduced the two phase Stefan problem to the following general problem:

$$v'(t) + w(t) + \mathcal{C}(t)u(t) = \tilde{f}(t), \ t > 0, \tag{10.0.11}$$

$$v(t) \in \mathcal{A}(t)u(t), \ \ w(t) \in \mathcal{B}(t)u(t), \ t > 0, \tag{10.0.12}$$

$$v(0) = v_0. \tag{10.0.13}$$

In the next sections we shall investigate the existence, uniqueness, and continuous dependence of the solution of (10.0.11)-(10.0.13). Also the existence of a periodic solution of (10.0.11)-(10.0.12) will be established. As a consequence of those investigations, we shall see under which conditions the two-phase Stefan problem (10.0.1)-(10.0.6) has a generalized solution

$$(u, v, \mathbf{j}) \in L^2\big(0, T; H^1(\Omega)\big) \times H^1\big(0, T; H^{-1}(\Omega)\big) \times L^2\big(0, T; L^2(\Omega)\big)^n$$

in Sobolev's sense.

Of course, the two-phase Stefan problem is only one application of the abstract problem (10.0.11)-(10.0.13). For other models, see [DiBSh]. Our main assumption will be that the operators $\mathcal{A}(t)$ and $\mathcal{B}(t)$ are maximal monotone such that $\mathcal{A}(t) + \mathcal{B}(t)$ is uniformly coercive for all $t \in [0, T]$. Hence the common character of the possible applications is that they are nonlinear *parabolic* problems with time dependent coefficients.

10.1 Existence of solution

In this section we shall give sufficient conditions under which problem (10.0.11)-(10.0.13) has a solution. Many authors have studied such a problem in the case where $\mathcal{A}$ and $\mathcal{B}$ do not depend on time. Let us mention three important works in chronological order by O. Grange and F. Mignot [GraMi], V. Barbu [Barbu4], and E. Di Benedetto and R.E. Showalter [DiBSh]. In [GraMi] and [Barbu4] the operators $\mathcal{A}$ and $\mathcal{B}$ are assumed to be bounded subdifferentials of some convex functions. In [DiBSh] $\mathcal{A}$ is a subdifferential and $\mathcal{B}$ is just a maximal monotone operator, but in turn they are both assumed to be affinely bounded. The main method in [GraMi] is time-discretization and a priori estimates which by compactness assumptions guarantee that the discrete approximations converge to a solution of the problem. In [Barbu4] a more general problem, namely a Volterra integral equation

$$v(t) + a * w(t) = \int_0^t \tilde{f}(s)\, ds + v_0, \ t > 0, \tag{10.1.1}$$

has been studied. Here, a is a real-valued continuous function with $a(0) > 0$ and $*$ denotes the convolution of two functions. The idea is to approximate (10.1.1) by an integro-differential equation, whose solvability is known, to derive a priori estimates and then show that the solutions of approximate problems converge to a solution of (10.1.1). In the calculations the properties of the Yosida approximation of the maximal monotone operator have been applied. As an advantage, the existence of the solution for integral equations is also established. However, in this approach an extra condition, the so-called angle condition $(w(t), v(t))_H \geq -1$, has to be assumed. In [DiBSh] the original problem is approximated by a problem where $\mathcal{B}$ is replaced by its Yosida approximation and $\mathcal{A}$ by a more coercive operator. Again, the solutions of approximate problems turn out to be bounded and hence to converge toward a solution of (10.0.11)-(10.0.13). The approach of [GraMi] has been applied in [Saguez], where the third operator $\mathcal{C}(t)$ is introduced, and later in [BeDuS], where $\mathcal{A}$ depends on time. The approach of Barbu has been applied in [Hokk3] and [Hokk4]. We shall here apply the approach of Di Benedetto and Showalter. For further details, see [Hokk1], [Hokk2], [Hokk5], and, of course, [DiBSh].

Let us make the following hypotheses.

We are given positive numbers a, M, and T, an increasing continuous function $g : \mathbb{R} \to \mathbb{R}$, a real reflexive Banach space W, a real Hilbert space V, and a linear compact injection $i \colon V \to W$ such that iV is dense in W. We denote by $\langle \cdot, \cdot \rangle$ the dual pairing between V and V^*, and by $i^* \colon W^* \to V^*$ the adjoint of $i \colon V \to W$. The Riesz map $V \to V^*$ is denoted by $\mathcal{R}$. The function $\phi \colon [0, T] \times W \to \mathbb{R}$ satisfies:

(A.1) For each $t \in [0, T]$, $\phi(t, \cdot)$ is continuous and convex.

(A.2) For each $x, y \in W$, the function $t \mapsto \phi(t, x)$ is differentiable a.e. and

$$\left|\phi_t(t, x)\right| \leq M\left(1 + \|x\|_W^2\right),$$
$$\left|\phi_t(t, x) - \phi_t(t, y)\right| \leq g\left(\|x\|_W + \|y\|_W\right)\|x - y\|_W$$

for a.a. $t \in (0, T)$.

(A.3) For each $t \in [0, T]$ and $(x, y) \in \partial\phi(t, \cdot)$, $\|y\|_{W^*} \leq g(\|x\|_W)$.

Define $\mathcal{A}(t) = i^* \partial\phi(t, \cdot)i$, $t \in [0, T]$, and its realization

$$\mathcal{A} = \left\{(x, y) \in L^2(0, T; V) \times L^2(0, T; V^*) \mid\right.$$
$$\left.(x(t), y(t)) \in \mathcal{A}(t) \text{ for a.a. } t \in (0, T)\right\}.$$

(A.4) If $x \in H^1(0, T; V)$, $y \in H^1(0, T; V^*)$, and $y \in \mathcal{A}x$, then

$$\langle x'(t), y'(t)\rangle \geq -M(1 + \|x'(t)\|_V\|x(t)\|_V) \text{ for a.a. } t \in (0, T).$$

The family of operators $\{\mathcal{B}(t) \subset V \times V^* \mid t \in [0, T]\}$ satisfies:

(B.1) $\mathcal{B}(t)$ is maximal monotone in V such that $\|y\|_{V^*} \leq M(1 + \|x\|_V)$, for each $(x, y) \in \mathcal{B}(t)$ and for a.a. $t \in (0, T)$.

(B.2) For each $\alpha > 0$ and $z \in V^*$, the function $t \mapsto \left(\mathcal{R} + \alpha\mathcal{B}(t)\right)^{-1} z$ is measurable.

The function $\mathcal{C}: [0, T] \times V \to V^*$ satisfies:

(C.1) For each $z \in V$, the function $t \mapsto \mathcal{C}(t)z$ is measurable.

(C.2) For each $x, y, z \in V$ and for a.a. $t \in (0, T)$, $\mathcal{C}(t)0 = 0$ and
$$|\langle x, \mathcal{C}(t)y - \mathcal{C}(t)z \rangle| \leq M\|ix\|_W\|y - z\|_V.$$

(C.3) If $x_n, x \in L^2(0, T; V), \|x_n\|_{L^2(0,T;V)} \leq M$, for each $n \in \mathbb{N}^*$, and $x_n(t) \to x(t)$ in W uniformly on $[0, T]$, as $n \to \infty$, then $\mathcal{C}x_n \to \mathcal{C}x$ weakly in $L^2(0, T; V^*)$, as $n \to \infty$, on a subsequence (we denote $\mathcal{C}x(t) = \mathcal{C}(t)x(t)$).

We are now prepared to state the first existence result. In fact, we shall use it as a lemma in proving existence for problems with degenerate operator $\mathcal{A}$. Observe that we do not state uniqueness, since the solution is not usually unique.

THEOREM 10.1.1

Assume (A.1)-(A.4), (B.1)-(B.2), and (C.1)-(C.3). Then, for each $\tilde{f} \in L^2(0, T; V^*)$ and $(u_0, \tilde{v}_0) \in \mathcal{A}(0)$, there exist functions $u \in H^1(0, T; V)$, $v \in H^1(0, T; V^*)$, and $w \in L^2(0, T; V^*)$ such that

$$\mathcal{R}u'(t) + v'(t) + w(t) + \mathcal{C}(t)u(t) = \tilde{f}(t), \tag{10.1.2}$$

$$v(t) \in \mathcal{A}(t)u(t), \ w(t) \in \mathcal{B}(t)u(t) \ \text{for a.a.} \ t \in (0, T), \tag{10.1.3}$$

$$\mathcal{R}u(0) + v(0) = \mathcal{R}u_0 + \tilde{v}_0. \tag{10.1.4}$$

PROOF   Let us first transform problem (10.1.2)-(10.1.4) into an equivalent Hilbert space problem, since we shall heavily use its inner product. We identify V and V^* by the Riesz map $\mathcal{R}: V \to V^*$ and denote

$$A(t) = \mathcal{R}^{-1}\mathcal{A}(t), \ A = \mathcal{R}^{-1}\mathcal{A}, \ B(t) = \mathcal{R}^{-1}\mathcal{B}(t), \ B = \mathcal{R}^{-1}\mathcal{B}$$
$$C(t) = \mathcal{R}^{-1}\mathcal{C}(t), \ C = \mathcal{R}^{-1}\mathcal{C}, \ f = \mathcal{R}^{-1}\tilde{f}, \ v_0 = \mathcal{R}^{-1}\tilde{v}_0.$$

Using these operators, we see that problem (10.1.2)-(10.1.4) is equivalent to the following one:

$$u'(t) + v'(t) + w(t) + C(t)u(t) = f(t), \tag{10.1.5}$$

$$v(t) \in A(t)u(t), \ w(t) \in B(t)u(t) \ \text{for a.a.} \ t \in (0, T), \tag{10.1.6}$$

$$u(0) + v(0) = u_0 + v_0. \tag{10.1.7}$$

Let $\lambda > 0$ and consider the following problem, which seems to approximate (10.1.5)-(10.1.7).

$$u'_\lambda(t) + v'_\lambda(t) + B_\lambda(t)u_\lambda(t) + C(t)u_\lambda(t) = f(t), \qquad (10.1.8)$$

$$v_\lambda(t) \in A(t)u_\lambda(t) \ \text{ for a.a. } t \in (0,T), \qquad (10.1.9)$$

$$u_\lambda(0) + v_\lambda(0) = u_0 + v_0, \qquad (10.1.10)$$

where $B_\lambda(t)$ is the Yosida approximate of $B(t)$ (if such exists). The idea of the proof is to indicate first that this approximate problem has a solution, then to establish that those solutions form a bounded sequence, whose subsequence turns out to converge toward a solution of problem (10.1.5)-(10.1.7). Let us first state and prove some lemmas concerning the Hilbert space operators A, B, C.

LEMMA 10.1.1

For each $t \in (0,T)$, the operators $A(t)$ and $B(t)$ are maximal monotone and everywhere defined in V.

PROOF Let $t \in (0,T)$ be arbitrary. First we observe that there exists a $(x_0, y_0) \in \partial\phi(t, \cdot) \subset W \times W^*$, since $\phi(t, \cdot)$ is a proper lower semicontinuous function on W (cf. Theorems 1.2.8 and 1.2.12). By (A.3) and the definition of the subdifferential

$$\phi(t, x) \leq \langle x_0 - x, y\rangle_{W \times W^*} + \phi(t, x_0) \leq g\big(\|x\|_W\big)\|x - x_0\|_W + \phi(t, x_0),$$

whenever $(x, y) \in \partial\phi(t, \cdot)$ and $x \in iV$. Since iV is dense in W, the domain of $\phi(t, \cdot)$ is the whole W. Thus it is continuous everywhere (cf. Theorem 1.2.10), in particular at $0 \in V$. By Theorem 1.2.17 $\mathcal{A}(t) \subset V \times V^*$ is the subdifferential of $\phi(t, i\cdot): V \to \mathbb{R}$. So it is maximal monotone, which implies both monotonicity and $R\big(\mathcal{A}(t) + \mathcal{R}\big) = V^*$ (see Minty's theorem). Thus $A(t)$ is monotone in V and $R\big(A(t) + I\big) = V$, whence it follows that $A(t)$ is maximal monotone in V. Minty's theorem also implies that $B(t)$ is maximal monotone. Since both are bounded operators, they are everywhere defined. ∎

LEMMA 10.1.2

Let $z \in H^1(0,T;V)$ and $(y, z) \in A$. Then, the mapping $t \mapsto \phi^*\big(t, z(t)\big)$, where $\phi^*(t, \cdot)$ is the conjugate of $\phi(t; \cdot)$, belongs to $W^{1,1}(0,T)$ and

$$\frac{d}{dt}\phi^*\big(t, z(t)\big) = -\phi_t\big(t, y(t)\big) + \big(y(t), z'(t)\big)_V \ \text{ for a.a. } t \in (0,T).$$

PROOF   Denote

$$F_h(t) = \frac{1}{h}\Big(\phi^*\big(t, z(t+h)\big) - \phi^*\big(t, z(t)\big)\Big) + \phi_t\big(t, iy(t)\big) - \big(y(t), z'(t)\big)_H.$$

Since $\phi\big(t, y(t)\big) + \phi^*\big(t, z(t)\big) = \big(y(t), z(t)\big)_V$, then $\phi^*\big(\cdot, z(\cdot)\big) \in L^1(0, T)$. From (A.2) and the definition of the subdifferential we see that, for a.a. $t, t+h \in (0, T)$,

$$\big(y(t), z(t+h) - z(t)\big)_V - \int_t^{t+h} \phi_t\big(s, y(s)\big)\, ds \leq$$

$$\leq \phi^*\big(t+h, z(t+h)\big) - \phi^*\big(t, z(t)\big) \leq$$

$$\leq \big(y(t+h), z(t+h) - z(t)\big)_V - \int_t^{t+h} \phi_t\big(s, y(t+h)\big)\, ds.$$

Thus $\int_0^{T-h} |F_h(t)|\, dt \to 0$, as $h \to 0^+$. So, we obtain the desired result.
∎

LEMMA 10.1.3
Let $z \in H^1(0, T; V)$ and $\alpha > 0$. Then, $t \mapsto \big(I + \alpha A(t)\big)^{-1} z(t)$ belongs to $H^1(0, T; V)$.

PROOF   Let $t, s \in (0, T)$, $s \leq t$. Define

$$\psi : V \to \mathbb{R}, \quad \psi(t, x) = \frac{1}{2}\|x\|_V^2 + \alpha\phi(t, x).$$

Then, $\partial\psi(t, \cdot) = I + \alpha A(t)$ and, by Theorem 1.2.12, $\big(I + \alpha A(t)\big)^{-1} = \partial\psi^*(t, \cdot)$. As a resolvent, it is single-valued. By Theorem 1.2.13 it is the Gâteaux differential of $\psi^*(t, \cdot)$. Hence,

$$\left\|\big(I + \alpha A(t)\big)^{-1} z(s) - \big(I + \alpha A(s)\big)^{-1} z(s)\right\|_V =$$

$$= \sup_{\|y\|_V \leq 1} \lim_{\epsilon \to 0^+} \frac{1}{\epsilon} \int_s^t \left(\frac{d}{d\tau}\psi^*\big(\tau, z(s) + \epsilon y\big) - \frac{d}{d\tau}\psi^*\big(\tau, z(s)\big)\right) d\tau \leq$$

$$\leq \sup_{\|y\|_V \leq 1} \lim_{\epsilon \to 0^+} \frac{1}{\epsilon} g\big(2\|z(t)\|_W + \epsilon\|y\|_W\big)\|\epsilon y\|_V (t - s) = M(t - s),$$

since Lemma 10.1.2 also holds for ψ. Indeed, $\psi_t = \phi_t$ is locally Lipschitzian and $\big(I + \alpha A(t)\big)^{-1}$ is a contraction. Using $z \in H^1(0, T; V)$ and Theorem 1.1.2, we get $\big(I + \alpha A(\cdot)\big)^{-1} z(\cdot) \in H^1(0, T; V)$.   ∎

LEMMA 10.1.4

For each $\lambda > 0$, the operators A,

$$B = \{(u,v) \in L^2(0,T;V)^2 \mid (u(t),v(t)) \in B(t) \ \text{for a.a.} \ t \in (0,T)\},$$

$$B_\lambda = \{(u,v) \in L^2(0,T;V)^2 \mid (u(t),v(t)) \in B_\lambda(t) \ \text{for a.a.} \ t \in (0,T)\}$$

are maximal monotone and everywhere defined in $L^2(0,T;V)$.

PROOF   Let $(x_1,y_1) \in B$ be such that $(x_1 - x, y_1 - y)_{L^2(0,T;V)} \geq 0$ for all $(x,y) \in B$ whenever there exists a measurable set

$$E_{x,y} = \left\{ t \in (0,T) \mid (x_1(t) - x(t), y_1(t) - y(t))_V > 0 \right\}.$$

Since $D(B(t)) = V$, we may set $x = x_1$ outside $E_{x,y}$, whence it follows that the measure of $E_{x,y}$ is zero. Hence $(x_1(t) - x(t), y_1(t) - y(t))_V \geq 0$ for all $(x(t),y(t)) \in B(t)$ a.e. on $(0,T)$. Thus B is maximal monotone.

The maximal monotonicity of A and B_λ can be proved similarly. All these maximal monotone operators are everywhere defined, since they are bounded (see the proof of Lemma 10.1.1). ∎

Let us continue the proof of Theorem 10.1.1. By $M_i, i \in \mathbb{N}^*$, we mean positive constants that are independent of parameters λ, and we drop the subscripts on M.

For every $t \in [0,T]$, the mapping

$$\hat{J}_\lambda(t) = \left(B_\lambda(t) + C(t)\right)\left(I + A(t)\right)^{-1} : V \to V$$

is Lipschitzian with constant $1/\lambda + M$ (see Theorem 1.4.4 and (C.2)). By Lemma 10.1.4 and (C.1), $\hat{J}_\lambda(\cdot)x$ is measurable for all $x \in V$. Using (B.1), (C.2), and Lemma 10.1.3, we see that these mappings are integrable. By Lemma 1.5.3, the equation

$$y_\lambda' + \hat{J}_\lambda(t)y_\lambda(t) = f(t) \ \text{for a.a.} \ t \in (0,T)$$

has a solution $y_\lambda \in H^1(0,T;V)$ satisfying $y_\lambda(0) = u_0 + v_0$. Set $u_\lambda = (I+A)^{-1}y_\lambda$ and $v_\lambda = y_\lambda - u_\lambda$. By Lemma 10.1.3, $u_\lambda, v_\lambda \in H^1(0,T;V)$ and they satisfy (10.1.8)-(10.1.10).

LEMMA 10.1.5

There exists a constant $M > 0$, which satisfies for every $\lambda \in (0,1)$:

$$\|u_\lambda\|_{L^\infty(0,T;V)} \leq M, \quad \|v_\lambda\|_{L^\infty(0,T;W^*)} \leq M, \tag{10.1.11}$$

$$\|J_\lambda(\cdot)u_\lambda(\cdot)\|_{L^\infty(0,T;V)} \leq M, \quad \|B_\lambda u_\lambda\|_{L^\infty(0,T;V)} \leq M, \tag{10.1.12}$$

$$\|u_\lambda'\|_{L^2(0,T;V)} \leq M, \quad \|v_\lambda'\|_{L^2(0,T;V)} \leq M, \tag{10.1.13}$$

where $J_\lambda(t)$ is the resolvent of $B(t)$, $t \in (0,T)$.

PROOF   From (10.1.8) one obtains that

$$(u'_\lambda, f)_V = \|u'_\lambda\|^2_V + (u'_\lambda, v'_\lambda)_V + (u'_\lambda, B_\lambda u_\lambda)_V + (u'_\lambda, Cu_\lambda)_V, \quad (10.1.14)$$

whence it follows, by (A.4), (B.1), and (C.2), that

$$\|u'_\lambda(t)\|^2_V \leq M\left(1 + \|u_\lambda(t)\|^2_V + \|f(t)\|^2_V\right) \text{ for a.a. } t \in (0,T). \quad (10.1.15)$$

Hence

$$\|u_\lambda(t)\|^2_V = \|u_0\|^2_V + \int_0^t 2\big(u_\lambda(s), u'_\lambda(s)\big)_V \, ds \leq M \int_0^t \|u_\lambda(s)\|^2_V \, ds + M$$

for all $t \in [0,T]$. Now, the first estimate in (10.1.11) follows by Gronwall's inequality. Since $\partial\phi(t,\cdot)$ is uniformly bounded in $W \times W^*$ (see (A.3)), we obtain the second inequality in (10.1.11). On the other hand, using (B.1), we get (10.1.12). Now, the first inequality in (10.1.13) is implied by (10.1.15). Finally, the second inequality in (10.1.13) can be derived from (10.1.8) and (C.2).   ∎

LEMMA 10.1.6
There exist $u, v \in H^1(0,T;V)$ and $w \in L^2(0,T;V)$ such that

$$u_\lambda \to u, u'_\lambda \to u', \ v'_\lambda \to v', \ B_\lambda u_\lambda \to w \text{ weakly in } L^2(0,T;V), \quad (10.1.16)$$

$$u_\lambda(t) \to u(t) \text{ weakly in } V \text{ for all } t \in [0,T], \quad (10.1.17)$$

$$v_\lambda(t) \to v(t) \text{ in } V \text{ uniformly on } [0,T], \quad (10.1.18)$$

$$u_\lambda(t) \to u(t) \text{ in } W \text{ uniformly on } [0,T], \quad (10.1.19)$$

$$Cu_\lambda \to Cu \text{ weakly in } L^2(0,T;V), \text{ as } \lambda \to 0^+, \quad (10.1.20)$$

all on a subsequence.

PROOF   Since $L^2(0,T;V)$ is a Hilbert space and the corresponding sequences are bounded, there exist functions $u, u^*, v^*, w \in L^2(0,T;V)$ such that

$$u_\lambda \to u, u'_\lambda \to u^*, v'_\lambda \to v^*, \ B_\lambda u_\lambda \to w \text{ weakly in } L^2(0,T;V), \quad (10.1.21)$$

as $\lambda \to 0^+$, on a subsequence. The embedding $W^* \subset V^*$ is compact. Hence, the set $\cup_{\lambda \in (0,1)} R(v_\lambda(\cdot))$ is included in a compact set of V. Due to the boundedness of (v'_λ), the family $\{v_\lambda(\cdot) \mid \lambda \in (0,1)\}$ is equicontinuous. Hence Ascoli's theorem (see Theorem 1.1.4) guarantees that there exists a $v \in C([0,T];V)$ satisfying the uniform limit (10.1.18), on a subsequence.

Let $x \in C_0^\infty\big((0,T);V\big)$. By (10.1.21) and (10.1.18)

$$(v^*, x)_{L^2(0,T;V)} \leftarrow (v'_\lambda, x)_{L^2(0,T;V)} =$$
$$= -(v_\lambda, x')_{L^2(0,T;V)} \rightarrow -(v, x')_{L^2(0,T;V)},$$
$$(u^*, x)_{L^2(0,T;V)} \leftarrow (u'_\lambda, x)_{L^2(0,T;V)} =$$
$$= -(u_\lambda, x')_{L^2(0,T;V)} \rightarrow -(u, x')_{L^2(0,T;V)},$$

as $\lambda \rightarrow 0^+$. Thus $u^* = u'$ and $v^* = v'$ in $L^2(0,T;V)$, and so we obtain (10.1.16) and $u, v \in W^{1,2}(0,T;V) = H^1(0,T;V)$.

Since (u_λ) and (u'_λ) are bounded and $V \subset W$ compactly, we can apply Ascoli's theorem again. So, there exists a $\hat{u} \in C\big([0,T];W\big)$ such that

$$iu_\lambda(t) \rightarrow \hat{u}(t) \text{ in } W \text{ uniformly on } [0,T], \text{ as } \lambda \rightarrow 0^+,$$

on a subsequence. By (10.1.16)

$$\int_0^t \langle \hat{u}(s), x \rangle_{W \times W^*} \, ds = \lim_{\lambda \rightarrow 0^+} \int_0^t \langle iu_\lambda(s), x \rangle_{W \times W^*} \, ds = \int_0^t \langle u(s), i^*x \rangle \, ds$$

for all $x \in W^*$, $t \in [0,T]$. Hence $\hat{u}(t) = iu(t)$ for all $t \in [0,T]$. Thus we obtain (10.1.19), which implies (10.1.20) by (C.3).

Let $\epsilon > 0$, $x \in V^*$, and $t \in [0,T]$. Since i^*W^* is dense in V^*, there exists a $x_\epsilon \in W^*$ such that

$$\|x - i^*x_\epsilon\|_{V^*} \leq \frac{\epsilon}{M + \|u(t)\|_V}.$$

Using (10.1.19) and (10.1.11), we get

$$\limsup_{\lambda \rightarrow 0^+} \big| \langle u_\lambda(t), x \rangle - \langle u(t), x \rangle \big| \leq \limsup_{\lambda \rightarrow 0^+} \big| \langle u_\lambda(t), x - i^*x_\epsilon \rangle \big| +$$
$$+ \limsup_{\lambda \rightarrow 0^+} \big| \langle u_\lambda(t), x_\epsilon \rangle_{W \times W^*} - \langle u(t), x \rangle \big| \leq \epsilon + \epsilon.$$

This gives the limit (10.1.17), since $\epsilon > 0$ is arbitrary.   ∎

Let $t \in [0,T]$ and $x \in V$. By (10.1.8),

$$\big(u_\lambda(t) + v_\lambda(t), x\big)_V + \int_0^t \big(B_\lambda(s)u_\lambda(s), x\big)_V \, ds +$$

$$+ \int_0^t \big(C(s)u_\lambda(s), x\big)_V \, ds = \int_0^t \big(f(s), x\big)_V \, ds + (u_0 + v_0, x)_V.$$

Using Lemma 10.1.6 and the fact that the weak convergence in $L^2(0,T;V)$ implies the weak convergence in $L^2(0,t;V)$, we get

$$\big(u(t) + v(t), x\big)_V + \int_0^t \big(w(s) + C(s)u(s), x\big)_V \, ds = (u_0 + v_0, x)_V.$$

Since t and x are arbitrary, we obtain the differential equation (10.1.5) and the initial condition (10.1.7).

It remains to prove (10.1.6). Since the inverse A^{-1} is maximal monotone in $L^2(0,T;V)$ (see Lemma 10.1.4), $v_\lambda \to v$ strongly, $u_\lambda \to u$ weakly, and $(v_\lambda, u_\lambda) \in A^{-1}$, we can apply the demiclosedness result of maximal monotone operators (see Theorem 1.2.3, (c)). Thus $(v,u) \in A^{-1}$, i.e., $v \in Au$, which is the first relation in (10.1.6). For the second one, i.e., $w \in Bu$, we must work more.

We have $B_\lambda u_\lambda \to w$ and $J_\lambda u_\lambda = u_\lambda - \lambda B_\lambda u_\lambda \to u$ weakly in $L^2(0,T;V)$, as $\lambda \to 0^+$, and $B_\lambda u_\lambda \in BJ_\lambda u_\lambda$ (see Theorem 1.2.4). By the demiclosedness property of maximal monotone operators (see Theorem 1.2.3, (d)), it suffices to show that

$$\liminf_{\lambda \to 0^+} (B_\lambda u_\lambda, J_\lambda u_\lambda)_{L^2(0,T;V)} \leq (w,u)_{L^2(0,T;V)}. \tag{10.1.22}$$

However, by (10.1.8), $J_\lambda = I - \lambda B_\lambda$, Lemma 10.1.5, and (C.2),

$$(B_\lambda u_\lambda, J_\lambda u_\lambda)_{L^2(0,T;V)} - (B_\lambda u_\lambda, u)_{L^2(0,T;V)} = \lambda \|B_\lambda u_\lambda\|^2_{L^2(0,T;V)} +$$
$$+(f - u'_\lambda - v'_\lambda - Cu_\lambda, u_\lambda - u)_{L^2(0,T;V)} \leq M\|u - u_\lambda\|_{L^2(0,T;W)} +$$
$$\leq +M + (f, u_\lambda - u)_{L^2(0,T;V)} - (u'_\lambda + v'_\lambda, u_\lambda - u)_{L^2(0,T;V)}.$$

In light of Lemma 10.1.6, this implies (10.1.22) if

$$\liminf_{\lambda \to 0^+} (u'_\lambda + v'_\lambda, u_\lambda - u)_{L^2(0,T;V)} \geq 0. \tag{10.1.23}$$

Define $\psi\colon [0,T] \times V \to \mathbb{R}$, $\psi(t,x) = \frac{1}{2}\|x\|^2_V + \phi(t, ix)$. By integrating by parts and Theorem 1.2.18, we get

$$(u'_\lambda + v'_\lambda, u_\lambda)_{L^2(0,T;V)} = \big(u_\lambda(T) + v_\lambda(T), u_\lambda(T)\big)_V - (u_0 + v_0, u_0)_V -$$
$$-(u_\lambda + v_\lambda, u'_\lambda)_{L^2(0,T;V)} = \psi^*\big(T, u_\lambda(T) + v_\lambda(T)\big) -$$
$$-\psi^*(0, u_0 + v_0) + \int_0^T \phi_t\big(t, u_\lambda(t)\big)\, dt \tag{10.1.24}$$

and, similarly,

$$(u' + v', u)_{L^2(0,T;V)} = \psi^*\big(T, u(T) + v(T)\big) -$$
$$-\psi^*(0, u_0 + v_0) + \int_0^T \phi_t\big(t, u(t)\big)\, dt. \tag{10.1.25}$$

Since $\psi^*(T, \cdot)$ is weakly lower semicontinuous, Lemma 10.1.6 yields

$$\psi^*\big(T, u(T) + v(T)\big) \leq \liminf_{\lambda \to 0^+} \psi^*\big(T, u_\lambda(T) + v_\lambda(T)\big).$$

Hence (10.1.24) and (10.1.25) imply that

$$\liminf_{\lambda \to 0^+} (u'_\lambda + v'_\lambda, u_\lambda - u)_{L^2(0,T;V)} \geq$$

$$\geq \liminf_{\lambda \to 0^+} \int_0^T \phi_t\big(t, u_\lambda(t)\big)\, dt - \int_0^T \phi_t\big(t, u(t)\big)\, dt.$$

By (A.2) and the boundedness of (u_λ), we may apply the Lebesgue-Fatou lemma. Thus we obtain (10.1.23), and Theorem 10.1.1 is now completely proved. ∎

Let us state more hypotheses for the main theorem of this section.

(A.5) For each $(x_1, y_1), (x_2, y_2) \in \mathcal{A}(t)$ and $t \in [0, T]$,

$$\langle x_1 - x_2, y_1 - y_2 \rangle \geq a\|ix_1 - ix_2\|_W^2.$$

(A.6) There exists a $p \in [1, 2)$ such that

$$|\phi_t(t, x)| \leq M\big(1 + \|x\|_W^p\big) \text{ for a.a. } t \in (0, T) \text{ and for all } x \in W.$$

(A.7) For a.e. $t \in (0, T)$, $\|y\|_{W^*} \leq M\big(1 + \|x\|_W\big)$, for all $(x, y) \in \partial\phi(t, \cdot)i$.

(B.3) For a.a. $t \in (0, T)$ and each $(x, y) \in \mathcal{B}(t)$,

$$\langle x, y \rangle + M\|ix\|_W^2 \geq a\|x\|_V^2 - M.$$

(C.4) The uniform convergence in (C.3) is replaced by a.e.-convergence.

THEOREM 10.1.2
Assume (A.1)-(A.7), (B.1)-(B.3), (C.1)-(C.2), and (C.4). Then, for each $\tilde{f} \in L^2(0, T; V^*)$ and $\tilde{v}_0 \in R\big(\mathcal{A}(0)\big)$, there exist functions $u \in L^2(0, T; V)$, $w \in L^2(0, T; V^*)$, and $v \in H^1(0, T; V^*)$ that satisfy (10.0.11)-(10.0.13).

PROOF   We consider the following problem, which is equivalent to (10.0.11)-(10.0.13):

$$v'(t) + w(t) + C(t)u(t) = f(t), \tag{10.1.26}$$

$$v(t) \in A(t)u(t), w(t) \in B(t)u(t) \text{ for a.a. } t \in (0, T) \tag{10.1.27}$$

$$v(0) = v_0. \tag{10.1.28}$$

We choose $u_0 \in A(0)^{-1}v_0$, $\mu \in (0, 1]$ and write:

$$\mu u'_\mu(t) + v'_\mu(t) + w_\mu(t) + C(t)u_\mu(t) = f(t), \tag{10.1.29}$$

$$v_\mu(t) \in A(t)u_\mu(t), w_\mu(t) \in B(t)u_\mu(t) \text{ for a.a. } t \in (0, T) \tag{10.1.30}$$

$$\mu u_\mu(0) + v_\mu(0) = \mu u_0 + v_0. \tag{10.1.31}$$

Theorem 10.1.1 guarantees that problem (10.1.29)-(10.1.31) has a solution $(u_\mu, v_\mu, w_\mu) \in H^1(0,T;V)^2 \times L^2(0,T;V)$. Indeed, it suffices to replace $A(t)$, $B(t)$, $C(t)$, $f(t)$, and v_0 by $1/\mu \, A(t)$, $1/\mu \, B(t)$, etc.

LEMMA 10.1.7

There exists a constant $M > 0$ such that, for each $\mu \in (0,1]$,

$$\|u_\mu\|_{L^2(0,T;V)} \le M, \quad \sqrt{\mu}\|u_\mu\|_{L^\infty(0,T;V)} \le M, \qquad (10.1.32)$$

$$\|w_\mu\|_{L^2(0,T;V)} \le M, \quad \|v_\mu\|_{L^2(0,T;W^*)} \le M, \qquad (10.1.33)$$

$$\|\mu u'_\mu\|_{L^2(0,T;V)} \le M, \quad \|v'_\mu\|_{L^2(0,T;V)} \le M. \qquad (10.1.34)$$

PROOF From the equation (10.1.29) we obtain that

$$\mu(u'_\mu, u_\mu)_V + (v'_\mu, u_\mu)_V + (w_\mu, u_\mu)_V + (Cu_\mu, u_\mu)_V = (f, u_\mu)_V.$$

We integrate this over $[0,t]$ and apply (B.3) and (C.2). Then we get

$$\frac{1}{2}\mu\|u_\mu(t)\|_V^2 - \frac{1}{2}\mu\|u_0\|_V^2 + \big(v_\mu(t), u_\mu(t)\big)_V - (v_0, u_0)_V -$$

$$- \int_0^t \big(v_\mu(s), u'_\mu(s)\big)_V dt + \int_0^t a\|u_\mu(s)\|_V^2 \, ds \le \int_0^t M\|u_\mu(s)\|_W^2 \, ds +$$

$$+ \int_0^t \Big(M + M\|u_\mu(s)\|_V\|u_\mu(s)\|_W + \big(f(s), u_\mu(s)\big)_V\Big) ds.$$

By Theorem 1.2.18, (A.2), and the definition of the subdifferential,

$$\int_0^t \big(v_\mu(s), u'_\mu(s)\big)_V ds = \int_0^t \Big(\frac{d}{ds}\phi\big(s, u_\mu(s)\big) - \phi_t\big(s, u_\mu(s)\big)\Big) ds \le$$

$$\le \big(u_\mu(t) - 0, v_\mu(t)\big)_V + \phi(t,0) - \phi(0, u_0) - \int_0^t \phi_t\big(s, u_\mu(s)\big) ds.$$

Using (A.2), we obtain from this and the previous inequality that

$$\mu\|u_\mu(t)\|_V^2 + \int_0^t a\|u_\mu(s)\|_V^2 \, ds \le \mu\|u_0\|_V^2 + 2(v_0, u_0)_V + M -$$

$$-2\phi(0, u_0) + M\int_0^t \|u_\mu(s)\|_W^2 \, ds \text{ for all } t \in [0,T]. \qquad (10.1.35)$$

Since $\phi(0, u_0) > -\infty$, we obtain that

$$\mu\|u_\mu(t)\|_V^2 + \int_0^t \|u_\mu(s)\|_V^2 \, ds \le M\int_0^t \|u_\mu(s)\|_W^2 \, ds + M \qquad (10.1.36)$$

for all $t \in [0,T]$.

Multiplying (10.1.29) by v_μ, it follows that

$$\mu(u_\mu', v_\mu)_V + (v_\mu', v_\mu)_V + (w_\mu, v_\mu)_V + (Cu_\mu, v_\mu)_V = (f, v_\mu)_V.$$

We integrate this over $[0, t]$ and apply Theorem 1.2.18, (B.1), and (C.2). Then,

$$\mu\phi\big(t, u_\mu(t)\big) - \mu\phi(0, u_0) - \mu \int_0^t \phi_t\big(s, u_\mu(s)\big)\, ds +$$

$$+ \frac{1}{2}\|v_\mu(t)\|_V^2 - \frac{1}{2}\|v_0\|_V^2 \leq \frac{1}{2}\int_0^t \|v_\mu(s)\|_V^2\, ds +$$

$$+ 6M^2 \int_0^t \|u_\mu(s)\|_V^2\, ds + M \quad \text{for all } t \in [0, T]. \qquad (10.1.37)$$

From (A.2) we obtain that, for each $x \in V$ and $t \in [0, T]$,

$$\phi(t, x) = \int_0^t \phi_t(s, x)\, ds + \phi(0, x) \geq -M(1 + \|x\|_W^2) + \phi(0, x). \qquad (10.1.38)$$

By Lemma 10.1.1, $A(0)$ is everywhere defined and thus there exists $(0, v^*) \in A(0)$. By the definition of the subdifferential and Theorem 1.2.17,

$$\phi\big(0, u_\mu(t)\big) - \phi(0, 0) \geq \big(u_\mu(t), v^*\big)_V \geq -M - \|u_\mu(t)\|_W^2.$$

Now, using (10.1.38), we get

$$\phi\big(t, u_\mu(t)\big) - \phi(0, u_0) \geq -M - M\|u_\mu(t)\|_W^2,$$

since $\phi(0, \cdot)$ is everywhere finite. Thus (10.1.37), (A.2), and (10.1.36) yield

$$\frac{1}{2}\|v_\mu(t)\|_V^2 \leq \mu M + M \int_0^t \|u_\mu(t)\|_W^2\, ds + M +$$

$$+ \mu \int_0^t M(1 + \|u_\mu(t)\|_W^2)\, ds + \frac{1}{2}\|v_0\|_V^2 +$$

$$+ \frac{1}{2}\int_0^t \|v_\mu(s)\|_V^2\, ds + 6M^2\Big(M \int_0^t \|u_\mu(s)\|_W^2 + M\Big)$$

for all $t \in [0, T]$. Gronwall's inequality gives now that

$$\|v_\mu(t)\|_V^2 \leq M \int_0^t \|u_\mu(s)\|_W^2\, ds + M \quad \text{for all } t \in [0, T]. \qquad (10.1.39)$$

Let $\hat{v}^*(t) \in A(t)0$ for all $t \in [0, T]$. By (A.5),

$$\|v_\mu(t) - \hat{v}^*(t)\|_V = \sup_{\xi \in V, \xi \neq 0} \frac{\big(\xi, v_\mu(t) - \hat{v}^*(t)\big)_V}{\|\xi\|_V} \geq \frac{a\|u_\mu(t) - 0\|_W^2}{\|u_\mu(t)\|_V}.$$

Hence, using (A.3), we obtain that

$$\|u_\mu(t)\|_W^2 \le \frac{1}{M_*}\|u_\mu(t)\|_V^2 + \frac{2M_*}{a^2}\|v_\mu(t)\|_V^2 + M$$

for all $M_* > 0$. Hence the inequalities (10.1.36) and (10.1.39) imply that

$$\int_0^t \|u_\mu(s)\|_V^2 \, ds \le M \int_0^t \Big(\frac{1}{M_*}\|u_\mu(s)\|_V^2 + \frac{2M_*}{a^2}\|v_\mu(s)\|_V^2 + M\Big)\, ds +$$

$$+M \le \frac{M}{M_*}\int_0^t \|u_\mu(s)\|_V^2 \, ds + \frac{2MM_*}{a^2}\int_0^t \int_0^s \|u_\mu(\tau)\|_V^2 \, d\tau \, ds + M.$$

We choose $M_* = 2M$. Then,

$$\int_0^t \|u_\mu(s)\|_V^2 \, ds \le M \int_0^t \int_0^s \|u_\mu(\tau)\|_V^2 \, d\tau \, ds + M \qquad (10.1.40)$$

for all $t \in [0, T]$. By Gronwall's inequality, we obtain the first inequality in (10.1.32). Now, the second inequality in (10.1.32) is implied by (10.1.36). The estimates (10.1.33) are entailed by (B.1) and (A.7).

We multiply (10.1.29) by v'_μ. Then

$$(\mu u'_\mu, v'_\mu)_V + \|v'_\mu\|_V^2 + (w_\mu, v'_\mu)_V + (Cu_\mu, v'_\mu)_V = (f, v'_\mu)_V,$$

which, together with (A.4), (B.2), and (C.2), gives that

$$\|v'_\mu\|_V^2 \le \frac{1}{4}\|\mu u_\mu(t)\|_V^2 + M\Big(1 + \|u_\mu(t)\|_V^2 + \|f(t)\|_V^2\Big) \quad (10.1.41)$$

for a.a. $t \in (0, T)$.

Using (B.2) and (C.2) to (10.1.29), we obtain that

$$\|\mu u'_\mu\|_V \le \|v'_\mu\|_V + M\big(1 + \|u_\mu\| + \|f(t)\|_V\big).$$

This implies, for a.a. $t \in (0, T)$, that

$$\|\mu u'_\mu(t)\|_V^2 \le 2\|v'_\mu(t)\|_V^2 + M\Big(1 + \|u_\mu(t)\|_V^2 + \|f(t)\|_V^2\Big). \qquad (10.1.42)$$

From (10.1.41)-(10.1.42), we obtain the second estimate in (10.1.34). The first estimate in (10.1.34) follows now from (10.1.42). ∎

LEMMA 10.1.8
There exist $u, w \in L^2(0, T; V)$ and $v \in H^1(0, T; V)$ such that

$$u_\mu \to u, \ v'_\mu \to v', \ \mu u'_\mu \to 0 \ \text{weakly in } L^2(0, T; V), \quad (10.1.43)$$

$$v_\mu \to v \ \text{in } L^2(0, T; V), \qquad (10.1.44)$$

$$w_\mu \to w, \ Cu_\mu \to Cu \ \text{weakly in } L^2(0, T; V), \qquad (10.1.45)$$

$$v_\mu(t) \to v(t) \ \text{weakly in } V \ \text{for all } t \in [0, T], \qquad (10.1.46)$$

$$u_\mu(t) \to u(t) \ \text{in } W \ \text{for a.a. } t \in (0, T), \qquad (10.1.47)$$

as $\mu \to 0^+$, on a subsequence.

PROOF   Since the sequences (u_μ), (v'_μ), $(\mu u'_\mu)$, and (w_μ) are bounded in the Hilbert space $L^2(0, T; V)$, there exist $u, v^*, u^*, w \in L^2(0, T; V)$ such that

$$u_\mu \to u, \ v'_\mu \to v^*, \ \mu u'_\mu \to u^*, \ w_\mu \to w \text{ weakly in } L^2(0, T; V), \quad (10.1.48)$$

as $\mu \to 0^+$, on a subsequence.

Since $i: V \to W$ is compact and we have Lemma 10.1.7, we can apply Theorem 1.1.5. Hence $\{\mathcal{R}v_\mu \mid \mu > 0\}$ is a relatively compact set in $L^2(0, T; V^*)$. Thus there exists a $v \in L^2(0, T; V)$ satisfying (10.1.44).

Let $x \in C_0^\infty((0, T); V)$. By (10.1.48), (10.1.44), and (10.1.32),

$$(v, x')_{L^2(0,T;V)} \leftarrow (v_\mu, x')_{L^2(0,T;V)} =$$
$$= -(v'_\mu, x)_{L^2(0,T;V)} \to -(v^*, x)_{L^2(0,T;V)},$$
$$\left|(u^*, x)_{L^2(0,T;V)}\right| \leftarrow \left|(\mu u'_\mu, x)_{L^2(0,T;V)}\right| \leq$$
$$\leq \mu \|u_\mu\|_{L^\infty(0,T;V)} M \|x'\|_{L^2(0,T;V)} \to 0, \text{ as } \mu \to 0^+.$$

Thus $v^* = v'$ and $u^* = 0$ in $L^2(0, T; V)$ and so $v \in H^1(0, T; V)$.

Since L^2-convergence implies a pointwise convergence on a subsequence, it follows from (10.1.44) that there exists a $t' \in (0, T)$ such that $v_\mu(t') \to v(t')$ in V, on a subsequence. Hence the weak limit of (v'_μ) gives the pointwise weak limit (10.1.46).

Since $u_\mu \to u$ weakly, $v_\mu \to v$ strongly, and A is maximal monotone in $L^2(0, T; V)$, Theorem 1.2.3 gives that $(u, v) \in A$. Then (A.5) implies that

$$\|u_\mu - u\|^2_{L^2(0,T;W)} \leq \frac{1}{a} \int_0^T \left(u_\mu(t) - u(t), v_\mu(t) - v(t)\right)_V dt \to 0,$$

as $\mu \to 0^+$. Thus there is a subsequence of (u_μ) that satisfies the pointwise limit (10.1.47). Then, finally, the limit $Cu_\mu \to Cu$ weakly follows from (C.4). $\blacksquare$

Let $x \in V$ and $t \in [0, T]$. From (10.1.29) we obtain

$$\langle \mu u_\mu(t), x \rangle + \langle v_\mu(t), x \rangle_V + \int_0^t \langle w_\mu(s) + C(s)u_\mu(s), x \rangle_V ds$$

$$= \int_0^t \langle f(s), x \rangle_V ds + \langle \mu u_0 + v_0, x \rangle_V, \quad (10.1.49)$$

whence by Lemma 10.1.8 it follows that

$$\langle v(t), x \rangle_V + \int_0^t \langle w(s) + C(s)u(s), x \rangle_V ds = \int_0^t \langle f(s), x \rangle_V ds + \langle v_0, x \rangle_V.$$

Now, we obtain the differential equation (10.1.26) and the initial condition (10.1.28). We have already proved $v \in Au$, so it remains to prove only $w \in Bu$.

Due to the demiclosedness of maximal monotone operators (see Theorem 1.2.3), it suffices to show that

$$\liminf_{\mu \to 0^+}(w_\mu, u_\mu)_{L^2(0,T;V)} \leq (w, u)_{L^2(0,T;V)}. \tag{10.1.50}$$

The inequality

$$\liminf_{\mu \to 0^+}(\mu u'_\mu + v'_\mu, u_\mu)_{L^2(0,T;V)} \geq (v', u)_{L^2(0,T;V)} \tag{10.1.51}$$

implies (10.1.50), since if it were valid we could calculate using (10.1.5), (10.1.29), (C.4), and Lemma 10.1.8:

$$(w, u)_{L^2(0,T;V)} = (f - v' - Cu, u)_{L^2(0,T;V)} \geq \lim_{\mu \to 0^+}(f, u_\mu)_{L^2(0,T;V)} -$$

$$- \liminf_{\mu \to 0^+}(\mu u'_\mu + v'_\mu, u_\mu)_{L^2(0,T;V)} + \lim_{\mu \to 0^+}(-Cu_\mu, u)_{L^2(0,T;V)} =$$

$$= \limsup_{\mu \to 0^+}\left((f - \mu u'_\mu - v'_\mu - Cu_\mu, u_\mu)_{L^2(0,T;V)} + \right.$$

$$\left. +(Cu_\mu, u_\mu - u)_{L^2(0,T;V)}\right) \geq$$

$$\geq \limsup_{\mu \to 0^+}\left((w_\mu, u_\mu)_{L^2(0,T;V)} - M\|u_\mu\|_{L^2(0,T;V)}\|u_\mu - u\|_{L^2(0,T;W)}\right) \geq$$

$$\geq \liminf_{\mu \to 0^+}(w_\mu, u_\mu)_{L^2(0,T;V)} + 0.$$

We integrate by parts and apply Theorems 1.2.12 and 1.2.18:

$$(\mu u'_\mu + v'_\mu, u_\mu)_{L^2(0,T;V)} = \frac{1}{2}\mu\|u_\mu(T)\|_V^2 - \frac{1}{2}\mu\|u_0\|_V^2 +$$

$$+\phi^*\big(T, v_\mu(T)\big) - \phi^*(0, v_0) + \int_0^T \phi_t\big(t, u_\mu(t)\big)\,dt. \tag{10.1.52}$$

Using (A.2), Egorov's theorem, (A.6), and the Lebesgue-Fatou Lemma we can prove that

$$\liminf_{\mu \to 0^+}\int_0^T \phi_t\big(t, u_\mu(t)\big)\,ds \geq \int_0^T \phi_t\big(t, u(t)\big)\,dt. \tag{10.1.53}$$

Now, we obtain from (10.1.52)-(10.1.53) and the lower semicontinuity of $\phi^*(T, \cdot)$ that

$$\liminf_{\mu \to 0^+}(\mu u'_\mu + v'_\mu, u_\mu)_{L^2(0,T;V)} \geq 0 + \liminf_{\mu \to 0^+}\phi^*\big(T, v_\mu(T)\big) - \phi^*(0, v_0) +$$

$$+ \liminf_{\mu \to 0^+}\int_0^T \phi_t\big(t, u_\mu(t)\big)\,dt \geq \phi^*\big(T, v(T)\big) - \phi^*(0, v_0) +$$

$$+ \int_0^T \phi_t\big(t, x(t)\big)\,dt = (v', u)_{L^2(0,T;V)}.$$

We have now shown (10.1.51), and so Theorem 10.1.2 is completely proved. ∎

Let us state more hypotheses for the degenerated case, where the operators $\partial\phi(t,\cdot) \subset W \times W^*$ are not strongly monotone.

(A.8) If a sequence (x_n) of elements of $L^2(0,T;V)$ converges weakly in $L^2(0,T;V)$ toward a point $x \in L^2(0,T;V)$, then

$$\liminf_{n\to\infty} \int_0^T \phi_t\big(t, ix_n(t)\big)\, dt \geq \int_0^T \phi_t\big(t, ix(t)\big)\, dt,$$

on a subsequence.

(B.4) For each $t \in [0,T]$ and $(x,y) \in \mathcal{B}(t)$,

$$\langle x, y \rangle \geq a\|x\|_V^2 - M.$$

(C.5) There is a $p \in [0,2)$ such that

$$\langle x, \mathcal{C}(t)x \rangle \geq -M\|x\|_V^p - M \ \text{ for all } x \in V,\ t \in [0,T].$$

(C.6) If a sequence (x_n) of elements of $L^2(0,T;V)$ converges weakly in $L^2(0,T;V)$ toward a point $x \in L^2(0,T;V)$, then $(\mathcal{C}x_n)$ converges weakly toward $\mathcal{C}x$ in $L^2(0,T;V^*)$, on a subsequence.

THEOREM 10.1.3
Assume (A.1)-(A.2), (A.4), (A.6)-(A.8), (B.1)-(B.2), (B.4), (C.1)-(C.2), and (C.5)-(C.6). Then, for each $\tilde{f} \in L^2(0,T;V^*)$ and $\tilde{v}_0 \in R\big(\mathcal{A}(0)\big)$, there exist functions $u \in L^2(0,T;V)$, $w \in L^2(0,T;V^*)$, and $v \in H^1(0,T;V^*)$ that satisfy (10.0.11)-(10.0.13).

PROOF　Since (C.6) is stronger than (C.3), we may apply Theorem 10.1.1 and conclude that the problem (10.1.29)-(10.1.31) with $\mu \in (0,1]$ and $u_0 \in A(0)^{-1}v_0$ has a solution $(u_\mu, v_\mu, w_\mu) \in H^1(0,T;V)^2 \times L^2(0,T;V)$.

LEMMA 10.1.9
There exists a constant $M > 0$ such that the estimates (10.1.32)-(10.1.34) are satisfied for all $\mu \in (0,1]$.

PROOF　From the equation (10.1.29) we obtain that

$$\mu(u_\mu', u_\mu)_V + (v_\mu', u_\mu)_V + (w_\mu, u_\mu)_V + (Cu_\mu, u_\mu)_V = (f, u_\mu)_V.$$

We integrate this over $[0, t]$ and apply (B.4) and (C.5). Then we get

$$\frac{1}{2}\mu\|u_\mu(t)\|_V^2 - \frac{1}{2}\mu\|u_0\|_V^2 + \big(v_\mu(t), u_\mu(t)\big)_V - (v_0, u_0)_V -$$

$$- \int_0^t \big(v_\mu(s), u_\mu'(s)\big)_V \, dt + \int_0^t a\|u_\mu(s)\|_V^2 \, ds \le$$

$$\le \int_0^t M\left(\|u_\mu(s)\|_V^p + M + \big(f(s), u_\mu(s)\big)_V\right) ds \le$$

$$\le \frac{a}{2}\int_0^t \|u_\mu(s)\|_V^2 \, ds + M + M\int_0^T \|f(s)\|_V^2 \, ds,$$

where Young's inequality is also applied. By Theorem 1.2.18, (A.2), and the definition of the subdifferential,

$$\int_0^t \big(v_\mu(s), u_\mu'(s)\big)_V \, ds = \int_0^t \left(\frac{d}{ds}\phi\big(s, u_\mu(s)\big) - \phi_t\big(s, u_\mu(s)\big)\right) ds \le$$

$$\le \big(u_\mu(t) - 0, v_\mu(t)\big)_V + \phi(t, 0) - \phi(0, u_0) - \int_0^t \phi_t\big(s, u_\mu(s)\big) \, ds.$$

Using (A.6) and Young's inequality, we obtain from this and the previous inequality that

$$\mu\|u_\mu(t)\|_V^2 + \int_0^t a\|u_\mu(s)\|_V^2 \, ds \le \mu\|u_0\|_V^2 + 2(v_0, u_0)_V +$$

$$+ M - 2\phi(0, u_0) \quad \text{for all } t \in [0, T]. \tag{10.1.54}$$

Since $\phi(0, u_0) > -\infty$, we obtain that

$$\mu\|u_\mu(t)\|_V^2 + \int_0^t \|u_\mu(s)\|_V^2 \, ds \le M \quad \text{for all } t \in [0, T]. \tag{10.1.55}$$

Hence, (10.1.32) is valid. The estimates (10.1.33) are entailed by (B.1) and (A.7). The rest of the proof is the same as that of Lemma 10.1.7. ∎

LEMMA 10.1.10
There exist $u, w \in L^2(0, T; V)$ and $v \in H^1(0, T; V)$ such that (10.1.43)-(10.1.46) are satisfied, as $\mu \to 0^+$, on a subsequence.

PROOF The proof is exactly the same as that of Lemma 10.1.8, except for $Cu_\mu \to Cu$ weakly in $L^2(0, T; V)$. That follows directly from (C.6) and $u_\mu \to u$ weakly in $L^2(0, T; V)$. ∎

Let us continue the proof of Theorem 10.1.3. The differential equation (10.1.26) and the initial condition (10.1.28) are obtained exactly as in the proof of Theorem 10.1.2. The demiclosedness of A, $u_\mu \to u$ weakly, $v_\mu \to v$ strongly in $L^2(0,T;V)$ imply, again, that $v \in Au$.

Assume that (10.1.51) and, on a subsequence,

$$\liminf_{\mu \to 0^+} \int_0^T \left(u_\mu(t), C(t)u_\mu(t) \right)_V dt \geq \int_0^T \left(u(t), C(t)u(t) \right)_V dt \quad (10.1.56)$$

are valid. Then we calculate, using (10.1.5), (10.1.29), (C.6), and Lemma 10.1.10:

$$(w,u)_{L^2(0,T;V)} = (f - v' - Cu, u)_{L^2(0,T;V)} \geq \lim_{\mu \to 0^+}(f, u_\mu)_{L^2(0,T;V)} -$$

$$- \liminf_{\mu \to 0^+}(\mu u'_\mu + v'_\mu, u_\mu)_{L^2(0,T;V)} - \liminf_{\mu \to 0^+}(Cu_\mu, u_\mu)_{L^2(0,T;V)} =$$

$$= \limsup_{\mu \to 0^+}(f - \mu u'_\mu - v'_\mu, u_\mu)_{L^2(0,T;V)} +$$

$$+ \limsup_{\mu \to 0^+} -(Cu_\mu, u_\mu)_{L^2(0,T;V)} \geq$$

$$\geq \limsup_{\mu \to 0^+}(f - \mu u'_\mu - v'_\mu - Cu_\mu, u_\mu)_{L^2(0,T;V)} =$$

$$= \limsup_{\mu \to 0^+} (w_\mu, u_\mu)_{L^2(0,T;V)}.$$

Now, we proceed exactly as in the end of the proof of Theorem 10.1.2, except (10.1.53) is implied directly from (A.8). Thus (10.1.51) is proved if we can show (10.1.56).

Let $z \in V$, $t \in [0,T]$, and (x_n) be a sequence of elements of $L^2(0,T;V)$, which converges toward $x \in L^2(0,T;V)$ weakly in $L^2(0,T;V)$. By (C.2),

$$\sup_{y \in V, y \neq 0} \frac{\langle y, C(t)z \rangle}{\|iy\|_W} \leq \sup_{y \in V, y \neq 0} \frac{M\|iy\|_W \|z\|_V}{\|iy\|_W} = M\|z\|_V.$$

Thus, indeed, $C(t)x_n(t)$ and $\int_0^t C(s)x_n(s)\,ds$ are W^*-valued. Moreover,

$$\left\| \int_0^t C(s)x_n(s)\,ds \right\|_{W^*} \leq \int_0^t \|C(s)x_n(s)\|_{W^*}\,ds \leq$$

$$\leq M \int_0^t \|x_n(s)\|_V\,ds \leq M\|x_n\|_{L^2(0,T;V)} \leq M',$$

where $M' > 0$ is a constant, since a weakly converging sequence is bounded. We obtain also, for each $t' \in [0,T]$,

$$\left\| \int_{t'}^t C(s)x_n(s)\,ds \right\|_{V^*} \leq \int_{t'}^t \|C(s)x_n(s)\|_{W^*}\,ds \leq$$

$$\leq M \int_{t'}^t \|x_n(s)\|_V\,ds \leq \sqrt{|t - t'|}M\|x_n\|_{L^2(0,T;V)} \leq M'\sqrt{|t - t'|}.$$

Thus the set

$$\left\{\xi\colon [0,T] \to V^* \,\middle|\, \xi(t) = \int_0^t \mathcal{C}(s)x_n(s)\,ds, \ t \in [0,T]\right\}$$

is an equicontinuous family of mappings from $[0,T]$ into a compact metric space. By Ascoli's theorem, it has a subsequence that converges in V^* uniformly on $[0,T]$. By (C.6), the limit function is $t \mapsto \int_0^t \mathcal{C}(s)x(s)\,ds$. Hence

$$\int_0^t \mathcal{C}(s)x_n(s)\,ds \to \int_0^t \mathcal{C}(s)x(s)\,ds \text{ in } V, \text{ uniformly on } [0,T], \quad (10.1.57)$$

as $n \to \infty$, on a subsequence.

Let $\epsilon > 0$ be arbitrary. Choose $u_\mu^\epsilon, u^\epsilon \in C_0^\infty\big((0,T);V\big)$ such that

$$\|u_\mu^\epsilon - u_\mu\|_{L^2(0,T;V)} < \epsilon, \ \|u^\epsilon - u\|_{L^2(0,T;V)} < \epsilon, \qquad (10.1.58)$$

$$(u_\mu^\epsilon)' \to (u^\epsilon)' \text{ weakly in } L^2(0,T;V), \text{ as } \mu \to 0^+. \qquad (10.1.59)$$

Indeed, u_μ^ϵ, u^ϵ, can be the usual convolution approximation of u_μ, u, respectively:

$$u_\mu^\epsilon(t) = \int_0^T \rho_\epsilon(t-s)u_\mu(s)\,ds, \ u^\epsilon(t) = \int_0^T \rho_\epsilon(t-s)u(s)\,ds$$

for all $t \in [0,T]$. Here, $\rho_\epsilon \in C_0^\infty(\mathbb{R})$ is the appropriate mollifier. Now, for each $t \in [0,T]$ and $z \in V$,

$$\big(z, (u_\mu^\epsilon)'(t)\big)_V = \int_0^T \big(\rho_\epsilon'(t-s)z, u_\mu(s)\big)_V\,ds \to$$

$$\to \int_0^T \big(\rho_\epsilon'(t-s)z, u(s)\big)_V\,ds = \big(z, (u^\epsilon)'(t)\big)_V, \text{ as } \mu \to 0^+,$$

which is even a stronger convergence than (10.1.59). Using (C.2), (10.1.29), and (10.1.58), we calculate that

$$(u_\mu, Cu_\mu)_{L^2(0,T;V)} = (u_\mu^\epsilon, Cu_\mu^\epsilon)_{L^2(0,T;V)} +$$

$$+ (u_\mu - u_\mu^\epsilon, Cu_\mu^\epsilon)_{L^2(0,T;V)} + (u_\mu, Cu_\mu - Cu_\mu^\epsilon)_{L^2(0,T;V)} \geq$$

$$\geq - \int_0^T \left((u_\mu^\epsilon)'(t), \int_0^t Cu_\mu^\epsilon(s)\,ds\right)_V dt - \epsilon M,$$

where we have also integrated by parts. By (10.1.57)-(10.1.59),

$$\liminf_{\mu \to 0^+}(u_\mu, Cu_\mu)_{L^2(0,T;V)} \geq - \int_0^T \left((u^\epsilon)'(t), \int_0^t Cu^\epsilon(s)\,ds\right)_V dt - \epsilon M =$$

$$= (u^\epsilon, Cu^\epsilon)_{L^2(0,T;V)} - \epsilon M \geq (u, Cu)_{L^2(0,T;V)} - \epsilon M.$$

Since $\epsilon > 0$ is arbitrary, we obtain (10.1.56). The proof of Theorem 10.1.3 is now complete. ∎

REMARK 10.1.1   The most degenerate case is $\mathcal{A}(t) = 0$. So it is evident that (C.5) cannot be dropped. Observe that (A.8) can easily be checked by Theorem 1.2.19 if the functions $\phi_t(t, \cdot)$, $t \in [0, T]$, are all convex or concave. ∎

REMARK 10.1.2   Since the problem (10.1.5)-(10.1.7) is by its character a parabolic one, it has certain smoothening properties of the initial data $\tilde{v}_0$. Roughly, the existence can be proved if $\tilde{v}_0$ belongs to the closure of $\phi(0, \cdot)$ with respect to the topology of W^*. ∎

REMARK 10.1.3   The autonomous case, where $\mathcal{B}(t)$ is a coercive subdifferential and $\mathcal{A}(t)$ is just a maximal monotone operator, has been investigated in [DiBSh]. ∎

10.2   Uniqueness of solution

In this section we consider the uniqueness of the solutions of (10.0.11)-(10.0.13). First we present a counterexample of nonuniqueness and then state some sets of sufficient conditions for the uniqueness. For convenience, we formulate our uniqueness theorems for the equivalent Hilbert space problem (10.1.26)-(10.1.28). Let $T > 0$ be fixed and V be a real Hilbert space that is embedded into a real Banch space W by means of a linear injection $i: V \to W$.

Example 10.2.1
([DiBSh, p. 739]). Let $V = W = R$ and, for $t \in [0, 1]$,

$$A(t), B(t), C(t): \mathbb{R} \to \mathbb{R}, A(t)x = B(t)x = x + H(x - 1), C(t)x = 0,$$

where $H \subset \mathbb{R} \times \mathbb{R}$ is the Heaviside operator,

$$H(x) = \begin{cases} 1 & \text{if } x > 0, \\ [0, 1] & \text{if } x = 0, \\ 0 & \text{if } x < 0. \end{cases}$$

We consider the following system

$$v'(t) + w(t) = 0 \text{ for a.a. } t \in (0, 1), \ v(0) = 2, \tag{10.2.1}$$

$$v(t) \in u(t) + H\big(u(t) - 1\big), \quad w(t) \in u(t) + H\big(u(t) - 1\big). \quad (10.2.2)$$

Clearly, this problem is a special case of (10.0.11)-(10.0.13) with $T = 1$, $\tilde{f} = 0$, and $\tilde{v}_0 = 2$. We see that, for any $c \in [1/2, 1]$, the functions $u_c, v_c, w_c \colon [0, 1] \to \mathbb{R}$,

$$u_c(t) = 1, \quad v_c(t) = 2 - \frac{t}{c}, \quad w_c(t) = \frac{1}{c} \text{ if } t \in [0, c],$$
$$u_c(t) = v_c(t) = w_c(t) = \mathrm{e}^{\,c-t} \text{ if } t \in (c, 1],$$

are solutions of (10.2.1)-(10.2.2). The nonuniqueness is met in every element of the triple (u, v, w). Observe that both operators $A(t)$ and $B(t)$ are even strongly monotone, maximal monotone operators. In the standard form of Cauchy problems, (10.2.1)-(10.2.2) read as

$$v'(t) + B(t)A(t)^{-1}v(t) \ni 0 \text{ for a.a. } t \in (0, 1), \quad v(0) = 2,$$

where

$$B(t)A(t)^{-1}x = \begin{cases} [1, 2] & \text{if } x \in [1, 2], \\ x & \text{otherwise}, \end{cases}$$

which is not a monotone operator. $\quad \square$

A little bit less than monotonicity of $B(t)A(t)^{-1}$ suffices for the uniqueness.

THEOREM 10.2.1

Let $\eta \in L^1(0, T)$ be positive. Assume that the operators $A(t), B(t) \subset V \times V$, $t \in [0, T]$, satisfy:

(i) $A(t)$ or $B(t)$ is an injection.

(ii) For each $(x_1, y_1), (x_2, y_2) \in A(t)$, $(x_1, z_1), (x_2, z_2) \in B(t)$,

$$(y_1 - y_2, z_1 - z_2)_V \geq -\eta(t)\|y_1 - y_2\|_V^2.$$

If $f \in L^1(0, T; V)$ and $C(t) = 0$, then there exists at most one triple

$$(u, v, w) \in L^1(0, T; V) \times W^{1,1}(0, T; V) \times L^1(0, T; V,)$$

which satisfies (10.1.26)-(10.1.28).

PROOF Let (10.1.26)-(10.1.28) be satisfied by

$$(u_1, v_1, w_2), (u_2, v_2, w_2) \in L^1(0, T; V) \times W^{1,1}(0, T; V) \times L^1(0, T; V).$$

By the differential equation (10.1.26),

$$v_1'(t) - v_2'(t) + w_1(t) - w_2(t) = 0 \ \text{ for a.a. } t \in (0, T). \qquad (10.2.3)$$

We multiply this by $v_1 - v_2$, use (ii), and integrate over $[0, t]$. Then

$$\frac{1}{2}\left\|v_1(t) - v_2(t)\right\|_V^2 \le \frac{1}{2}\left\|v_1(0) - v_2(0)\right\|_V^2 + \int_0^t \eta(s)\left\|v_1(s) - v_2(s)\right\|_V^2 ds,$$

for all $t \in [0, T]$. By the initial condition (10.1.28) and Gronwall's inequality,

$$\|v_1(t) - v_2(t)\|_V^2 = 0 \ \text{ for all } t \in [0, T].$$

Thus $v_1 = v_2$. By (10.2.3), $w_1 = w_2$. Using (i) we obtain that $u_1 = u_2$.

∎

THEOREM 10.2.2
Assume the conditions of Theorem 10.2 except (i), and, in addition, that there exist mappings $C(t) : V \to V$, $t \in [0, T]$, satisfying:

(iii) For all $(x_1, y_1), (x_2, y_2) \in A(t)$, $z \in V$,

$$\left|(z, C(t)x_1 - C(t)x_2)_V\right| \le \eta(t)\|y_1 - y_2\|_V \|z\|_V;$$

(iv) $A(t)$, $B(t)$, or $C(t)$ is an injection.

If $f \in L^1(0, T; V)$, then there exists at most one triple

$$(u, v, w) \in L^1(0, T; V) \times W^{1,1}(0, T; V) \times L^1(0, T; V), \qquad (10.2.4)$$

which satisfies (10.1.26)-(10.1.28).

PROOF   Let (10.1.26)-(10.1.28) and (10.2.4) be satisfied by (u_1, v_1, w_2) and (u_2, v_2, w_2). Denote by $M_1, M_2, \ldots$ some positive constants and

$$u = u_1 - u_2, \ v = v_1 - v_2, \ w = w_1 - w_2,$$

$$c = Cu_1 - Cu_2, \ F(t) = \int_0^t w(s)\, ds \ \text{ for all } t \in [0, T].$$

By the differential equation (10.1.26) and the initial condition (10.1.28),

$$v(t) + F(t) + \int_0^t c(s)\, ds = 0 \ \text{ for all } t \in T. \qquad (10.2.5)$$

We multiply this by $F' = w$ and integrate over $[0,t]$. By (ii) and (iii),

$$\frac{1}{2}\|F(t)\|_V^2 = \int_0^t \Big(-\big(v(s),w(s)\big)_V + \big(c(s),F(s)\big)_V \Big) ds -$$

$$-\Big(\int_0^t c(s)\,ds, F(t)\Big)_V \leq \int_0^t \eta(s)\Big(\|v(s)\|_V^2 + \|v(s)\|_V\|F(s)\|_V\Big) ds +$$

$$+\frac{1}{4}\|F(t)\|_V^2 + M_1 \int_0^t \eta(s)\|v(s)\|_V^2\,ds \quad \text{for all } t \in [0,T].$$

Using Gronwall's inequality, we get

$$\|F(t)\|^2 \leq M_2 \int_0^t \eta(s)\|v(s)\|_V^2\,ds \quad \text{for all } t \in [0,T]. \tag{10.2.6}$$

We multiply (10.2.5) by v and use (iii). Then we get

$$\|v(t)\|_V^2 \leq 2\|F(t)\|_V^2 + M_3 \int_0^t \eta(s)\|v(s)\|_V^2\,ds \quad \text{for all } t \in [0,T].$$

Together with (10.2.6) and Gronwall's inequality this yields $v = 0$. By (10.2.6), $F = 0$, whence $w = 0$. Using (10.2.5), we get $Cu_1 = Cu_2$. Finally, by (i), $u = 0$. $\blacksquare$

In [DiBSh, p. 741] there is a simple counterexample of the nonuniqueness of the solution in the autonomous case where $A(t)$ is even a linear operator and $B(t)$ is a strictly monotone single-valued subdifferential. Essential in this example is that $A(t)$ is not symmetric. However, in our existence theory, we have assumed $A(t)$ to be a subdifferential, which means in the linear case that it is symmetric.

THEOREM 10.2.3

Let $M, a > 0$ be constants and $g \colon \mathbb{R} \to \mathbb{R}$ be continuous. Assume that the operators $A(t), B(t) \subset V \times V$ and the mappings $\phi(t,\cdot) \colon V \to \mathbb{R}_+$ and $C(t) \colon V \to V$ satisfy the following conditions for all $t \in [0,T]$.

(1) $\phi(t,\cdot)$ is convex, continuous, and $\phi(t,x) \leq M\|x\|_V^2$ for all $x \in V$.

(2) For each $x, y \in V$, $\phi(\cdot,x) \in H^1(0,T)$ and

$$\big|\phi_t(t,x) - \phi_t(t,y)\big| \leq g\big(\|x\|_V + \|y\|_V\big)\|x - y\|_V.$$

(3) $A(t) = \partial\phi(t,\cdot)$ and $A(t)$ is linear.

(4) $(x_1 - x_2, y_1 - y_2)_V \geq a\|ix_1 - ix_2\|_W^2$ for all $(x_1,y_1),(x_2,y_2) \in A(t)$.

(5) For all $(x_1, y_1), (x_2, y_2) \in B(t)$,

$$(x_1 - x_2, y_1 - y_2)_V \geq a\|x_1 - x_2\|_V^2 - M\|ix_1 - ix_2\|_W^2.$$

(6) $\left|(z, C(t)x - C(t)y)_V\right| \leq M\|iz\|_W\|x - y\|_V$ for all $x, y, z \in V$.

If $f \in L^1(0, T; V)$, then there is at most one triple that satisfies (10.1.26)-(10.1.28) and (10.2.4).

PROOF   We use the same notation as in the proof of Theorem 10.2.2. Multiplying (10.2.5) by F and using (6), we get

$$\left\|F(t)\right\|^2 \leq M_4\left\|v(t)\right\|_V^2 + M_4 \int_0^t \left\|u(s)\right\|_V^2 ds, \quad \text{for all } t \in [0, T]. \quad (10.2.7)$$

We differentiate (10.2.5), multiply it by u, and use (5)-(6). Then

$$\left(v'(t), u'(t)\right)_V + \frac{a}{2}\left\|u(t)\right\|_V^2 \leq M_5\left\|iu(t)\right\|_W^2 \quad \text{for a.a. } t \in (0, T). \quad (10.2.8)$$

By the linearity of $A(t)$, ϕ, u and v satisfy the conditions of Lemma 10.1.2. Thus

$$\frac{d}{dt}\phi^*\left(t, v(t)\right) = \left(u(t), v'(t)\right)_V - \phi_t\left(t, u(t)\right) \quad \text{for a.a. } t \in (0, T). \quad (10.2.9)$$

By (1), $\phi(t, 0) = 0$, whence $\phi_t(t, 0) = 0$ for a.a. $t \in (0, T)$. Using (10.2.8), (10.2.9), and (2), we get

$$\phi^*\left(t, v(t)\right) + \frac{a}{2} \int_0^t \left\|u(s)\right\|_V^2 ds \leq \phi^*(0, 0) +$$

$$+ \int_0^t \left(- \phi_t\left(s, u(s)\right) + M_5\left\|iu(s)\right\|_W^2\right) ds \leq$$

$$\leq \phi^*(0, 0) + M_6 \int_0^t \left\|iu(s)\right\|_W^2 ds \quad \text{for all } t \in [0, T]. \quad (10.2.10)$$

By (1) one also has

$$\phi^*(0, 0) = 0, \quad \phi^*\left(t, v(t)\right) \geq \frac{1}{4M}\left\|v(t)\right\|_V^2 \quad \text{for all } t \in [0, T].$$

Hence (10.2.10) yields

$$\left\|v(t)\right\|_V^2 + \frac{a}{2} \int_0^t \left\|u(s)\right\|_V^2 ds \leq M_6 \int_0^t \left\|iu(s)\right\|_W^2 ds \quad \text{for all } t \in [0, T],$$

whence it follows by (4) and Gronwall's inequality that $v = 0$. By (4) also $u = 0$ and so, by (10.2.7), $w = 0$.   ∎

REMARK 10.2.1 A uniqueness result similar to Theorem 10.2.3 can also be proved in the case where $B(t)$ is a symmetric linear operator (see [DiBSh, p. 740] and [Hokk3, p. 664]. Observe that the conditions required by these uniqueness results are rather strong. In general the solution is not unique. ∎

10.3 Continuous dependence of solution

We are motivated by the problem of finding an optimal control of the continuous casting process of steel. The solidification and cooling of the steel is described by the general equations (10.0.11)-(10.0.13). The cost function is of the type

$$J\colon L^2(U) \to (-\infty, \infty], J(\xi) = J(\xi; u, v, w) := \|u - u_d\|^2_{L^2(0,T;V)} +$$
$$+\|v - v_d\|^2_{L^2(0,T;W^*)} + \|w - w_d\|^2_{L^2(0,T;V^*)} + \|\xi\|_{L^2(0,T;U)},$$

where U is a Banach space of the values of the control variable ξ, (u, v, w) the solution of (10.0.11)-(10.0.13) corresponding to the control variable, and u_d, v_d, w_d are the desired values. For example, the cooling should be uniform enough in order to avoid flaws, stresses, cavities, etc. The control variable contains the initial temperature field and cooling rate on the boundary. The latter is related to the operator $B(t)$ by $h(x,t)$, appearing in the Neumann boundary conditions (10.0.5), since usually the boundary is cooled by a water spray. In the case of continuous casting, the solution (u, v, w) is unique, as can be shown by classical methods, but if that were not the case, we could consider the following cost function:

$$\xi \mapsto \sup \big\{ J(\xi; u, v, w) \mid (u, v, w) \text{ is a solution of } (10.0.11)\text{-}(10.0.13) \big\}.$$

Clearly, there is no hope that the mapping $\xi \mapsto J(\xi)$ could be convex. So, Theorem 1.2.9 does not give us the existence of the optimal control. However, the method of sequences, which we used in proving the existence of solution, is now applicable. We take a sequence (ξ_n) of control variables converging weakly toward ξ. So, we have a sequence of operators (B_n) and initial values $(\tilde{v}_0^n)$ that converge in some sense toward B and $\tilde{v}_0$ corresponding to ξ. For these objects instead of B and $\tilde{v}_0$, there exist solutions (u_n, v_n, w_n) of (10.0.11)-(10.0.13). We try to show that these solutions form a bounded sequence, which, by the compactness assumptions, has a weakly converging subsequence. By the closedness properties, the limit of this subsequence is a solution of (10.0.11)-(10.0.13) corresponding to ξ. In such a manner we can prove that the cost function is *weakly lower semicontinuous*. Since it is evidently coercive, we obtain the existence of an optimal control.

In this section we state and prove explicitly one continuous dependence result of this kind. For further details and extensions we refer to [Hokk2].

Let $n \in \mathbb{N}^*$. We consider the problems

$$v_n'(t) + w_n(t) + \mathcal{C}^n(t)u_n(t) = \tilde{f}_n(t), \qquad (10.3.1)$$

$$v_n(t) \in \mathcal{A}^n(t)u_n(t), \ w_n(t) \in \mathcal{B}^n(t)u_n(t), \ \text{ for a.a. } t \in (0,T), \quad (10.3.2)$$

$$v_n(0) = \tilde{v}_0^n, \qquad (10.3.3)$$

as compared to the problem

$$v'(t) + w(t) + \mathcal{C}^\infty(t)u(t) = \tilde{f}_\infty(t), \qquad (10.3.4)$$

$$v(t) \in \mathcal{A}^\infty(t)u(t), \ w(t) \in \mathcal{B}^\infty(t)u(t), \ \text{ for a.a. } t \in (0,T), \quad (10.3.5)$$

$$v(0) = \tilde{v}_0^\infty. \qquad (10.3.6)$$

Our results are analogous to a part of the Neveu-Trotter-Kato theorem for nonlinear semigroups [Brézis1, pp. 120, 102], by A. Pazy and H. Brézis, i.e., for the continuous dependence of the solution of the Cauchy problem

$$u'(t) + Au(t) \ni f(t) \ \text{ for a.a. } t \in (0,T), \ u(0) = u_0$$

on A, f, and u_0, where A is a maximal monotone operator in a real Hilbert space H. The key condition is the *resolvent convergence*, i.e.,

$$(I + A^n)^{-1}x \to (I + A)^{-1}x \ \text{ for all } x \in H, \ \text{ as } n \to \infty,$$

by which the exact meaning of the limit of a sequence of maximal monotone operators (A^n) has been clarified.

We begin with the following generalization of the demiclosedness result of maximal monotone operators.

LEMMA 10.3.1

Let X be a real reflexive Banach space with continuous duality mapping F, $G^n \subset X \times X^*$ be monotone operators, $G \subset X \times X^*$ be a maximal monotone operator, and $(x_n, y_n) \in G^n$ for all $n \in \mathbb{N}^*$ such that, as $n \to \infty$,

$$x_n \to x \ \text{weakly in } X, \ y_n \to y \ \text{weakly in } X^*,$$

$$(F + G^n)^{-1}z \to (F + G)^{-1}z \ \text{in } X \ \text{for all } z \in X^*,$$

$$\liminf_{n \to \infty} \langle x_n, y_n \rangle_{X \times X^*} \le \langle x, y \rangle_{X \times X^*}.$$

Then $(x, y) \in G$.

PROOF   For each $(\eta, \xi) \in G^n$, we have $\langle x_n - \eta, y_n - \xi \rangle_{X \times X^*} \ge 0$. Thus

$$\langle x_n - \eta, y_n - \xi \rangle_{X \times X^*} \ge 0, \ \eta = (F + G^n)^{-1}\xi', \ \xi = \xi' - F\eta$$

for all $\xi' \in X^*$. Passing to the limit as $n \to \infty$, we obtain that

$$\langle x - \eta, y - \xi \rangle_{X \times X^*} \geq 0, \ \eta = (F + G)^{-1}\xi', \ \xi = \xi' - F\eta \ \text{ for all } \xi' \in X^*.$$

Hence $\langle x - \eta, y - \xi \rangle_{X \times X^*} \geq 0$ for all $\eta \in D(G), \xi \in G\eta$. Thus $y \in Gx$. ∎

Let us state some hypotheses, which are very similar to those of the first section of this chapter. Let the objects $a, M, T \in (0, \infty)$, $g \in C(\mathbb{R})$, W, V, $\mathcal{R}$, and $i : V \mapsto W$ be as there. We denote $\mathbb{N}^\infty = \{1, 2, \ldots, \infty\}$.

For each $n \in \mathbb{N}^\infty$, the functions $\phi_n \colon [0, T] \times W \mapsto \mathbb{R}$ satisfy:

(A-1) For each $t \in [0, T]$, the function $\phi_n(t, \cdot) \colon W \to \mathbb{R}$ is convex and continuous.

(A-2) For each $x, y \in W$ and for a.a. $t \in (0, T)$, the function $t \mapsto \phi_n(t, x)$ is differentiable and

$$\phi_n(0, 0) \leq M, \ |\phi_{n,t}(t, x)| \leq M(1 + \|x\|_W^2),$$
$$|\phi_{n,t}(t, x) - \phi_{n,t}(t, y)| \leq g(\|x\|_W + \|y\|_W)\|x - y\|_W.$$

(A-3) For a.e. $t \in (0, T)$ and $(x, y) \in \partial\phi_n(t, \cdot)$, $\|y\|_{W^*} \leq M + M\|x\|_W$.

For each $n \in \mathbb{N}^\infty$ and $t \in [0, T]$, the operators $\mathcal{A}_W^n(t) = \partial\phi_n(t, \cdot)$ and $\mathcal{A}^n(t) = i^*\partial\phi_n(t, \cdot)i$ satisfy:

(A-4) For each $(x_1, y_1), (x_2, y_2) \in \mathcal{A}_W^n(t)$,

$$\langle x_1 - x_2, y_1 - y_2 \rangle_{W \times W^*} \geq a\|x_1 - x_2\|_W^2.$$

(A-5) For each $x \in W^*$, $\mathcal{A}_W^n(t)^{-1}x \to \mathcal{A}_W^\infty(t)^{-1}x$ in W, as $n \to \infty$.

(A-6) $\liminf_{n \to \infty} \phi_n^*(T, x) \geq \phi_\infty^*(T, x)$ for all $x \in W^*$.

(A-7) For each $x \in V$ and $b, c \in [0, T]$ such that $c > b$,

$$\liminf_{n \to \infty} \big(\phi_n(c, x) - \phi_n(b, x)\big) \geq \phi_\infty(c, x) - \phi_\infty(b, x).$$

The maximal monotone operators $\mathcal{B}^n(t) \subset V \times V^*$, $n \in \mathbb{N}^\infty$, $t \in [0, T]$, satisfy:

(B-1) For each $(x, y) \in \mathcal{B}^n(t)$, $\|y\|_{V^*} \leq M(1 + \|x\|_V)$.

(B-2) For each $\alpha \in (0, \infty)$ and $z \in V^*$, the function $t \mapsto \big(\mathcal{R} + \alpha\mathcal{B}^n(t)\big)^{-1}z$ is measurable in V.

(B-3) For each $(x, y) \in \mathcal{B}^n(t)$, $\langle x, y \rangle \geq a\|x\|_V^2 - M\|ix\|_W^2 - M$.

(B-4) For each $x \in V^*$, $(\mathcal{R}+\mathcal{B}^n(t))^{-1}x \to (\mathcal{R}+\mathcal{B}^\infty(t))^{-1}x$ in V, as $n \to \infty$.

The operators $\mathcal{C}^n(t) : [0,T] \times V \mapsto V^*$ satisfy, for each $n \in \mathbb{N}^\infty$:

(C-1) For each $z \in V^*$, the function $t \mapsto \mathcal{C}^n(t)z$ is measurable in V^*.

(C-2) For each $x, y, z \in V$ and for a.e. $t \in (0,T)$, $\mathcal{C}^n(t)0 = 0$ and

$$|\langle x, \mathcal{C}^n(t)y - \mathcal{C}^n(t)z \rangle| \leq M \|ix\|_W \|y - z\|_V.$$

(C-3) If $x_n \to x$ weakly in $L^2(0,T;V)$ and strongly in $L^2(0,T;W)$, as $n \to \infty$, then $\mathcal{C}^n x_n \to \mathcal{C}^\infty x$ weakly in $L^2(0,T;V^*)$, as $n \to \infty$, on a subsequence.

The functions $\tilde{f}_n \in L^2(0,T;V^*)$ and the points $\tilde{v}_0^n$, $n \in \mathbb{N}^\infty$, satisfy:

$$\tilde{f}_n \to \tilde{f}_\infty \text{ in } L^2(0,T;V^*), \tag{10.3.7}$$

$$\tilde{v}_0^n \to \tilde{v}_0^\infty \text{ weakly in } V^*, \text{ as } n \to \infty, \tag{10.3.8}$$

$$\phi_n^*(0,\tilde{v}_0^n) \leq M, \ \limsup_{n\to\infty} \phi_n^*(0,\tilde{v}_0^n) \leq \phi_\infty^*(0,\tilde{v}_0^\infty). \tag{10.3.9}$$

THEOREM 10.3.1

Assume (A-1)-(A-7), (B-1)-(B-4), (C-1)-(C-3), (10.3.7)-(10.3.9). If the problems (10.3.1)-(10.3.3) have a solution

$$(u_n, v_n, w_n) \in L^2(0,T;V) \times H^1(0,T;V^*) \times L^2(0,T;V^*)$$

for all $n \in \mathbb{N}^*$, then the problem (10.3.4)-(10.3.6) has a solution

$$(u, v, w) \in L^2(0,T;V) \times \left(H^1(0,T;V^*) \cap L^\infty(0,T;W^*) \right) \times L^2(0,T;V^*),$$

which is the limit of a subsequence of $((u_n, v_n, w_n))$, as $n \to \infty$, in the sense of

$$u_n \to u \text{ weakly in } L^2(0,T;V), \tag{10.3.10}$$

$$u_n(t) \to u(t) \text{ in } W \text{ for a.a. } t \in (0,T), \tag{10.3.11}$$

$$v_n(t) \to v(t) \text{ in } V^* \text{ for all } t \in [0,T], \tag{10.3.12}$$

$$v_n \to v \text{ weakly in } L^2(0,T;W^*), \tag{10.3.13}$$

$$v_n' \to v', \ w_n \to w, \ \mathcal{C}^n u_n \to \mathcal{C}^\infty u \text{ weakly in } L^2(0,T;V^*). \tag{10.3.14}$$

PROOF   Again, for convenience, we investigate the equivalent Hilbert space problems. Let $n \in \mathbb{N}^\infty$ and denote $A^n(t) = \mathcal{R}^{-1}\mathcal{A}^n(t)$, $B^n(t) = \mathcal{R}^{-1}\mathcal{B}^n(t)$, $C^n(t) = \mathcal{R}^{-1}\mathcal{C}^n(t)$, $f_n = \mathcal{R}^{-1}\tilde{f}_n$, and $v_0^n = \mathcal{R}^{-1}\tilde{v}_0^n$. The problem

$$v_n'(t) + w_n(t) + C^n(t)u_n(t) = f_n(t), \tag{10.3.15}$$

$$v_n(t) \in A^n(t)u_n(t), \ w_n(t) \in B^n(t)u_n(t), \ \text{for a.a. } t \in (0,T), \tag{10.3.16}$$

$$v_n(0) = v_0^n, \tag{10.3.17}$$

is equivalent to (10.3.1)-(10.3.3), and

$$v'(t) + w(t) + C^\infty(t)u(t) = f_\infty(t) \ \text{ for a.a. } t \in (0,T), \qquad (10.3.18)$$

$$v(t) \in A^\infty(t)u(t), \ w(t) \in B^\infty(t)u(t) \ \text{ for a.a. } t \in (0,T), \qquad (10.3.19)$$

$$v(0) = v_0^\infty, \qquad (10.3.20)$$

is equivalent to (10.3.4)-(10.3.6). By $M_i > 0$ we mean constants that are independent of the parameter $n \in \mathbb{N}^\infty$. We drop the last subscripts on M.

Since ϕ_n, A^n and B^n, $n \in \mathbb{N}^\infty$,.satisfy the conditions of Lemmas 10.1.1 and 10.1.2, we have the following two lemmas.

LEMMA 10.3.2

Let $n \in \mathbb{N}^\infty$. The operators

$$A^n = \left\{(x,y) \in L^2(0,T;V)^2 \mid y(t) \in A^n(t)x(t), \ \text{ for a.a. } t \in (0,T)\right\},$$

$$B^n = \left\{(x,y) \in L^2(0,T;V)^2 \mid y(t) \in B^n(t)x(t), \ \text{ for a.a. } t \in (0,T)\right\}$$

are maximal monotone in $L^2(0,T;V)$.

LEMMA 10.3.3

Let $n \in \mathbb{N}^\infty$, $z \in H^1(0,T;V)$, and $(y,z) \in A^n$. Then the mapping $t \mapsto \phi_n^*\big(t,z(t)\big)$ belongs to $W^{1,1}(0,T)$ and

$$\frac{d}{dt}\phi^*\big(t,z(t)\big) = -\phi_{n,t}\big(t,y(t)\big) + \big(y(t),z'(t)\big)_V \ \text{ for a.a. } t \in (0,T).$$

LEMMA 10.3.4

For each $y \in L^2(0,T;V)$,

$$\left(I + B^n\right)^{-1}y \to \left(I + B^\infty\right)^{-1}y \ \text{ in } L^2(0,T;V), \ \text{ as } n \to \infty.$$

PROOF Let $y \in L^2(0,T;V)$. The measurability of the mapping $t \mapsto \big(I + B^n(t)\big)^{-1}y$ in V follows from (B-2) and the continuity of $\big(I + B^n(t)\big)^{-1} : V \to V$. Moreover, by the boundedness of $B^n(t)$,

$$\left\|\big(I + B^n(t)\big)^{-1}y(t)\right\|_V \leq \left\|y(t) - (I + B^n(t))0\right\|_V \leq \left\|y(t)\right\|_V + M,$$

for a.a. $t \in (0,T)$. Thus, by the Lebesgue Dominated Convergence Theorem and (B-4),

$$\lim_{n\to\infty}\left\|\big(I + B^n\big)^{-1}y - \big(I + B^\infty\big)^{-1}y\right\|^2_{L^2(0,T;V)} = 0.$$

∎

LEMMA 10.3.5

There exists a constant $M_1 > 0$, independent of $n \in \mathbb{N}^\infty$, such that

$$\|u_n\|_{L^2(0,T;V)} \leq M_1, \quad \|u_n\|_{L^\infty(0,T;W)} \leq M_1, \qquad (10.3.21)$$

$$\|\mathcal{R}v_n\|_{L^\infty(0,T;W^*)} \leq M_1, \quad \|v_n'\|_{L^2(0,T;V)} \leq M_1, \qquad (10.3.22)$$

$$\|w_n\|_{L^2(0,T;V)} \leq M_1. \qquad (10.3.23)$$

PROOF   Using the boundedness of $\partial\phi_n(t,\cdot)$ and (A-2), we get

$$\phi_n(t,x) \leq \phi_n(t,0) + \langle ix - 0, y\rangle_{W\times W^*} \leq M\big(1 + \|ix\|_W^2\big)$$

for all $(x,y) \in \partial\phi_n(t,i\cdot)$, since $A^n(t)$ is everywhere defined. Hence,

$$\phi_n^*(t,y) = \sup_{\xi \in V}(\langle \xi, y\rangle - \phi_n(t,\xi)) \geq$$

$$\geq \sup_{\xi \in V}\left(\langle \xi, y\rangle_{W\times W^*} - M - M\|i\xi\|_W^2\right) \geq$$

$$\geq \frac{1}{4M}\|y\|_{W^*}^2 - M \ \text{ for all } y \in W^*, \qquad (10.3.24)$$

since iV is dense in W. We choose $v_n^*(t) \in \partial\phi_n(t,\cdot)0$ for a.a. $t \in (0,T)$. Then $\|v_n^*(t)\|_W \leq M$, by (A-3), and it follows from (A-4) that

$$\big\|\mathcal{R}v_n(t) - v^*(t)\big\|_{W^*} \geq \frac{\langle iu_n(t) - 0, v_n(t) - v^*(t)\rangle_{W\times W^*}}{\|iu_n(t) - 0\|_W} \geq a\big\|iu_n(t)\big\|_W.$$

Thus,

$$\big\|u_n(t)\big\|_W^2 \leq M\big\|v_n(t)\big\|_{W^*}^2 + M \ \text{ for a.a. } t \in (0,T). \qquad (10.3.25)$$

We multiply (10.3.15) by u_n. Then,

$$\big(u_n(t), v_n'(t)\big)_V + \big(u_n(t), w_n(t)\big)_V = \big(u_n(t), f_n(t) - C^n(t)u_n(t)\big)_V \qquad (10.3.26)$$

for a.a. $t \in (0,T)$. By integrating over $[0,t]$, Lemma 10.3.3 (B-3), (A-2), (C-2), and (10.3.9), we get

$$\phi_n^*\big(t, v_n(t)\big) + a\int_0^t \big\|u_n(s)\big\|_V^2\,ds \leq \phi_n^*(0, v_0^n) - \int_0^t \phi_{n,t}\big(s, u_n(s)\big)\,ds +$$

$$+ M\int_0^t \big\|u_n(s)\big\|_W^2\,ds + M + \int_0^t \big(u_n(s), f_n(s) - C^n(s)u_n(s)\big)_V\,ds \leq$$

$$\leq M + M\int_0^t \big\|iu_n(s)\big\|_W^2\,ds + \frac{a}{2}\int_0^t \|u_n(s)\|_V^2\,ds. \qquad (10.3.27)$$

Now, using Gronwall's inequality, (10.3.24), and (10.3.25), applied to $v_n(t)$, we obtain (10.3.21). By (A-3) and (B-1), $\mathcal{R}v_n$ and w_n are bounded. Finally, (C-2) and (10.3.15) imply the estimate for v_n'.   ∎

LEMMA 10.3.6

Let $x \in L^2(0,T;W^*)$ and $n \in \mathbb{N}^\infty$. Then $\mathcal{A}_W^n(\cdot)^{-1}x(\cdot) \in L^2(0,T;W)$ and

$$\mathcal{A}_W^n(\cdot)^{-1}x(\cdot) \to \mathcal{A}_W^\infty(\cdot)^{-1}x(\cdot) \text{ in } L^2(0,T;W), \text{ as } n \to \infty.$$

PROOF   By (A-4), $\mathcal{A}_W^n(t)^{-1}\colon W^* \to W$ is Lipschitzian with the constant $1/a$. As in Lemma 10.1.3, one can prove that $\mathcal{A}_W^n(\cdot)^{-1}z$ belongs to $H^1(0,T;W)$ for all $z \in W^*$. Thus $t \mapsto \mathcal{A}_W^n(t)^{-1}x(t)$ is measurable in W. It belongs to $L^2(0,T;W)$, since by (A-3) and $0 \in D\big(\mathcal{A}_W^n(t)\big)$,

$$\big\|\mathcal{A}_W^n(t)^{-1}x(t)\big\|_W \le \frac{1}{a}\big\|x(t) - \mathcal{A}_W^n(t)0\big\|_{W^*} \le \frac{1}{a}\big(\|x(t)\|_W + M\big)$$

for a.a. $t \in (0,T)$. Thus we may use the Lebesgue Dominated Convergence Theorem and (A-5).   ∎

Let us continue the proof of Theorem 10.3.1. Using Lemma 10.3.5 and the reflexivity of spaces $L^2(0,T;V)$ and $L^2(0,T;W^*)$, we obtain that there exist $u, \tilde{v}, w \in L^2(0,T;V)$ and $v \in L^2(0,T;W^*)$ such that

$$u_n \to u, v_n' \to \tilde{v}, \ w_n \to w \text{ weakly in } L^2(0,T;V), \qquad (10.3.28)$$

$$v_n \to v \text{ weakly in } L^2(0,T;W^*), \text{ as } n \to \infty. \qquad (10.3.29)$$

By Theorem 1.1.5, Lemma 10.3.5, and the compactness of $i^*\colon W^* \to V^*$, there exists a $\hat{v} \in L^2(0,T;V)$ satisfying

$$v_n \to \hat{v} \text{ in } L^2(0,T;V), \text{ as } n \to \infty, \qquad (10.3.30)$$

on a subsequence. Let $x \in C_0^\infty\big((0,T);V\big)$ and $y \in L^2(0,T;V)$. Then

$$(x', \hat{v})_{L^2(0,T;V)} \leftarrow (x', v_n)_{L^2(0,T;V)} =$$
$$= -(x, v_n')_{L^2(0,T;V)} \to -(x, \tilde{v})_{L^2(0,T;V^*)},$$
$$(y, \hat{v})_{L^2(0,T;V)} \leftarrow (y, v_n)_{L^2(0,T;V)} = \langle iy, v_n \rangle_{L^2(0,T;W) \times L^2(0,T;W^*)} \to$$
$$\to \langle y, i^*\tilde{v} \rangle, \text{ as } n \to \infty.$$

Hence we can denote $\hat{v} = v$, since i^* is an injection. Thus $v \in H^1(0,T;V)$ and $\tilde{v} = v'$.

By (10.3.30), on a subsequence,

$$v_n(t) \to v(t) \text{ in } V, \text{ for a.a. } t \in (0,T), \text{ as } n \to \infty.$$

Let $t \in [0,T]$ and $\epsilon > 0$ be arbitrary. Then there exists a $t_\epsilon' \in (t-\epsilon, t+\epsilon)$, for which $v_n(t_\epsilon') \to v_n(t_\epsilon')$ in V, as $n \to \infty$. Hence, by Lemma 10.3.5,

$$\limsup_{n \to \infty} \big\|v_n(t) - v(t)\big\|_V \le$$

$$\le \limsup_{n \to \infty} \Big\|v_n(t_\epsilon') - v(t_\epsilon') + \int_{t_\epsilon'}^{t} \big(v_n'(s) - v(s)\big)\, ds \Big\|_V \le 0 + \sqrt{\epsilon}.$$

Thus (10.3.10), (10.3.12), (10.3.13), and the first two limits in (10.3.14) are satisfied.

Let us prove next that $v(t) \in A^\infty(t)u(t)$, i.e.,

$$\mathcal{R}v(t) \in \mathcal{A}_W^\infty(t)u(t) \text{ for a.a. } t \in (0, T).$$

Since A^n is monotone and $v_n \in A^n$,

$$(u_n - \xi, v_n - \eta)_{L^2(0,T;V)} \geq 0 \text{ for all } (\xi, \eta) \in A^n.$$

Hence we have for every $\xi \in L^2(0, T; W^*)$ with $\eta(t) = \mathcal{A}_W^n(t)^{-1}\xi(t)$ a.e. on $(0, T)$,

$$(u_n, v_n)_{L^2(0,T;V)} - (u_n, \xi)_{L^2(0,T;V)} - \langle \eta, v_n - \xi \rangle_{L^2(0,T;W) \times L^2(0,T;W^*)} \geq 0.$$

We take $n \to \infty$. It follows by Lemma 10.3.6, (10.3.10), (10.3.29), and (10.3.30) that

$$(u, v - \xi)_{L^2(0,T;V)} - \langle \eta, v - \xi \rangle_{L^2(0,T;W) \times L^2(0,T;W^*)} \geq 0$$

for all $\xi \in L^2(0, T; W^*)$ where $\eta(t) = \mathcal{A}_W^\infty(t)^{-1}\xi(t)$ a.e. on $(0, T)$. Thus

$$(u - \eta, v - \xi)_{L^2(0,T;V)} \geq 0 \text{ for all } (\eta, \xi) \in A^\infty.$$

Since A^∞ is maximal monotone, $v \in A^\infty u$.

Now, we can calculate, using (A-4), (10.3.21), (10.3.30), and (10.3.29), that

$$a \int_0^T \|u_n(t) - \mathcal{A}_W^n(t)^{-1}\mathcal{R}v(t)\|_W^2 \, dt \leq$$

$$\leq \langle u_n - \mathcal{A}_W^n(\cdot)^{-1}\mathcal{R}v, v_n - v \rangle_{L^2(0,T;W) \times L^2(0,T;W^*)} =$$

$$= (u_n, v_n - v)_{L^2(0,T;V)} -$$

$$-\langle \mathcal{A}_W^n(\cdot)^{-1}\mathcal{R}v(t), v_n - v \rangle_{L^2(0,T;W) \times L^2(0,T;W^*)} \to 0,$$

as $n \to \infty$. Thus, at least on a subsequence,

$$\|u_n(t) - \mathcal{A}_W^n(t)\mathcal{R}v(t)\|_W \to 0, \text{ as } n \to \infty,$$

for a.a. $t \in (0, T)$. Since $v \in A^\infty u$, we obtain from (A-5) that

$$\|u_n(t) - u(t)\|_W \leq \|u_n(t) - \mathcal{A}_W^n(t)\mathcal{R}v(t)\|_W +$$

$$+\|\mathcal{A}_W^n(t)^{-1}\mathcal{R}v(t) - u(t)\|_W \to \|\mathcal{A}_W^\infty(t)^{-1}\mathcal{R}v(t) - u(t)\|_W = 0$$

for a.a. $t \in (0, T)$, as $n \to \infty$. Thus (10.3.11) is valid. The last limit in (10.3.14) follows now from (C-3). So, all limits have been proved true.

Let $x \in V$ and $t \in [0, T]$. By (10.3.15) and (10.3.17),

$$\left(x, v_n(t) - v_0^n\right)_V + \int_0^t \left(x, w_n(s) + C^n(s)u_n(s)\right)_V ds = \int_0^t \left(x, f_n(s)\right)_V ds,$$

from which we obtain, as $n \to \infty$, that

$$\left(x, v_n(t) - v_0^\infty\right)_V + \int_0^t \left(x, w(s) + C^\infty(s)u(s)\right)_V ds = \int_0^t \left(x, f_\infty(s)\right)_V ds.$$

The initial condition (10.3.6) follows by setting $t = 0$, and the differential equation (10.3.4) by differentiating with respect to t. It remains only to prove $w \in Bu$.

In light of the generalization of the demiclosedness result (see Lemma 10.3.1) it suffices to prove that

$$\limsup_{n \to \infty}(u_n, w_n)_{L^2(0,T;V)} \le (u, w)_{L^2(0,T;V)}. \tag{10.3.31}$$

Starting from (10.3.15) we calculate

$$\limsup_{n \to \infty}(u_n, w_n)_{L^2(0,T;V)} \le \limsup_{n \to \infty}(u_n, f_n - v_n' - C^n u_n)_{L^2(0,T;V)} \le$$

$$\le \limsup_{n \to \infty}(u_n, f_n)_{L^2(0,T;V)} + \limsup_{n \to \infty} -(u_n, v_n')_{L^2(0,T;V)} +$$

$$+ \limsup_{n \to \infty} -(u, C^n u_n)_{L^2(0,T;V)} + \limsup_{n \to \infty}(u - u_n, C^n u_n)_{L^2(0,T;V)} =$$

$$= (u, f - C^\infty u)_{L^2(0,T;V)} - \liminf_{n \to \infty}(u_n, v_n')_{L^2(0,T;V)}, \tag{10.3.32}$$

using $u_n \to u$, $C^n u_n \to C^\infty u$ weakly, and $f_n \to f$ strongly in $L^2(0, T; V)$; the term with $u_n - u$ is zero by (C-2), (10.3.21), (10.3.11), and by the Lebesgue Dominated Convergence Theorem. Using Lemma 10.3.3 we calculate:

$$\liminf_{n \to \infty}(u_n, v_n')_{L^2(0,T;V)} =$$

$$= \liminf_{n \to \infty} \int_0^T \left(\frac{d}{dt}\phi_n^*\big(t, v_n(t)\big) + \phi_{n,t}\big(t, u_n(t)\big)\right) dt \ge$$

$$\ge \liminf_{n \to \infty} \phi_n^*\big(T, v_n(T)\big) + \liminf_{n \to \infty} -\phi_n^*(0, v_0^n) +$$

$$+ \liminf_{n \to \infty} \int_0^T \phi_{n,t}\big(t, u_n(t)\big) dt \ge$$

$$\ge \phi_\infty^*\big(T, v(T)\big) - \phi_\infty^*(0, v_0^\infty) + \int_0^T \phi_{\infty,t}\big(t, u(t)\big) dt, \tag{10.3.33}$$

where the second limit follows from (10.3.9), and the first limit is calculated as follows. By (A-6) and the definition of the subdifferential,

$$\liminf_{n \to \infty} \phi_n^*\big(T, v_n(T)\big) =$$

$$= \liminf_{n\to\infty} \left(\phi_n^*\big(T, v(T)\big) + \phi_n^*\big(T, v_n(T)\big) - \phi_n^*\big(T, v(T)\big) \right) \geq$$

$$\geq \phi_\infty^*\big(T, v(T)\big) + \liminf_{n\to\infty} \langle \mathcal{A}_W^n(T)^{-1} v(T), v_n(T) - v(T) \rangle_{W\times W^*} \geq$$

$$\geq \phi_\infty^*\big(T, v(T)\big) + \liminf_{n\to\infty} \langle u(T), v_n(T) - v(T) \rangle +$$

$$+ \liminf_{n\to\infty} \langle \mathcal{A}_W^n(T)^{-1} v(T) - \mathcal{A}_W^\infty(T) v(T), v_n(T) - v(T) \rangle_{W\times W^*} =$$

$$= \phi_\infty^*\big(T, v(T)\big),$$

where (A-5), the boundedness of $\|v_n(T)\|_{W^*}$, and (10.3.12) were used. In order to calculate the last limit in (10.3.33), let $\epsilon > 0$ be arbitrary. There exists a step function $y_\epsilon \in L^\infty(0,T;V)$ such that

$$\|u - y_\epsilon\|_{L^2(0,T;W)} < \epsilon, \quad \|y_\epsilon\|_{L^\infty(0,T;W)} \leq M.$$

By (A-2) and (10.3.22),

$$\liminf_{n\to\infty} \int_0^T \phi_{n,t}\big(t, u_n(t)\big)\, dt \geq \liminf_{n\to\infty} \int_0^T \phi_{n,t}\big(t, y_\epsilon(t)\big)\, dt - M\epsilon \geq$$

$$\geq \int_0^T \phi_{\infty,t}\big(t, y_\epsilon(t)\big)\, dt - M\epsilon \geq \int_0^T \phi_{\infty,t}\big(t, y(t)\big)\, dt - M\epsilon,$$

where (A-7) and (10.3.11) were used. Since $\epsilon > 0$ is arbitrary, the conclusion follows. By applying again Lemma 10.3.3 and $v_0^\infty = v(0)$ to (10.3.33), we conclude that

$$\liminf_{n\to\infty} (u_n, v_n')_{L^2(0,T;V)} \geq (u, v')_{L^2(0,T;V)},$$

which combined with (10.3.32) yields (10.3.31).

Theorem 3.1 is completely proved.   ∎

REMARK 10.3.1   Observe that the resolvent convergence is assumed only for the operators $\mathcal{B}^n(t)$, since by their strong monotonicity the operators $\mathcal{A}_W^n(t)$ have Lipschitz continuous inverses. For a case of degenerate $\mathcal{A}_W^n(t)$, see [Hokk2].   ∎

10.4   Existence of periodic solutions

In this section we discuss briefly the following periodic problem:

$$v'(t) + w(t) + \mathcal{C}(t)u(t) = \tilde{f}(t), \tag{10.4.1}$$

$$v(t) \in \mathcal{A}(t)u(t), \quad w(t) \in \mathcal{B}(t)u(t) \quad \text{for a.a. } t \in (0,T), \tag{10.4.2}$$

$$v(0) = v(T). \tag{10.4.3}$$

The standard method to study periodic problems is to apply the appropriate fixed point theorem to the Poincaré mapping $P\colon V^* \to V^*$, $Px = v(T)$, where v is the solution of (10.4.1)-(10.4.2) with $v(0) = x$. However, this Poincaré mapping is generally not a contraction, nor is it single valued, convex valued, or compact. So the contraction principle, Schauder's principle, and Kakutani's fixed point theorem are all inapplicable. Instead, we can again make use of the method of approximating solutions: we approximate (10.4.1)-(10.4.3) by a more regular problem with four regularization parameters. This regularized problem is equivalent to the problem

$$y'(t) + \mathcal{G}y(t) \ni g\big(t, y(t)\big) \ \text{ for a.a. } t \in (0, T), \ y(0) = y(T), \qquad (10.4.4)$$

where maximal monotone operator $\mathcal{G} \subset V \times V^*$ generates a compact semigroup and $g(t, \cdot)\colon V \to V^*$ are Lipschitzian. The Poincaré mapping is single valued and compact. So one can infer from Schauder's fixed point theorem that (10.4.4) has a solution. The periodicity is restored when the first regularization parameter is chosen big enough and when the others tend to zero, successively. By means of this procedure the following theorem can be proved. For the proof and further details we refer to [Hokk5].

THEOREM 10.4.1

Assume the conditions of Theorem 10.1.3 and, in addition,

$$\tilde{f} \in L^2(0, T; V^*) \ \text{ and } \ \phi(0, x) = \phi(T, x) \ \text{ for all } x \in W.$$

Then there exists a triple

$$(u, v, w) \in L^2(0, T; V) \times \big(H^1(0, T; V^*) \cap L^2(0, T; W^*)\big) \times L^2(0, T; V^*),$$

which satisfies (10.4.1)-(10.4.3).

References

[Agmon] S. Agmon, *Lectures in Elliptic Boundary Value Problems*, Van Nostrand, New York, 1965.

[Barbu4] V. Barbu, Existence for nonlinear Volterra equations in Hilbert spaces, *SIAM J. Math. Anal.*, 10 (1979), pp. 552–569.

[BarPr] V. Barbu & Th. Precupanu, *Convexity and Optimization in Banach Spaces*, Editura Academiei and Reidel, Bucharest and Dordrecht, 1986.

[Brézis1] H. Brézis, *Opérateurs maximaux monotones et semi-groupes de contractions dans les espaces de Hilbert*, North-Holland, Amsterdam, 1973.

[BeDuS] A. Bermudez, J. Durany & C. Saguez, An existence theorem for an implicit nonlinear evolution equation, *Collect. Math.*, 35, no. 1 (1984), pp. 19–34.

[DiBSh] E. Di Benedetto & R.E. Showalter, Implicit degenerate evolution equations and applications, *SIAM J. Math. Anal.*, 12:5 (1981), pp. 731–751.

[GraMi] O. Grange & F. Mignot, Sur la résolution d'une équation et d'une inéquation paraboliques non-linéaires, *J. Funct. Anal.*, 11 (1972), pp. 77–92.

[Hokk1] V.-M. Hokkanen, An implicit non-linear time dependent equation has a solution, *J. Math. Anal. Appl.*, 161:1 (1991), pp. 117–141.

[Hokk2] V.-M. Hokkanen, Continuous dependence for an implicit nonlinear equation, *J. Diff. Eq.*, 110:1 (1994), pp. 67–85.

[Hokk3] V.-M. Hokkanen, On nonlinear Volterra equations in Hilbert spaces, *Diff. Int. Eq.*, 5:3 (1992), pp. 647–669.

[Hokk4] V.-M. Hokkanen, Existence for nonlinear time dependent Volterra equations in Hilbert spaces, *An. Ştiinţ. Univ. Al. I. Cuza Iaşi Secţ. I a Mat.*, 38 (1992), no. 1, pp. 29–49.

[Hokk5] V.-M. Hokkanen, Existence of a periodic solution for implicit nonlinear equations, *Diff. Int. Eq.*, 9:4 (1996), pp. 745–760.

[Nečas] J. Nečas, *Les méthodes directes en théorie des équations elliptiques*, Masson Éditeur and Academia Éditeurs, Paris and Prague, 1967.

[Saguez] C. Saguez, *Contrôle optimal de systèmes à frontière libre*, Ph.D. thesis, Compiegne University of Technology, 1980.

Index